P9-DCA-829

# Designing the Molecular World

# Designing the Molecular World

## Chemistry at the Frontier

*Philip Ball*

Princeton University Press

Princeton, New Jersey

Published by Princeton University Press, 41 William Street, Princeton, New Jersey 08540
In the United Kingdom: Princeton University Press, Chichester, West Sussex



**Library of Congress Cataloging-in-Publication Data**
Ball, Philip, 1962-
Designing the molecular world: chemistry at the frontier/Philip Ball.
p. cm.
Includes bibliographical references and index.
ISBN 0-691-00058-1
1. Chemistry. I. Title.
QD31.2.B35 1994
540—dc20

93-38151
CIP

This book has been composed in Palatino

Princeton University Press books are printed on acid-free paper and meet the guidelines for permanence and durability of the Committee on Production Guidelines for Book Longevity of the Council on Library Resources

Printed in the United Kingdom

Editorial and production services Fisher Duncan, 10 Barley Mow Passage, London W4 4PH, UK

10 9 8 7 6 5 4 3 2 1

# Contents

## *PART II: NEW PRODUCTS, NEW FUNCTIONS*



## *PART III: CHEMISTRY AS A PROCESS*

# Acknowledgements

If this book achieves its goals, it will be because of the indispensible help that I have received. In particular, I would like to acknowledge valuable suggestions, advice and assistance from Harry Kroto, Ahmed Zewail, Charles Knobler, Mark Davis, Toyoichi Tanaka, Julius Rebek, Stefan Muller, Fleming Crim, Stephen Scott, Norman Herron, and Ilya Prigogine. My colleagues at *Nature*, particularly Laura Garwin, have provided a great deal of the stimulus that prompted me to attempt something so foolhardy in the first place, and those at Princeton University Press, in particular Emily Wilkinson and Malcolm Litchfield, persuaded me that it was worth persevering with. I am very grateful to Sue Fox and Steve Sullivan for assistance with the illustrations. Somewhere in the distant past, I owe considerable debts to Colin McCarthy, Kit Heasman and Peter Walker. And Julia has been wonderfully patient in the face of many late nights and early mornings.

Philip Ball
*London, September 1993*

# Designing the Molecular World

*Introduction*

# *Engineering the Elements*

*He who understands nothing but chemistry doesn't even understand chemistry.*

Georg Christoph Lichtenberg

How to Avoid Science

A good way in a science lesson is to wait until some old fashioned poison like sulphurick acid etc. turns up. As per ushual science master, who not forward-looking, sa: No boy is to touch the contents of the tube.

Make up tube which look the same and place alongside acid. Master begins lesson drone drone drone. Sudenly you spring to feet with grate cry: 'Sir Sir I can't stand it any longer!'

Drink coloured water and collapse to be carried out as if dead. n.b. if you make a mistake with this one you are still carried out as if dead and you *are.*

Geoffrey Willans and Ronald Searle
*Down With Skool!*

Even that most aberrant of schoolboys Nigel Molesworth would have to admit that there are times when it pays to have a little knowledge of chemistry. It remains, however, one of the least glamorous of sciences. Physicists, by comparison, are to be found pondering the deepest mysteries of the Universe: Where did everything come from? What will happen to it all? What is matter? What is time? Physics represents science at its most abstract, and also on its grandest scale, as gigantic telescopes search the heavens for the echoes of creation and particle accelerators miles in diameter smash subatomic particles into each other in order to glean clues about what the world is made of. The questions tackled by biologists, meanwhile, are the matters of life and death – it is for them to take up arms against the thousand natural diseases that flesh is heir to, or to strive towards understanding how we evolved from sea-bound blobs of jelly. Geologists brave the awesome fury of volcanoes and earthquakes; oceanographers plumb the hidden depths of the world. What do chemists do? Well, they make paint, among other things.

One might expect to find nothing more of interest in the practice of making paint than in watching it dry. But there is, as I hope to convince you later, a subtlety and cleverness to the art. If that still seems a prospect wanting in enticement, let me mention that we will also see what paint has in common with living cells and soap bubbles, with muscle tissues and plastics. The tiny corner of chemistry in which paint is contrived holds unguessed surprises, and supplies as good an illustration as any other of the way in which an understanding of the chemical nature of substances helps us to control the shape and form of our world. For the truth is that, while many of the other sciences are associated with mysteries of an awe-inspiring scale, chemistry is the science of everyday experience, of how plants grow and how snowflakes form and how a flame burns.

Yet chemistry has acquired the image of a mundane pursuit; and it must be said that some blame resides with chemists themselves, many of whom seem resigned to accept a perception of their research as worthy but dull. It is true that chemists are hampered from the outset by low expectations. (According to the fossilized wit of Oxford, the chemist (invariably male) is a dour clod with long hair and dirty hands – a formidable beer-swigger perhaps, but a social gorilla.) Yet chemists themselves often insist on a humility that borders on insecurity. They will say at conferences, "I don't claim to understand these results – I leave that to the physicists. All I did was make the materials."

I have no crusade in mind, however. Rather, what this book aims to do is to present a selection of some of the things that a chemist today may find her- or himself engaged in studying. If by doing so it succeeds in demonstrating simply that the new chemistry is no longer a matter of test tubes and bad smells (although both may be encountered along the way), that is fine enough. For this demonstration we will need to take a cursory glance not only at some of the basic principles of chemistry but also at a pot-pourri of ideas from disciplines as diverse as genetics, climatology, electronics and the study of chaos. Yet this is most certainly not a textbook: it will not cover chemistry comprehensively, nor will it provide a rigorous scientific description of the phenomena that will be discussed. Simply, I hope to show that in order to discover a sense of wonder about the world, it is not always necessary to look to the stars or to the theory of evolution; one can look instead at the washing-up liquid, the leaves on a tree or the catalytic convertors in our cars.

In 1950 the distinguished American chemist Linus Pauling said "Chemistry is a young science." It is true that chemistry of a sort was practised in Ancient China, in

Babylonia and beyond, but you could see his point. At that time only a few decades had passed since we had come to understand the constitution of the atom, chemistry's building block; and Dmitri Mendeleev's Periodic Table of chemical elements was just 81 years old, with several of the gaps only recently filled. But almost half a century later, does chemistry still retain any of its youthful vigor?

Much of chemistry today is becoming motivated and guided by principles dramatically different from those that informed Pauling's comment. The new chemistry pays scant regard to the disciplines into which the topic has been traditionally divided. At college, chemistry still is often taught in three distinct chunks: physical, organic and inorganic. But few are the chemists today who claim firm allegiance to a single one of these branches; rather, novel concepts and classifications are emerging through which researchers define their work. I shall give here an incomplete list of some of these; we will find these ideas cropping up many times in the subsequent chapters, often lending a common thread to studies that otherwise appear disparate. If you can, bear them in mind in what follows.

*Materials*: There may be many who lament the dawn of the plastic age. It has, however, demonstrated in unambiguous terms that we are no longer forced to manage as best we can with the materials that the natural world provides – we can design new ones that better suit our purposes. Plastics now have a seemingly limitless variety of properties: they show tensile strengths comparable to steel, they can dissolve in water or be eaten by microbes, conduct electricity, change color or contract and flex like muscles. Plastics generally consist of carbon-based chain-like "polymer" molecules; polymers based on silicon and oxygen, meanwhile, serve as the precursors to new kinds of ceramic materials, "artificial rocks" that promise new limits to hardness and strength.

The explosion of interest in materials science in recent years has gained tremendously from the realization that an understanding of the structure of materials at the molecular level can lead to the design of properties useful at the engineering level. We can now control the growth of materials atom by atom, opening up new possibilities in semiconductor microelectronics for example, or allowing the possibility of mimicking the impressive design of natural substances like bone and shell. And as our ability to control the microscopic structure of materials improves, chemistry continues occasionally to produce materials with unforeseen surprises in store, such as the carbon cages known as fullerenes or the metal alloys called quasicrystals.

*Electronics*: Did I say plastics that conduct electricity? Yes, not only do they exist but they are already being used in electronic devices. A broad range of synthetic chemical compounds are now known that possess metal-like electrical conductivities, and some even show the remarkable property of superconductivity – conductivity without resistance. Magnets too can now be made without a metal in sight, based on carbon- and nitrogen-containing molecules more like those found in the organic world. An entire electronics industry is beginning to look feasible that has no need for metals or conventional semiconductors such as silicon. For some, the ultimate dream is to build circuits from individual molecules, using conducting molecular wires to link up atomic-scale components into incredibly compact "molecular devices."

While one approach to molecular electronics is to make conventional microelectronic devices from unconventional materials, a still more daring suggestion is to set aside the familiar diodes and transistors and look for inspiration from nature. Photosynthesis, for instance, involves the passing from molecule to molecule of tiny electric currents within the cells of living organisms, while other biomolecules regulate currents as if they were themselves miniature electronic devices. Gaining an understanding of how these natural devices work will open doors to a kind of "organic" electronics.

*Self-assembly:* If, as hinted above, we want to build molecular structures one molecule at a time, we will need much more precision and speed at manipulation than is available to today's engineers of the microworld. But there is an alternative to the laborious process of molecule-by-molecule construction: get the molecules to assemble themselves. This might seem like expecting a house to suddenly leap together from a pile of bricks, but molecules are much more versatile than bricks. Soap molecules, for instance, can aggregate spontaneously into all manner of complex structures, including sheets, layered stacks and artificial cell-like membranes. Other organic molecules show the ability to organize themselves into the variety of orderly arrays that we recognize as liquid crystals.

The better we understand the way that molecules interact, the more able we will be to design them so that they assemble themselves into these intricate structures. Here again there is much to be learned from nature, which abounds with molecules that can recognize and team up with others in very specific and organized ways. Both in nature and in the laboratory, molecular "recognition" and self-assembly can lead to the possibility of molecules that put together copies of themselves from their component parts, or in other words to . . .

*Replication:* One of the primary attributes of a living organism is that it be able to make replicas of itself. There is nothing about this ability that requires a motivating intelligence, however; chemistry alone can do the job. The discovery, in 1953, of the structure of DNA led the way to an understanding of how chemical replication is possible. The replicating molecule acts as a template on which a copy is assembled; and this assembly process involves "complementarity" – a pairing up of structural elements – so that the molecule provides a scaffolding for construction of its replica.

It is now clear that molecules don't have to be anywhere near as complex as DNA in order to be able to replicate – small molecules and molecular assemblies have been devised that can do this in a test tube. In some sense, these molecules represent the first step towards a kind of artificial life. But as the raw materials provided to these synthetic replicators are generally not far removed from the end product, they are not so much building copies from scratch as simply speeding up the rate at which the final stages of replication take place: genuine synthetic life is still a long way off. However, the discovery in 1982 that DNA's relative RNA can perform the trick of replication all by itself (that is, without an army of helper molecules such as DNA requires) may provide a vital clue to our understanding of how life evolved through chemistry alone.

*Specificity:* Chemical reactions can be notoriously messy affairs, leaving one with the unwelcome task of extracting one's intended product from a whole host of substances produced in side reactions. That this simply does not happen in the biochemistry of the body, where each reaction generally gives just the one desired product, suggests that we need not resign ourselves to such a state of affairs in our own clumsy attempts at chemical synthesis. And indeed, by exploiting the principles of molecular recognition found in biology, these efforts are becoming progressively less maladroit. We are learning how to make chemistry specific.

It is the class of molecules called enzymes that is responsible for the remarkable specificity of biochemical processes. Despite a still far from complete understanding of how enzymes function, synthetic molecules have been designed that can mimic many of their attributes. The chemical industry, meanwhile, is learning how to exploit the exquisite chemical control that enzymes display by setting them to work in "bioreactors," biologically based chemical plants which produce complex pharmaceutical products that are otherwise beyond our wit to synthesize. And petrochemical companies are finding that the minerals known as zeolites can function as rudimentary "solid-state enzymes" to provide useful compounds from crude oil.

*Seeing at the atomic scale:* The process of chemical change happens in the twinkling of an eye. During the course of a chemical reaction, the interaction of two molecules may occupy no more than a trillionth of a second. In the past this has posed tremendous difficulties for attempts to discover exactly what goes on when molecules get together, but there now exist ways to capture these incredibly brief events "on film." Lasers that pump out thousands of discrete light pulses during the time it takes for individual molecules to interact allow one to capture snapshots of molecular motions frozen in time. We can now watch molecules as they tumble, collide and become transformed into new arrangements of atoms.

Microscopes, meanwhile, are letting us see matter at the scale of individual atoms. These abandon the use of light and employ electrons instead to obtain images of objects so small that many millions would fit on a pinhead. The regularly packed lattice of atoms in a crystal, the orderly stacks of molecules in a liquid crystal film or the double helix of DNA – all have been revealed by this new brand of microscope.

*Nonequilibrium:* The many complex shapes found in the natural world, ranging from snowflakes to the roots and fronds of plants, have long represented a source of fascination and bafflement alike to natural scientists. But one of the astonishing discoveries of recent years is that complicated patterns do not necessarily require a highly controlled process of formation; rather, they can arise spontaneously in systems that appear to be wildly out of control. Systems that are far from attaining any sort of equilibrium need not descend into disorder but may, under appropriate conditions, organize themselves into large-scale patterns that may be at once very intricate and beautifully symmetric. The "forbidden crystals" known as quasicrystals provide one such example; others display so-called "fractal" properties, appearing identical regardless of how closely one looks at them.

Systems far from equilibrium frequently exhibit dynamic, moving patterns which persist even though the system is constantly changing. Nonequilibrium chemical reactions produce propagating chemical waves, like spiralling whirlpools or ripples radiating from a splash in a pond. Oscillating, periodic behavior in nonequilibrium systems is a common precursor to the onset of complete unpredictability – that is, to chaos. The hallmarks of chaos have now been identified in several chemical reactions.

*Mesoscale chemistry:* Our understanding of chemical processes is now fairly well advanced at both the macroscopic scale – that at which we can see and touch – and the microscopic or molecular scale. But the region in between – the mesoscopic scale, by which typically we mean sizes ranging from thousands of atoms to those of living cells – contains much uncharted territory. Will assemblies of a thousand or so molecules behave like a lump of bulk material or still much like individual molecules? The answer often turns out to be neither: entirely new properties may be observed at these scales.

Our new-found ability to induce self-assembly of molecules into large structures such as artificial membranes or ordered liquid crystalline arrays has opened up this middle ground for investigation. We can also condense atoms from a vapor into clusters of any desired size, from just three or four atoms to many thousands, and thereby follow the way that properties change as the system develops from a molecular object into a piece of bulk solid. This evolution sometimes gets stuck at anomalously stable "magic numbers" of atoms, the reasons for which are still incompletely understood. One example of particular interest is provided by clusters of carbon atoms, which have the ability to arrange themselves into hollow cages of very specific sizes. These carbon cages are providing entirely new directions for research in chemistry, electronics and materials science.

*Energy conversion:* Many chemical reactions produce energy, usually in the form of heat. We have been able to exploit this fact to our benefit ever since mankind tamed fire; but it is not a little remarkable that our principal means of energy generation today continues to involve a chemical process as crude and inefficient as combustion. The more direct conversion of chemical energy into electrical energy is carried out by batteries, but these are not cheap or powerful enough to meet a significant part of the world's demand for power. Nevertheless, new kinds of battery are now being developed that promise to bring novel applications: as power sources in cars or on space satellites, for instance. Extremely small, compact and lightweight batteries provide efficient, safe and convenient energy supplies in all manner of situations that do not require vast output power.

We receive millions of megawatts of energy for free every day by courtesy of the Sun, but have few efficient means of capturing this energy and converting it to more useful forms. Solar cells are chemistry's answer: they employ materials that absorb light and store it away in the form of chemical energy or channel it directly into electricity. Modern solar cells are now taking cues from nature's own version, the photosynthetic reaction centers in plants.

*Sensors:* The ability to detect, quickly and efficiently, the presence or absence of specific chemicals can be a matter of life or death. Leaks of toxic gases, monitoring of glucose or of anesthetics in the bloodstream, testing for harmful compounds in foods – all require reliable and sensitive sensing devices. Many chemical sensors rely on electrochemical principles, whereby the relevant chemical species induces a change in electrical current or voltage at an electrode. Sensors of this kind which display a highly specific response to certain biochemicals are today being developed by exploiting the molecular recognition capabilities of natural enzymes. Polymer science, meanwhile, is able to supply plastic membranes which can be made selectively permeable to one kind of molecule but impenetrable to others.

In some specialized situations, the ultimate in detection sensitivity – detecting single molecules – is now possible. This outperforms even the capabilities of our own primary chemical sensor, the nose's olfactory system. Sensing via spectroscopy – the interaction of molecules with light – conveys the advantage that the substances do not have to encounter the sensing device physically, but rather can be very distant. In this way, chemical compounds can be monitored in the remote atmosphere or in interstellar space and the atmospheres of stars.

*The environment:* Humans have been discharging chemical wastes into the rivers, oceans, soil and air for as long as we have been on the planet. But now that the consequences of these actions are finally coming home to roost, we are being forced to take an unprecedented interest in the chemical composition of our environment. Pollution from Europe shows up in Arctic snow; flue gases from power-generating plants fall back to the ground in the form of acidic rain; gases previously thought too inert to pose a hazard are now causing erosion of the ozone layer. And the product of combusted carbon compounds, carbon dioxide, threatens to turn the planet into a sweltering greenhouse.

The chemical processes responsible for these environmental hazards are now becoming well understood, but their effects on the planet's ecology and climate are harder to predict. There are clues to be had, however, from studying the way in which changes in the atmosphere's chemistry, induced by purely natural processes, have warmed or cooled the planet in the past. Scientists are studying the composition of ancient air trapped in ice bubbles, and of sedimentary rocks deposited long ago on the ocean floor, in attempts to understand the links between atmospheric chemistry and climate change.

Others, meanwhile, trace out the paths by which metals are cycled in the atmosphere and oceans to gain insights into the transport of pollutants. And researchers are laboring to find safer replacements for the substances that are endangering or littering the planet:

alternatives to ozone-destroying CFCs, for instance, or plastics that can be broken down by bacteria.

I have divided this book into three parts. The first four chapters are concerned with some of the traditional aspects of chemical research – structure and bonding, thermodynamics and kinetics, spectroscopy and crystallography (Chapters 1 to 4, respectively). They will illustrate, I hope, that this tradition is a changing one, whereby established tools and concepts are being adapted to meet new aims and challenges. Some themes in science become obsolete and fall by the wayside once they have served their purpose, but in these four areas, at least, new discoveries and technological advances have guaranteed a valuable role for "traditional" approaches for decades to come.

Of the three chapters that follow in Part II, only the theme of Chapter 7 (colloid chemistry) might have meant anything to researchers of the 1950s, and even then its relevance would have borne little similarity to that of today. We will see in these chapters how advances in understanding at the molecular level are leading to entirely new ways of looking at chemical properties and reactions and are helping to bridge the divide between chemistry and disciplines such as molecular biology, electronics and materials science. In short, we will look at some of the new *functions* of chemical research.

In the final part I will discuss some aspects of what I call "chemistry as a process." That is to say, I will be less concerned with chemical change in terms of the products and mechanics of chemical reactions and interactions, and more with the consequences of these processes at a higher level. Life itself is one such consequence, having arisen from chemistry on the early Earth (Chapter 8); complexity of growth and form in the natural world must also evolve somehow from simple chemical processes (Chapter 9); and many of the important changes in our atmosphere, environment and climate (Chapter 10) have their origin in chemical transformations.

In the course of talking about these matters, I have exhausted the space that some chemists might wish to have seen devoted to other topics. Most notably, it is hard to find an excuse for saying relatively little about polymer science and electrochemistry. I can but ask for forbearance here; in the bibliography, however, you will find a few pointers to sources that might serve to plug these holes.

*Part I*

# The Changing Tradition

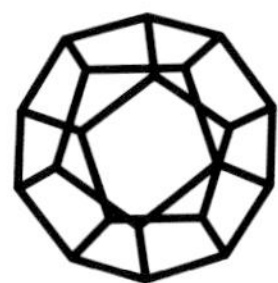

# 1

# How It All Fits Together

## The architecture of molecules

*The domain in which chemical synthesis exercises its creative power is vaster than that of nature herself.*

Marcellin Berthelot

In 1989, chemists working at Harvard University in Massachusetts brewed up a horribly lethal concoction called palytoxin – one of the most poisonous natural chemicals known and the most toxic ever to be synthesized artificially. No sinister motives lie behind this accomplishment, however. The Harvard chemists set their sights on palytoxin simply because building it "from scratch" represented such an extraordinary challenge.

A glance at Figure 1.1 might persuade you of the enormity of the task. This illustration shows the structure of the palytoxin molecule – the balls represent atoms, while the sticks linking them represent chemical bonds. (If you find the concepts of an atom, a molecule and a chemical bond unfamiliar or vague, don't despair; all will be explained shortly. But even without this understanding, you can appreciate that putting together such a complex object is no mean undertaking.) There is no important use for palytoxin. Its synthesis was a little like those recitals from memory of the Holy Bible or of the number pi to a million decimal places: a pure demonstration of technical prowess. Yet by tackling difficult tasks such as this, chemists are likely to discover new ways to solve problems that crop up in the synthesis of the complicated molecules needed by industries and by medical science.

Building molecules is big business, and understandably so. For although nature has provided us with a tremendous selection of substances with which to construct

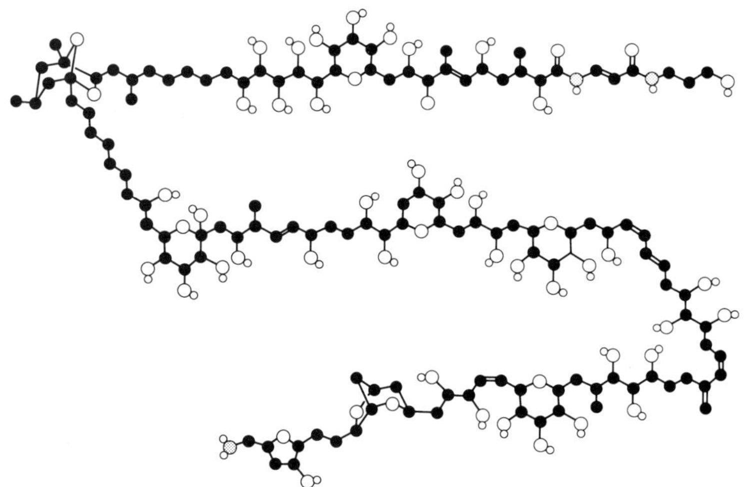

**Figure 1.1** *The molecular structure of palytoxin, simultaneously one of the most complicated and the most toxic compounds ever synthesized. The black circles represent carbon atoms, the large white circles oxygen, the gray circles nitrogen and the small white circles hydrogen. For clarity, hydrogen atoms attached to carbon are not shown.*

civilizations and improve the lives of its citizens, their range and abundance appears insufficient to meet our every need. The great variety of complex substances found in the living world, particularly in plants, has proved to be immensely valuable to physicians throughout the ages, but there are ailments for which natural cures are rare, ineffective or nonexistent. A great many chemists are therefore engaged in the enterprise of creating purely artificial substances that provide cheaper or more potent alternatives, or which can fill the gaps. The pharmaceutical industry is just one of the spheres in which artificial or synthetic substances are called for, but it is probably the example *par excellence* because the substances that it requires are often extremely sophisticated and accordingly hard to make.

In later chapters we will encounter some of the simpler synthetic molecules created via the techniques of modern chemistry. In general these are constructed from smaller molecules which are joined together or rearranged in chemical reactions. I don't propose to look at these techniques of synthesis in any detail – they are often ingenious, but in all honesty they don't hold much intrinsic interest for the nonchemist. There is more fun to be had, I feel, from looking at the behavior and the properties of the molecules that come out in the end. In this chapter, nevertheless, I do intend to take a close look at the synthesis of one particular molecule, which rejoices in the baroque name of buckminsterfullerene. Not only is this substance remarkable for all sorts of reasons, but the story of how it was identified and created is also well worth recounting. It shows how important scientific advances can come about in unexpected ways, and gives one of the best illustrations of why the often mundane task of molecule-building can occasionally inspire in its practitioners the most feverish excitement. The buckminsterfullerene story shows chemical research at its most colorful.

You will forgive me, I hope, for saving the story for the chapter's climax. In order to understand it better, we will first need to know a little more about this business of molecule-building. Not the least of the pertinent questions is that of what a molecule actually is. What are chemists trying to convey when they draw a picture like Figure 1.1? What in reality are those balls and sticks?

## The stuff of the Universe

### *Why the world really is an illusion*

It has become popular in recent years to draw analogies between modern physics and Eastern philosophies such as Taoism and Buddhism. Although this is a little like comparing two books because their covers are the same color, there *is* a sense in which modern science seems to suggest, like Taoism, that the physical world is but an illusion: it states that, very much contrary to appearances, the most solid of objects is nearly all empty space. If we compressed the Earth, for instance, down to a size in which all this empty space was eliminated, it would fit quite comfortably inside a (presumably extraterrestrial) football stadium. In fact, physicists are now having to ask themselves just how much of that football-field-sized lump of matter may also be empty space, but by that stage what we mean by "space" and "matter" is getting a little unclear.

Surely this qualifies as a remarkable illusion! We are sitting or standing on almost nothing but empty space. We *are* little more than empty space. Yet this book feels solid enough, and the mostly-empty-space of our fingers does not penetrate the mostly-empty-space of the pages. In this, as in many other ways, modern physics seems to be at odds with our everyday intuition. As I suggested in the Introduction, it is chemistry that acts as the go-between. At one end of the scale, chemistry can accept and utilize the description of the world provided by fundamental physics; while at the other, it gives us a very rational and self-consistent description of the way that we perceive matter to behave.

The crucial link-up between these two worlds is made at the level of the atom. For the most part, chemistry treats atoms as if they were tiny yet solid balls of matter which stick together in various arrangements to form the substances of which the everyday world is composed. The phenomena that we experience, be they the glowing of a candle's flame, the growth of a crystal, the browning of toast under the grill or the development of a human being from a single cell, can be described largely in terms of rearrangements in the patterns of bonding between these billiard-ball atoms.

But why, if they are mostly empty space, can chemists regard atoms as though they were as solid as billiard balls (that is, as solid as billiard balls *appear* to be)? What is an atom really like?

## *Order amongst the elements*

The Greek philosophers assumed that all matter was composed of just a few different components mixed in varying proportions. These basic ingredients of matter, called elements, were thought to be fourfold: earth, air, fire and water. (Aristotle posited a fifth element, the aether, as a component of the heavenly bodies, while Chinese alchemists proposed a fivefold group of elements: earth, fire, water, wood and metal.)

By the seventeenth century, natural philosophers had come to recognize that, while many substances could indeed be broken down into apparently more fundamental ones, the four-element picture was inadequate. Not only were the basic, irreducible substances very different from earth, air, fire and water, but there were certainly more than four. Many of the elements turned out to be metals, such as copper, iron, tin and lead. Several others were gases, including hydrogen, nitrogen and oxygen. A few were nonmetallic solids, like carbon (which was found in two elemental forms, diamond and graphite) and silicon. Substances that contain more than one element were named compounds.

Chemists have a shorthand notation for the elements, in which each is represented by a one- or two-letter symbol. Most of these are easy to decipher – hydrogen is represented by H, for instance, oxygen by O, nitrogen by N, nickel by Ni and aluminum by Al. Some are more cryptic, since they originate from a time when the elements were called by different names. Iron, for instance, is denoted by Fe, from its Latin form *ferrum*.

In the nineteenth century the French chemist Joseph Louis Proust and the Englishman John Dalton showed that the ratios of elements in a compound remained the same regardless of how the compounds were prepared. Proust enshrined this observation in a general rule which he called the law of definite proportions. The law can be rationalized by asserting that a compound consists of discrete atoms linked into clusters, called molecules, each of which contains a fixed number of atoms of each element. The idea that matter is composed of indivisible units was first posited by the Greek philosopher Leucippus in the fifth century BC; his student Democritus called these fragments *atomos*, meaning "unbreakable." But only thanks to Proust and Dalton was the atomistic hypothesis truly scientific, in the sense of helping logically to rationalize observed phenomena rather than comprising an *a priori* axiom.

The distinction between elements, atoms and molecules is an important one to get straight. If I talk, for example, about the element oxygen, atoms of oxygen and molecules of oxygen, I mean something different in each case. By an "element" I mean simply the substance, without any reference to an atomistic model; an atom is the smallest indivisible unit of an element; and a molecule is a cluster of atoms joined by chemical bonds.

It is rare to find, under normal conditions (which is to say, at temperatures in the region of room temperature), atoms on their own: usually they will be linked together with others in molecules with a well-defined composition, such as those of water (where one oxygen atom is linked to two hydrogen atoms) or in the oxygen and nitrogen gases which are the principal components of air (where the individual oxygen and nitrogen atoms are joined in pairs) (Figure 1.2*a*). Chemists present the composition

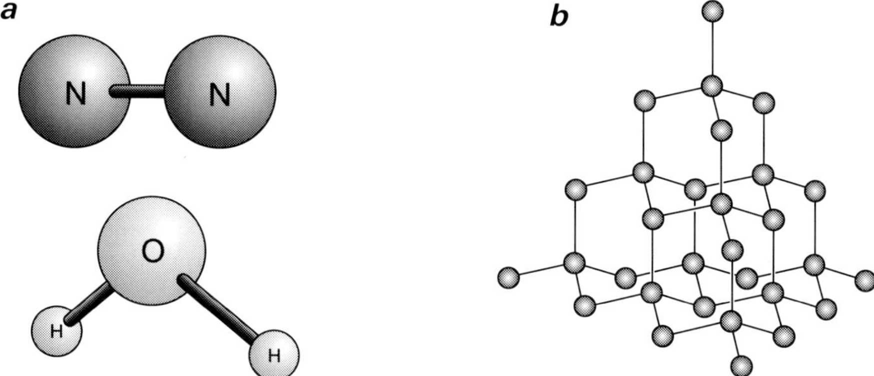


**Figure 1.2** *Molecules of nitrogen ($N_2$) and water ($H_2O$) (a) and the structure of diamond (b), in which carbon atoms are linked in a continuous crystalline framework.*

of molecules as a "chemical formula" which lists the atoms it contains (in their abbreviated forms). Subscripts denote the number of atoms of each respective element, so that the water molecule is $H_2O$ and the nitrogen molecule is $N_2$.

In some substances the constituent atoms are not joined into small molecules but are instead linked or stacked together in vast, continuous networks. This is the case in solids such as diamond (Figure 1.2*b*) or metals. There is no reason in principle why we could not consider an atomic network such as diamond to be a single, huge molecule, but it is not generally very instructive to do so. So when I use the term molecule, I will usually be referring to a discrete assembly of atoms of microscopic size, typically containing an easily countable number of atoms. I should mention, however, that we will encounter some molecules that are approaching a middle ground, consisting of perhaps several thousand or even several million atoms.

By the mid-nineteenth century, dozens of different elements had been identified. On the basis of the atomistic model, it was possible to assign each of these elements an atomic weight, which was defined relative to the weight of a hydrogen atom. The actual weight of an atom was a minute quantity and far from easy to measure; but the relative weights of elements were more easily determined. The Italian chemist Amedeo Avogadro suggested in 1811 that equal volumes of two gases at the same temperature and pressure contained equal numbers of atoms (or more precisely, of molecules). The atomic weight of oxygen was therefore the ratio of weights of equal volumes of oxygen and hydrogen gas (this comes out at a value of almost exactly 16).

It was also clear that certain groups of elements had similar chemical properties. The metals sodium, potassium, rubidium and cesium, for example, all react vigorously with water to liberate hydrogen gas. Fluorine and chlorine are both corrosive gases, while helium, neon and argon are all highly inert. The Russian chemist Dmitri Ivanovich

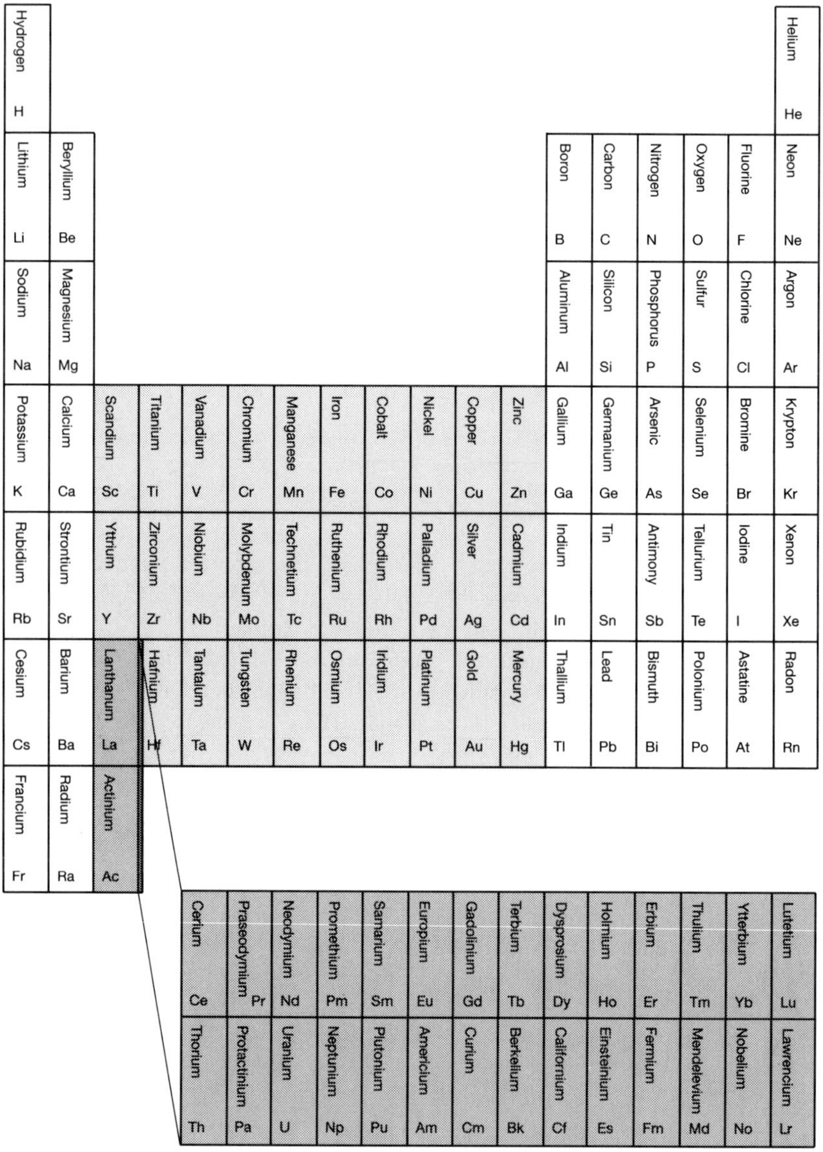

| Hydrogen H | | | | | | | | | | | | | | | | | Helium He |
|---|---|---|---|---|---|---|---|---|---|---|---|---|---|---|---|---|---|
| Lithium Li | Beryllium Be | | | | | | | | | | | Boron B | Carbon C | Nitrogen N | Oxygen O | Fluorine F | Neon Ne |
| Sodium Na | Magnesium Mg | | | | | | | | | | | Aluminum Al | Silicon Si | Phosphorus P | Sulfur S | Chlorine Cl | Argon Ar |
| Potassium K | Calcium Ca | Scandium Sc | Titanium Ti | Vanadium V | Chromium Cr | Manganese Mn | Iron Fe | Cobalt Co | Nickel Ni | Copper Cu | Zinc Zn | Gallium Ga | Germanium Ge | Arsenic As | Selenium Se | Bromine Br | Krypton Kr |
| Rubidium Rb | Strontium Sr | Yttrium Y | Zirconium Zr | Niobium Nb | Molybdenum Mo | Technetium Tc | Ruthenium Ru | Rhodium Rh | Palladium Pd | Silver Ag | Cadmium Cd | Indium In | Tin Sn | Antimony Sb | Tellurium Te | Iodine I | Xenon Xe |
| Cesium Cs | Barium Ba | Lanthanum La | Hafnium Hf | Tantalum Ta | Tungsten W | Rhenium Re | Osmium Os | Iridium Ir | Platinum Pt | Gold Au | Mercury Hg | Thallium Tl | Lead Pb | Bismuth Bi | Polonium Po | Astatine At | Radon Rn |
| Francium Fr | Radium Ra | Actinium Ac | | | | | | | | | | | | | | | |

| Cerium Ce | Praseodymium Pr | Neodymium Nd | Promethium Pm | Samarium Sm | Europium Eu | Gadolinium Gd | Terbium Tb | Dysprosium Dy | Holmium Ho | Erbium Er | Thulium Tm | Ytterbium Yb | Lutetium Lu |
|---|---|---|---|---|---|---|---|---|---|---|---|---|---|
| Thorium Th | Protactinium Pa | Uranium U | Neptunium Np | Plutonium Pu | Americium Am | Curium Cm | Berkelium Bk | Californium Cf | Einsteinium Es | Fermium Fm | Mendelevium Md | Nobelium No | Lawrencium Lr |

**Figure 1.3** *The Periodic Table of the elements, first formulated by Dmitri Mendeleev in 1869, brings some coherence to the profusion of natural elements. The atomic number (the number of protons) of adjacent elements increases by one from left to right. Those elements that appear in the same vertical column tend to have similar chemical properties. The elements in the central light gray area are the transition metals. Between lanthanum and hafnium, and beyond actinium, lie series of elements called the lanthanides and actinides respectively; these series are shown separately below the main table. Several unstable, radioactive elements have been created artificially that lie beyond lawrencium.*

Mendeleev showed that, when the elements then known were listed in order of increasing atomic weight, certain chemical properties cropped up at regular intervals. By chopping the list into rows and placing them one below another, one could obtain a table of elements in which these periodic similarities recurred down each column. Mendeleev presented his Periodic Table in 1869 as a speculative way of classifying the elements. He had no explanation, however, for why there should be these regularities. Moreover, in order to make the pattern work, Mendeleev had to leave some gaps where he assumed that an element was missing which had yet to be discovered. As chemists began to find in the ensuing decades that newly discovered elements slotted neatly into the predicted places in Mendeleev's table, they concluded that it was telling them something rather fundamental about the nature of atoms. But a lot more needed to be known about atomic structure before the patterns in the Periodic Table could be explained. Today the table has no more gaps (Figure 1.3), although physicists occasionally manage to add an extra very heavy and unstable element on to the end of the list.

## *Atomic anatomy*

The innermost secrets of atomic structure began to be unraveled in the early twentieth century. On observing that alpha particles (produced in the radioactive decay of radon gas) would mostly pass straight through gold foil without apparently encountering any obstacles, the physicist Ernest Rutherford proposed in 1916 that atoms are mostly empty space. He suggested that nearly all of their mass is concentrated in a tiny central nucleus, which bears a positive charge. (In rare events, alpha particles would encounter these dense nuclei in Rutherford's gold foil, and bounce straight back in the direction from which they had come.) In Rutherford's model, negatively charged particles called electrons circulate in orbit around the nucleus (Figure 1.4). Nuclei were shown soon after to contain two types of particle: protons and neutrons. Protons bear a positive charge of equal magnitude to the electron's negative charge, while neutrons are electrically neutral. A proton is, however, 1,837 times as heavy as an electron, and the neutron has a mass almost identical to the proton's.

The size of an atom is determined by the radii of the electrons' orbits, which are typically about 100,000 times that of the nucleus. It is tempting to explain why matter does not simply collapse in on itself on the basis that the electrostatic repulsion between electrons on different atoms (that is, the repulsion of electrical charges of the same sign) prevents them from overlapping. The real explanation, however, is more subtle, and regrettably I don't have space to go into it. Suffice it to say that the repulsion between electrons orbiting different atoms gives the atoms their apparent billiard-ball character.

A neutral atom has exactly the same number of electrons and protons; this number is called simply the atomic number. The atoms of different elements have different atomic numbers, and adjacent elements in Mendeleev's Periodic Table differ in atomic number by one. Carbon atoms, for instance, have six electrons and six protons; nitrogen atoms have seven of each, and oxygen atoms eight. Lead atoms have a grand total of 82. The

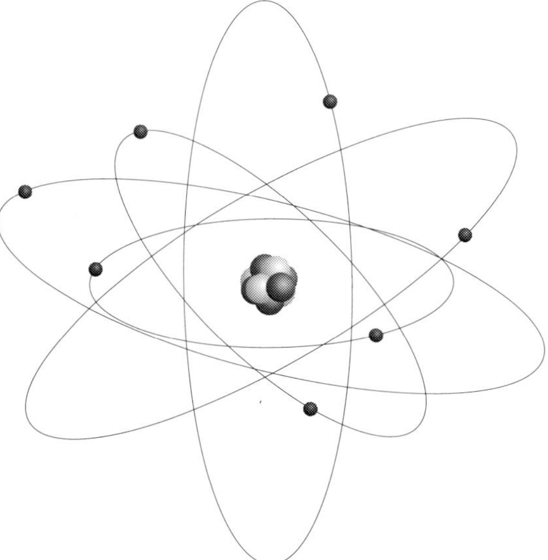

**Figure 1.4** *Ernest Rutherford proposed that atoms consist of a tiny, dense, positively charged nucleus orbited by negatively charged electrons.*

atomic number says nothing about the number of neutrons in each atom, however. For small atoms, the neutron count is roughly the same as the proton count – most carbon atoms have six neutrons, for example, and most nitrogens have seven – while heavier atoms tend to have a considerable excess of neutrons over protons. Most atoms of lead have 82 protons and 128 neutrons.

But I emphasize "most" here because the number of neutrons may vary in atoms of any given element. Some carbon atoms have seven neutrons, for instance, and some even eight. They remain atoms of carbon nonetheless, because the atomic number is the same. Atoms of an element that differ in their number of neutrons (and therefore in their overall mass, or "atomic mass") are called isotopes. The isotopes of hydrogen, popularly called "heavy hydrogen," are deuterium, which has a single neutron in the nucleus as well as a proton, and tritium, which has two neutrons and a proton.

## *The quantum atom*

It would be churlish to disparage Rutherford's "solar-system" model of the atom – it gives an idea of the relationship between the different subatomic components, and it also gives an intimation of how an atom can be mostly empty space. But it should not be taken too literally, because objects this small simply do not behave in the same way as objects the size of the Earth, or even the size of a billiard ball. This is perhaps the central message of quantum mechanics, the theory developed to describe objects at these microscopic scales.

Around the beginning of the twentieth century – even before Rutherford put forward his nuclear model of the atom – physicists began coming across unnerving intimations that there was something very wrong with their "classical" view of the world:

specifically, it appeared sometimes to make incorrect or even nonsensical predictions! The classical theory of electromagnetism formulated in the late nineteenth century by the Scotsman James Clerk Maxwell unified in a beautiful way a great deal of physical science, but unfortunately it also indicated that a hot body should radiate an infinite amount of heat, which was obviously absurd. And existing theories suggested that the speed of electrons kicked out of metals by shining light on them (a phenomenon known as the photoelectric effect) should depend on the intensity of light but not its color, whereas the opposite was found to be the case.

In 1902 the German physicist Max Planck set the stage for the new *Weltanschauung* of quantum theory by hypothesizing that a hot body radiates energy only in discrete packets, called quanta, each of which contains an amount of energy that depends on the wavelength of the radiation. Planck had no particular grounds for making this suggestion, other than the fact that it gave predictions that agreed with experiments. But in 1905 Albert Einstein showed that the same idea can explain the photoelectric effect, suggesting that energy quantization was not merely a mathematical trick but a feature of the real world.

The idea that energy is quantized – that it is transferred in discrete packets – was adopted by Niels Bohr in 1913 to explain the problematic fact that Rutherford's model of the atom contravened the laws of physics as they were then understood. According to the classical viewpoint, an electron orbiting a nucleus should radiate energy constantly until the orbit can no longer be sustained and the electron spirals into the nucleus. The atom would, in other words, be unstable. Bohr suggested that the electrons were restricted to specific orbits, each at a well-defined distance from the nucleus. This implied that the electrons' energies were quantized, and that while it stayed in given orbit an electron's energy remained at a fixed value. Electrons did not radiate energy continuously and spiral into the nucleus because they could increase or decrease their energy only in lumps of specific magnitude.

In Bohr's model of the atom, the allowed energy states of the electrons are like the rungs of a ladder; the spaces in between rungs represent forbidden energies (Figure 1.5).

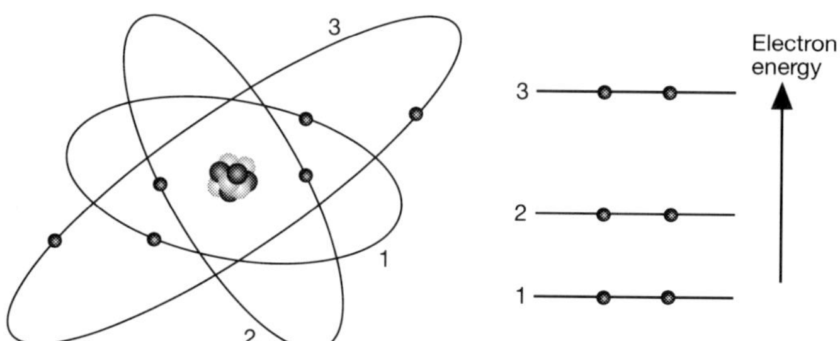


**Figure 1.5** *In Niels Bohr's model of the atom, electrons are allowed only specific, discrete energies. They can jump from one rung of the energy ladder to the next, by absorbing or emitting light, but they cannot possess energies in between the rungs.*

Each of these electron "energy levels" corresponds to a certain orbit around the nucleus, so that electrons in one energy level will follow one kind of trajectory whereas those in a different level will follow another. In the Bohr atom these trajectories were still pictured as circular orbits (although the German physicist Arnold Sommerfeld later introduced the possibility of elliptical orbits). But Werner Heisenberg demonstrated in the 1920s that quantum theory would not permit a description of such tiny particles as hard, well-defined objects following line-like trajectories. Rather, at these scales particles get smeared out in a way that prevents one from knowing, at any given time, both their exact position and their exact velocity (or more strictly, their momentum, which is just their velocity multiplied by their mass). In principle we can measure one property or the other as accurately as we like; but as this accuracy increases, the uncertainty with which we can know the value of the other property inevitably increases also. This is one way of expressing Heisenberg's famous Uncertainty Principle, which supplies one of the cornerstones of quantum mechanics. Because of this quantum-mechanical "smearing," one cannot talk about the precise positions of quantum particles but only about the *probability* of finding them at a given point in space.

As a consequence of the Uncertainty Principle, it is more appropriate to think of the electron orbits as corresponding to clouds of smeared-out charge surrounding the nucleus. Where the clouds are dense, there is a relatively high probability of finding the electron. To avoid misleading classical connotations, chemists call these clouds "orbitals" rather than orbits. For the first two energy levels of any atom, the orbitals are spherical – the electrons are more or less localized within a spherical region centered on the nucleus (Figure 1.6). These orbitals, called s orbitals, are therefore not so far removed from the picture of circular orbits, except for the peculiar fact that the electrons do not circulate around the nucleus at all but are instead moving along straight-line trajectories that pass right through it! But the orbitals of the third energy level, which are called p orbitals, are shaped rather like a dumbbell. To simplify again somewhat (although not dangerously), the electrons in these orbitals can be thought of as performing figures of eight through the nucleus. Some of the orbitals with greater energies have shapes that are merely larger version of these two; but others may have still more complicated shapes.

The electronic orbitals come in families or "shells," rather like those of a Russian doll. The first shell contains just one s orbital, denoted the 1s. The second shell has one spherical orbital (the 2s) and three p orbitals (the 2p's), which are arranged at right-angles to each other (Figure 1.6). The third shell has one s orbital (the 3s), three p orbitals (3p) and a group of five d orbitals (the 3d's). The pattern is that each successive shell comprises all the orbital types of the previous shell (but of greater size), plus a new group of orbitals that the previous shell does not possess. The energy of an electron depends both on the shell in which its orbital lies (successive shells correspond to successively higher energies) and, in general, on the nature of the orbital – that is, whether it is an s orbital, a p orbital, a d orbital and so on.

Another important tenet of quantum mechanics, called the Pauli exclusion principle after the Austrian-Swiss physicist Wolfgang Pauli, dictates that each orbital can

s = Sharp
p = principal
d = diffuse
f = fundamental

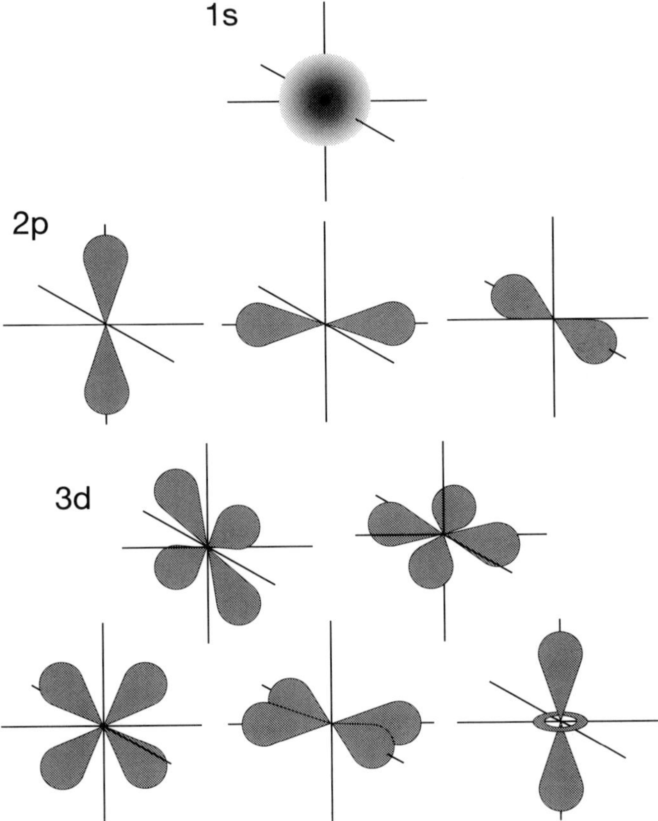


**Figure 1.6** *The smeared-out electron orbitals in the quantum atom are very different from the well-defined orbits of Rutherford's "classical" atom. The shaded regions are those in which there is the greatest probability of finding the electrons. The two orbitals of lowest energy (1s and 2s) are spherically symmetric, but the 2p orbitals have a dumbbell shape, and in the third electron shell are orbitals of double-dumbbell and ring-and-dumbbell shapes (3d).*

accommodate just two electrons. Taken together with the shell structure of the orbitals, this provides an explanation for the characteristics of the Periodic Table. The chemical behavior of an atom is determined largely by its outermost layer of electrons. These are sometimes, but not always, those in the orbitals of the outermost shell (the exceptions are the result of orbitals from lower shells "poking through," so that they too must be considered part of the outer layer). An atom's electrons can be considered to "fill up" the orbitals, two to an orbital, from the lowest in energy upwards. So the single electron of the hydrogen atom goes into the 1s orbital, and in helium this electron is joined by another (Figure 1.7). The next element in the table, lithium, has three electrons; two fill the 1s orbital, and the third must go into the second shell. But here it has two choices

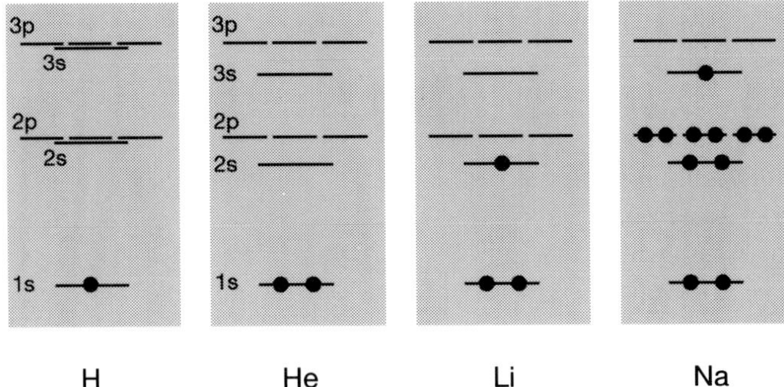


**Figure 1.7** *Electrons fill atomic orbitals two at a time, from those of lowest energy upwards. Thus hydrogen has a single 1s electron, helium has two 1s electrons, lithium has two 1s and one 2s, and so on. For hydrogen the energy of the electron levels depends only on the "shell number": the 2s and 2p orbitals lie at the same energy, as do the 3s, 3p and 3d. But for all other atoms this is no longer true. Hydrogen, lithium and sodium all have an outer electron shell containing a single s electron.*

– to go into the 2s or the 2p. For a hydrogen atom, these two orbitals have the same energy, but for any atom containing more than one electron this is no longer true: the 2s is lower in energy than the 2p. So the third electron in lithium goes into the 2s orbital (Figure 1.7). We then go on filling the second shell as the atomic number increases; carbon, for instance, has two electrons in the 1s orbital, two in the 2s, and two in 2p orbitals. By the time we get to neon, with eight electrons, the second shell is full – there are two electrons in the 2s orbital and six in the three 2p's. So for sodium, with nine electrons, the ninth must go into the third shell (that is, into the 3s). This means that both sodium and lithium have an outer shell consisting of one electron in an s orbital (2s for lithium, 3s for sodium) (Figure 1.7), with the result that they tend to react chemically in similar ways. Likewise the outer shell of chlorine is like a bigger version of fluorine's: both have two s electrons and five p electrons (which are in the second shell for fluorine and the third for chlorine). Bromine's outer shell is a yet bigger version of these two.

A new row in the Periodic Table begins each time a new shell starts to be filled. The sudden appearance of a new set of columns after calcium, corresponding to the so-called transition metals (Figure 1.3), signals the point at which the d orbitals begin to be filled. Similar excursions occur after the elements lanthanum and actinium (these are listed outside the main Table to prevent it from getting unmanageably wide), as a result of the filling of yet another new type of orbital in each case.

## Atomcraft – the structure of molecules

### *How the glue sticks*

Atoms are linked together in molecules by bonds, which were represented simplistically by sticks in Figure 1.1. Bonds between atoms are formed as a result of their sharing or redistributing electrons. The most usual arrangement is an egalitarian one: each atom donates one of its electrons to the union. Each of the electrons in this bond pair is then no longer restricted to an orbital around its own atom's nucleus, but can move around the nucleus of the other atom too. So the bonding electrons form a smeared-out cloud, or molecular orbital, around both nuclei. Bonds formed by the sharing of electrons between two or more nuclei are called covalent bonds.

When two hydrogen atoms combine to form the $H_2$ molecule, for instance, the spherical 1s orbitals on each atom combine to form a kind of rugby-ball-shaped molecular orbital (Figure 1.8). A rigorous, quantum-mechanical description of the bonding in the $H_2$ molecule, however, reveals something not intuitively obvious: the total number of orbitals must be "conserved." That is, the total number of molecular orbitals must be the same as the number of atomic orbitals that went into their formation. As two atomic orbitals go into forming the bond in $H_2$, two molecular orbitals must result. One is the bonding orbital in which the electron pair resides. The energy of the electrons in this orbital is lowered relative to that in the atomic orbitals, which is precisely why the atoms stay bound together instead of just drifting apart again – to separate them we have to supply the energy that the electrons have lost in forming the bond.

Where or what is the other orbital? Well, it exists sure enough, but we don't "see" it because it is empty: it contains no electrons. It is a kind of "potential" orbital, rather like a bank account that contains no money. An electron in this orbital would have a higher energy than those in the atomic orbitals (Figure 1.8). Putting electrons into this orbital will weaken the bond, since the total energy is then greater than that when electrons occupy just the molecular bonding orbital. The orbital is therefore called an antibonding orbital. If we force another electron on the $H_2$ molecule to make the ion $H_2^-$, the extra electron has to go into the antibonding orbital

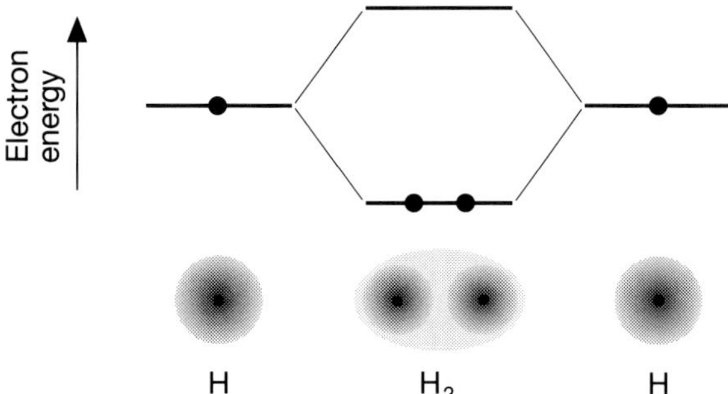


**Figure 1.8** *A covalent bond is formed when atoms share electrons. In effect, the shared electrons can orbit around both nuclei in "molecular" orbitals; more precisely, there is a significant probability of finding these electrons around either of the nuclei. In the hydrogen molecule ($H_2$) the two 1s orbitals on the individual atoms overlap to form an elongated molecular orbital encompassing both nuclei, with an energy lower than that of the constituent atomic orbitals. At the same time, an empty "antibonding" molecular orbital is created at a higher energy. The lowering of the electrons' total energy in forming the molecular orbitals is what holds the molecule together.*

because the bonding orbital is already full (like atomic orbitals, it can accommodate only two electrons). The bond between the two hydrogens in $H_2^-$ is therefore weaker than that in $H_2$. Electrons can be boosted up to these higher-energy empty orbitals by the absorption of light, a phenomenon discussed in Chapter 3.

Atoms can also become strongly bound together in another manner, which does not involve electron sharing but rather the *exchange* of electrons. That is to say, one atom donates an electron fully to another atom, such that this electron leaves the atomic orbital of the "donor" and is captured in an atomic orbital of the "acceptor." By losing a negatively charged electron, the donor atom acquires a net positive electric charge, while the acceptor atom now has one more electron than protons, and so is negatively charged. These electrically charged atoms are called ions, and by virtue of their opposite charges they are attracted to one another. So-called ionic bonding is prevalent in the compounds formed between metals and the nonmetallic elements at the extreme right of the Periodic Table. Rock salt, for instance, (the primary constituent of table salt) is an ionic compound of sodium and chlorine, in which each sodium atom can be considered to have given up an electron to each chlorine, creating positive sodium ions ($Na^+$) and negative chloride ions ($Cl^-$). Ionic compounds tend to be solids with relatively high melting points.

The number of electrons available to an atom for the purpose of forming bonds (either covalently or ionically) is usually only a small proportion of its total complement of electrons, in general comprising just those in the outer shell (or occasionally those in the outer two). Electrons in "deeper" shells are held too tightly to play an immediate role in chemical interactions. Thus each element tends to exhibit a well-defined bond-forming capacity which depends on the arrangement of its outermost electrons. Carbon, for example, has four electrons in its outer (second) shell, and so forms four bonds in most compounds. The number of covalent bonds that an atom will form does not, however, always correspond simply to the number of outer-shell electrons that it possesses. Nitrogen, for example, has five outer-shell electrons, and oxygen six; yet the former typically forms just three bonds, and the latter only two. In these atoms, pairs of electrons can be accommodated in orbitals that do not take part in bonding. These electron pairs, called lone pairs, reside in lobe-like orbitals that protrude from the atom and play an important part in determining the shape of nitrogen- and oxygen-containing molecules (Figure 1.9*a*). Lone pairs *can* become involved in bonding under some circumstances: they will allow their host atom to form an "extra" bond with a positive ion which has no free electrons of its own to contribute to the union. In this kind of bond, called a dative bond, both members of the bonding pair are therefore donated by just one atom. Nitrogen and oxygen atoms can often form dative bonds to hydrogen ions (which are just bare protons), forming positively charged "protonated" species (Figure 1.9*b*).

Despite the aforementioned propensity of carbon atoms to form four covalent bonds, in the ethylene molecule ($C_2H_4$) each of the two carbons is attached to just *three* atoms. The apparent deficiency in bonds to carbon is avoided, however, by the fact that the two carbon atoms are bound together by two bonds, or what is more properly

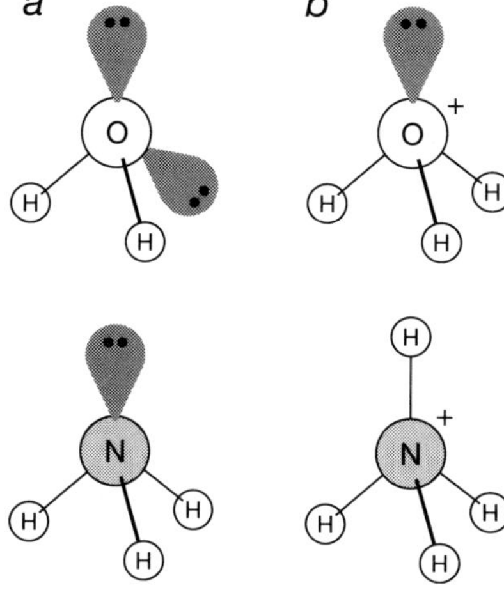


**Figure 1.9** *A pair of electrons in an outer-shell orbital that is not involved in covalent bonding represents a lone pair. The oxygen atom in water possesses two lone pairs, while the nitrogen atom in ammonia has just one pair (a). Lone pairs can become involved in bonding with positively charged ions, such as the hydrogen ion, which cannot themselves contribute electrons to the bond. When water or ammonia molecules form an additional bond to the $H^+$ ion, the result is a ''protonated'' molecular ion (b).*

known as a double bond. These bonds can be considered to have two distinct components: a regular "single" bond (known as a sigma bond) between the atoms in which the electron cloud is densest at the midpoint between the nuclei, and a so-called "pi bond," consisting of two separate, sausage-like electron clouds lying above and below the nuclei (Figure 1.10). Pi bonds are the results of overlap of dumbbell-like p orbitals lying adjacent to each other. One consequence of this structure is that the orbitals of the pi bond act like struts to stop the atoms at each end from rotating: the molecule stays more or less flat.

As one might expect, double bonds are stronger than single ones: to break the carbon–carbon link in ethylene requires more energy than is needed to do the same in ethane ($C_2H_6$), in which two $CH_3$ groups are linked by a single bond between carbons. The double bond is, however, considerably less than twice as strong; breaking open the pi component of the double bond is easier than breaking a single bond. For this reason, ethylene reacts with other compounds more readily than does ethane. Carbon compounds that contain pi bonds are said to be unsaturated, meaning that the carbon atoms, while having formed the requisite number of bonds, are not fully saturated in terms of their number of potential neighbors. Saturated carbon molecules, on the other hand, contain only single bonds. The polyunsaturates in oily foods are long, chain-like carbon-based molecules containing many double bonds between carbon atoms. "Saturating" the double bonds by adding hydrogen atoms to them (hydrogenation) yields polysaturates, which have higher melting points. Waxy polysaturated margarines are produced from liquid, unsaturated vegetable oils via this process of hydrogenation.

Double bonds by no means represent the greatest length to which atoms will go to ensure that they get their full complement of bonds: triple bonds are also possible. One such is found in acetylene ($C_2H_2$), in which the two carbons are bound to each other and to just one hydrogen apiece. A triple bond consists of a sigma bond plus two pi

## Drawing molecules

Chemists have several schemes for representing molecular structures. In general these are based on the principle of connecting symbols representing atoms of different elements via lines or sticks that represent chemical bonds. A single line corresponds to a single (sigma) bond, a double line to a double bond and a triple line to a triple bond:

$$H_2O\text{:}\quad H\text{—}O\text{—}H \qquad O_2\text{:}\quad O{=}O \qquad N_2\text{:}\quad N{\equiv}N$$

Sometimes it is convenient or useful for the positions of the atoms in the illustration to reflect their positions in three-dimensional space; but often this information is not essential. For example, although the three-dimensional structure of the ethane molecule looks like this:

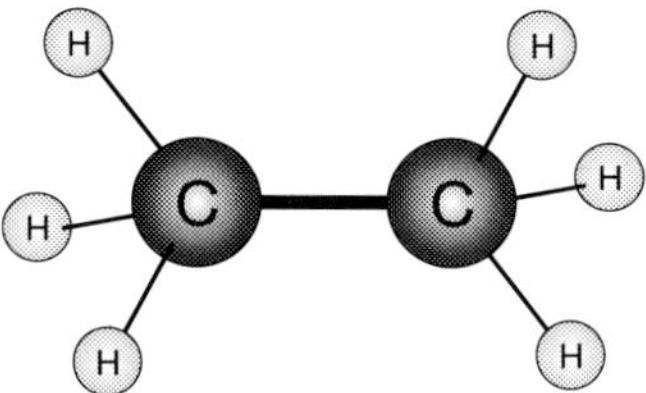


it is generally sufficient to ignore the tetrahedral arrangement of bonds around each carbon atom, and to depict the structure simply as:

```
    H  H
    |  |
H—C—C—H
    |  |
    H  H
```

Because so many molecules are built around a framework of carbon atoms, chemists often use a shorthand notation in which the carbon framework is depicted merely as lines, without the atoms being shown explicitly. The carbon atoms are understood to sit at the kinked vertices of the framework. Because they are very unreactive, the hydrogen atoms attached to carbon atoms are generally an unimportant part of the structure, and so within this shorthand scheme, these hydrogen atoms are omitted entirely for clarity. (Hydrogens attached to oxygen or nitrogen, on the other hand, often play an important structural or chemical role, so they *are* shown.) Thus, within this scheme, cyclohexane (Figure 1.12) and benzene (Figure 1.13) look like this:

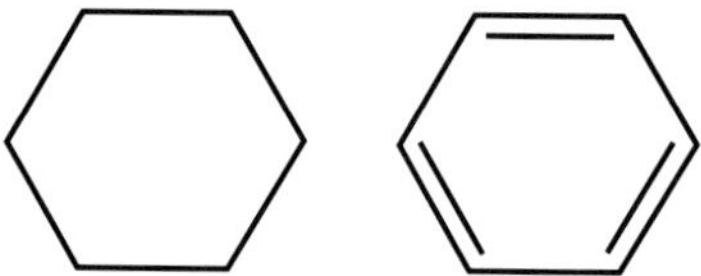

Although in reality the cyclohexane ring is puckered, this three-dimensional shape is ignored here.

While ball-and-stick or framework representations of molecules are very good for showing the way in which the atoms are connected, they bear as much relation to the *real* shape of molecules as a stick man does to a human being. Chemists are often more interested in the true, three-dimensional size and shape of molecules, so that they can understand the spatial constraints on the way that molecules interact. This is particularly important, for example, in determining how molecules might stack together with another in crystals. For these purposes, chemists use "space-filling models", in which the building blocks are designed to reflect the effective sizes of the constituent atoms. (As we have seen, atoms don't really have sharp, well-defined edges, but all the same it is possible to ascribe to them an effective radius based on the distance to which other atoms can easily approach.) In space-filling models the atoms are no longer complete spheres, since in a molecule their electron orbitals overlap. A space-filling model of benzene looks something like this:

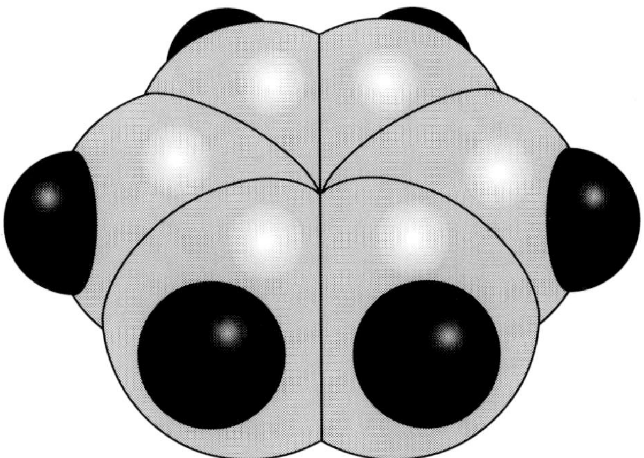

Here the large gray segments are carbon atoms, and the small black hemispheres are hydrogen.

In this book I will generally show each of the atoms in a molecule as balls. The exception is that, for clarity, I will omit hydrogen atoms attached to carbons in those molecules for which the structures are particularly complicated. When the hydrogens are omitted, I shall indicate as much. If the scale permits, I shall identify the atoms by their chemical symbols, but in large molecules I shall use the code employed in Figure 1.1:

| | | | |
|---|---|---|---|
| ● | Carbon | ○ | Oxygen |
| ◍ | Nitrogen | ○ | Hydrogen |

When it would be redundant or confusing to show the shape of molecules in atomic detail, I will employ more schematic ways of representing their structure – for example, by using the linear carbon-framework scheme or by depicting ring-shaped molecules as a featureless band or rod-like molecules as cylinders.

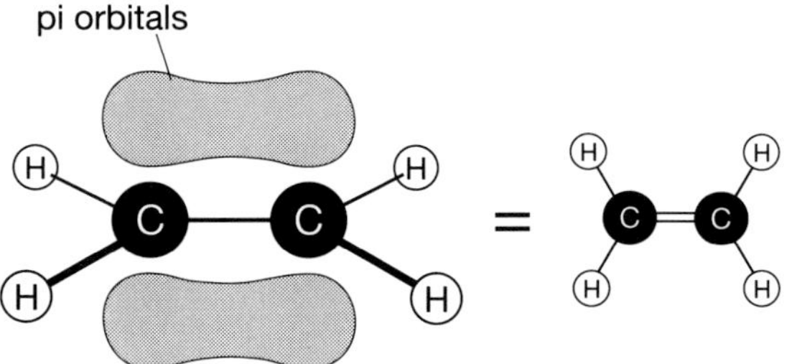


**Figure 1.10** *In the ethylene molecule, two carbon atoms are joined by a double bond. One component of the bond is formed from atomic orbitals on carbon which overlap in the region between the two nuclei; this part is called a sigma bond. The other component of the double bond is created by overlap of the two lobes of dumbbell-shaped 2p orbitals on carbon above and below the plane of the molecule, forming two sausage-like electron clouds. This is called a pi bond. In ball-and-stick diagrams, a double bond is represented by two "sticks."*

bonds, in which the sausage-like pi orbitals lie at right-angles to the other (Figure 1.11). Triple bonds are extremely strong, but in carbon compounds they are also very reactive. Acetylene's explosive nature is an indication of this – oxygen molecules will react with it very readily, bursting the triple bond open and releasing the energy bound up in it. This is what happens in the flame of an oxyacetylene blowtorch. But this reactivity is not necessarily the rule in other molecules. The nitrogen molecule ($N_2$), for instance, has a triple bond between its constituent atoms that is highly stable, which is why nitrogen gas is extremely inert.

In recent years it has become apparent that even higher-order multiple bonds are possible. The ultimate member of the two-carbon molecules, $C_2$, which contains a quadruple bond, is highly unstable and reactive, but relatively stable quadruple bonds have been identified between two metal atoms.

## *Full circle*

Carbon is the most versatile of atomic building blocks. Its ability to bond strongly to other carbon atoms (via single, double or triple bonds) gives rise to all manner of

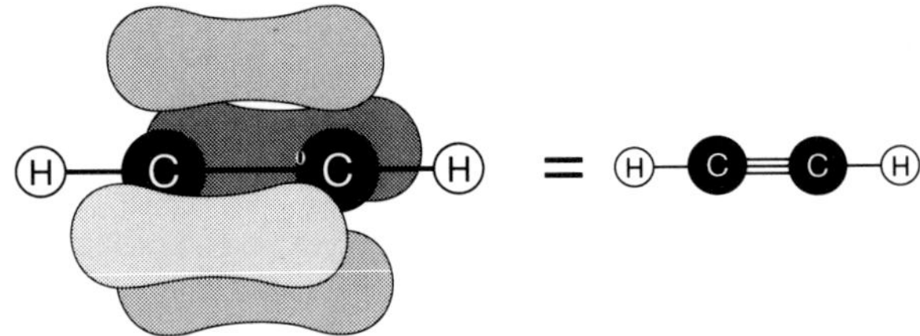


**Figure 1.11** *In the acetylene molecule, two carbon atoms are joined by a triple bond. This consists of a sigma bond and two pi bonds, at right angles to one another.*

molecular frameworks, some of which constitute the skeletons of the complex biochemicals, such as fats and steroids, found in living organisms. It should come as no surprise, therefore, that carbon is particularly valuable to synthetic chemists intent on designing molecules with new and unusual shapes. While these molecules have (as we shall see) a tendency to bring out the frivolous natures of their creators, the ultimate motivation for their synthesis may be far from whimsical. The design of molecules with peculiar geometries may be driven by the expectation that they will have useful chemical properties; equally, however, it may represent a dive into uncharted waters. An unusual shape may bestow on a molecule properties unguessed of, or provide insights into apparently unrelated areas of chemistry.

The basis of almost all of the work on unusual carbon molecules is the carbon ring. Hydrocarbons containing rings of five, six and (to a lesser degree) still more carbon atoms are found naturally in petroleum. The six-atom ring is the basis of cyclohexane, in which each carbon is joined to two others and to two hydrogen atoms. The most comfortable arrangement of bonds around each carbon is one that causes the ring to buckle (Figure 1.12).

In an important modern industrial process, a hydrogen atom is plucked from each carbon in cyclohexane to produce benzene ($C_6H_6$). The discovery that benzene (which is also found naturally in crude oil) is a carbon ring is usually attributed to the German chemist Friedrich August Kekulé, who reported it in 1865; but in fact another German, Johann Loschmidt, seems to have published the ring structure four years before. Legend has it that Kekulé's insight came to him in a dream, in which he had visions of a snake with its tail in its mouth. But as this story seems to have arisen 25 years after Kekulé's "discovery," it is hard to give it much credence. Indeed, some suggest that Kekulé's supposed insight may have actually derived from a glance at Loschmidt's book!

All the same, it is unlikely that the alternating arrangement of single and double bonds around the six-carbon ring will ever become known other than as the "Kekulé structure." There are two equivalent ways in which these bonds can be arranged (Figure 1.13); but if this equivalence is removed, for example by replacing two adjacent hydrogens by chlorine atoms (giving dichlorobenzene), one might expect two distinct structures for the molecule, one with a double bond between the chlorine-bearing carbons and the other with a single bond in this position. Experimental studies insist, however, that there is only one kind of dichlorobenzene. Kekulé proposed that the bonds in the benzene molecule oscillate rapidly back and forth between the two arrangements. The

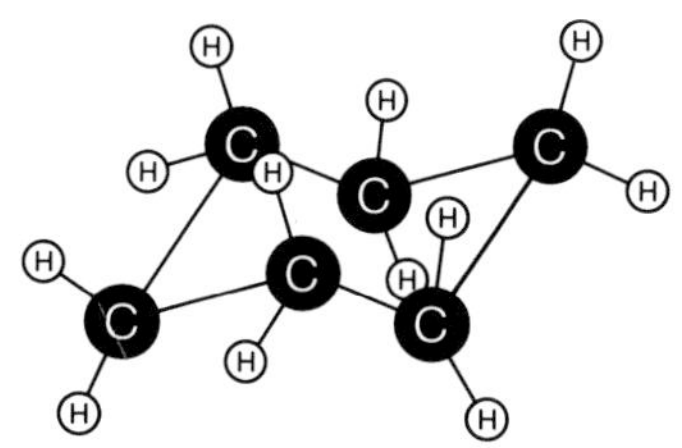


**Figure 1.12** *Cyclohexane is a hydrocarbon in which six carbon atoms are joined in a ring. The ring puckers to allow the most favorable arrangement of the four bonds around each carbon atom.*

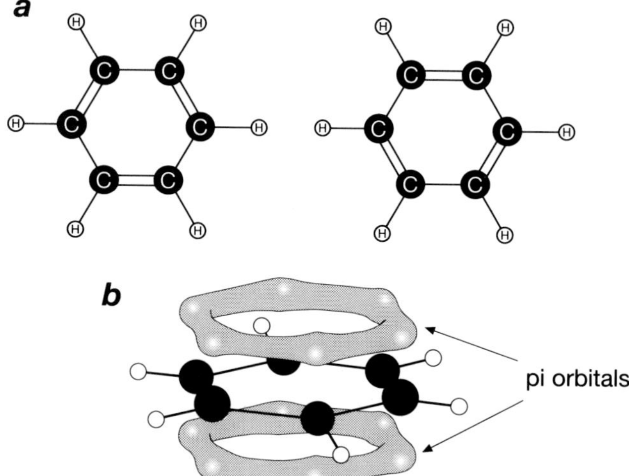


**Figure 1.13** *August Kekulé deduced that the benzene molecule contains a ring of six carbon atoms linked by alternating single and double bonds. In Kekulé's structure there are two possible arrangements of the bonds (a); but according to the modern theory of chemical bonding there is just one bonding arrangement, in which the pi electron clouds become smeared out into two "delocalized" ring-like orbitals above and below the plane of the molecule (b).*

modern view is that, rather than flipping in this manner, the pi bonds become smeared out into two continuous electron clouds running around the ring above and below (Figure 1.13). The electrons in these molecular orbitals can move readily from atom to atom all around the loop: they are said to be delocalized.

Benzene is the building block for a wide range of hydrocarbon molecules. Two can be linked side by side, for instance, to form naphthalene. Adding another benzene ring gives anthracene; or, if the third ring is added in a crooked fashion, phenanthrene (Figure 1.14). Both of these molecules, like benzene itself, are entirely flat, and the electrons in the pi orbitals are delocalized across all of the six-atom rings. There is an entire family of hydrocarbon compounds derived from benzene rings stuck together like hexagonal tiles; some of these compounds are found in coal, some are thought to be formed in the atmospheres of stars, and many are highly carcinogenic. An interesting molecule is obtained by continuing to add rings to phenanthrene in a crooked manner: the sixth will meet up with the first to form a larger ring, a kind of "superbenzene," called coronene because of its supposed resemblance to a (rather flat!) crown. If, however, we don't join the sixth ring to the first but leave them both with their hydrogen atoms instead, the molecule can no longer remain flat. It is forced to twist like a spring washer, and this helical structure has earned it the name helicene.

Is there any limit to the size of these benzene tilings? Apparently there is not – one can join them into flat sheets of any size, which must be edged with hydrogen atoms to ensure that each carbon forms four bonds. Graphite consists of vast carbon sheets stacked on top of one another like a sheaf of papers (Figure 1.15).

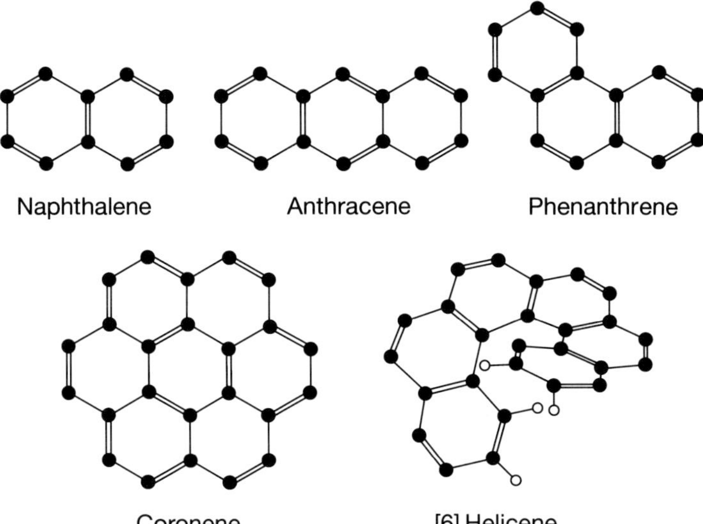


**Figure 1.14** *Benzene rings can be joined together edge-on to form a variety of molecules, known collectively as polycyclic aromatic hydrocarbons. The simplest is naphthalene; two possible arrangements of three rings correspond to anthracene and phenanthrene. More exotic variants include coronene and helicene. One can continue adding benzene rings to the six in [6]helicene to form larger, coiled helicenes. In each of these molecules the pi orbitals form extended delocalized orbitals that run continuously from one ring to the next. Hydrogen atoms not shown, except for those at each end in helicene.*

It is this sheet-like structure that makes graphite a good lubricant, as the sheets can slide over one another. Many of the other interesting properties of graphite derive from the fact that the pi orbitals on each benzene-like ring overlap to give the electrons the freedom to roam all over the surface. This makes graphite a reasonable electrical conductor (it is actually a so-called semiconductor, of which more is said in Chapter 6).

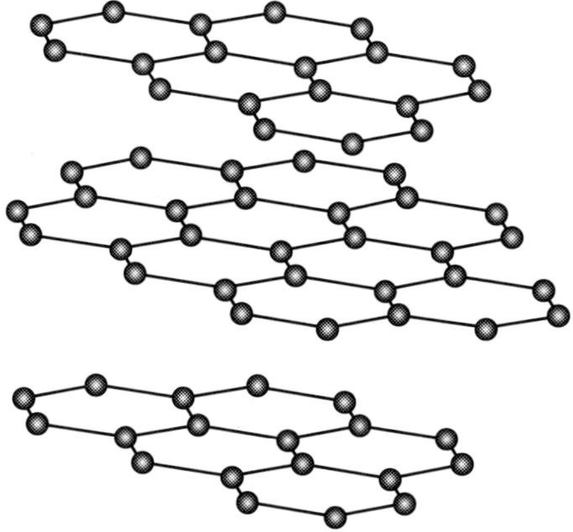

**Figure 1.15** *Graphite contains vast sheets (of which only small fragments are shown here) in which benzene-like carbon rings form a hexagonal network, like chicken wire. The sheets lie on top of one another, and are held together by weak interactions between the delocalized electron orbitals on adjacent sheets.*

## Building with carbon

The versatility of carbon as a structural element in natural compounds has encouraged chemists to attempt to construct all kinds of audacious molecular edifice. One suspects that the part which chemists relish most is not so much seeing their weird and wonderful molecules take shape, but thinking up names for them. It is perhaps unfair, though not entirely implausible, to imagine that some of these painstaking synthetic procedures have been devised solely because their architects have thought up an amusing name for the intended product!

Carbon rings are the central component of most of these molecules. To create just about any interesting kind of framework structure one needs more than one ring (making the molecule "polycyclic"). The simple example of a four-carbon-atom ring abutting a three-atom ring gives an indication of the way that the name game is played: it is called "housane," and Figure 1.16 shows why. The "-ane" ending reflects the fact that the hydrocarbon is saturated (contains no double or triple bonds), and therefore belongs to the class of compounds called alkanes; the number of bonds on each carbon is made up to four with hydrogen atoms. Both three- and four-carbon rings are said to be strained because the bonds between carbons have to be bent severely to get the ends to join up.

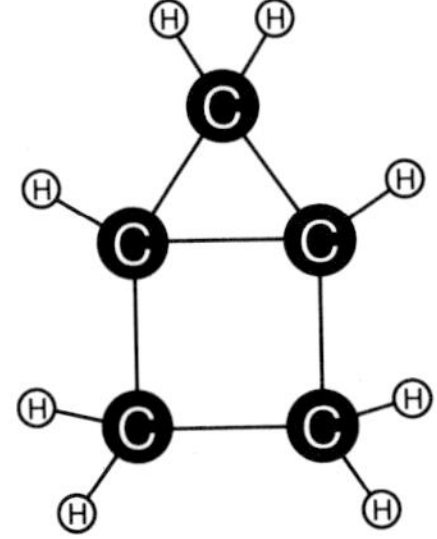


**Figure 1.16** *A four-membered and three-membered carbon ring joined together form the housane hydrocarbon.*

If three rings are joined along a common edge, the result is a kind of molecular propellor, therefore inevitably christened a "propellane." But when Jordan Bloomfield of the University of Oklahoma made the first propellane in 1966 (Figure 1.17*a*), the journal editors (a notoriously conservative breed) forced him to relegate the nickname (which he actually formulated as "propellerane") to a footnote. It was David Ginsberg of the Israel Institute of Technology who first got away with using "propellane" for his own variant (Figure 1.17*b*). Kenneth Wiberg of Yale University in Connecticut and colleagues later produced a very unlikely propellane in which three three-carbon rings were joined back to back (Figure 1.17*c*). Joining rings by corners rather than edges takes us from propellors to paddle wheels, which have been christened "rotanes" (Figure 1.17*d*).

The benzene ring provides a flat, disk-like unit with which some researchers have amused themselves by stacking them like dinner plates. The archetypal motif for these

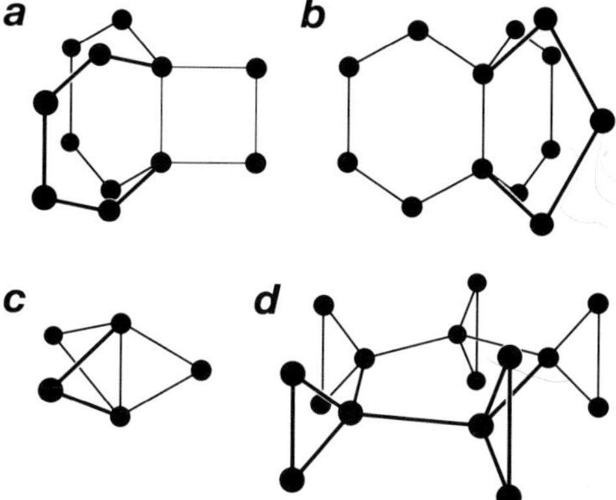


**Figure 1.17** *Carbon rings linked edge-on and by corners give rise to propellanes (a–c) and rotanes (d), respectively. Hydrogen atoms not shown.*

stacked molecules is cyclophane, in which two rings are clipped together by short hydrocarbon chains (Figure 1.18). Two or even three pairs of linking struts can be attached, in the latter case producing a molecule that looks like two spiders embracing. Synthesized for the first time in 1979 by Virgil Boekelheide of the University of Oregon, it goes by the name of superphane. An Oriental beauty can be discerned in the remarkable cyclophane made by Masao Nakazaki and colleagues of Osaka University (Figure 1.18); indeed, it reminded them of a traditional kind of Japanese lantern called a *chochin,* and in a break from the conventional suffix they adopted this name as it stood. These cyclophanes are not built for their attractiveness alone, however – some serve as the basic structural units of molecules that can mimic aspects of the behavior of the important natural compounds called enzymes.

More complicated polycyclic hydrocarbon networks can be made by joining rings along more than one edge.

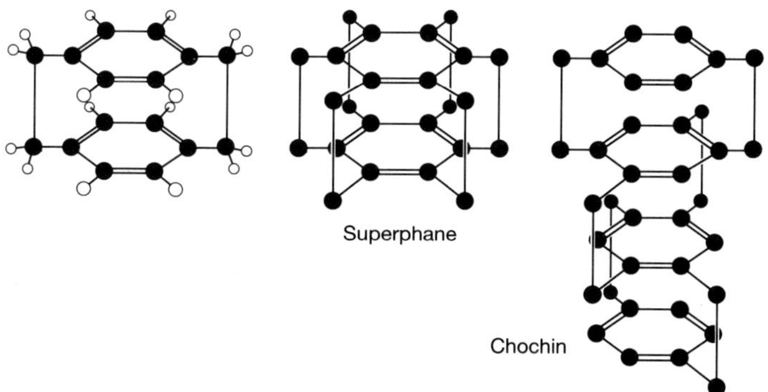


**Figure 1.18** *Cyclophanes are benzene rings stacked atop each other, linked by short hydrocarbon chains. Hydrogen atoms not shown in superphane and chochin.*

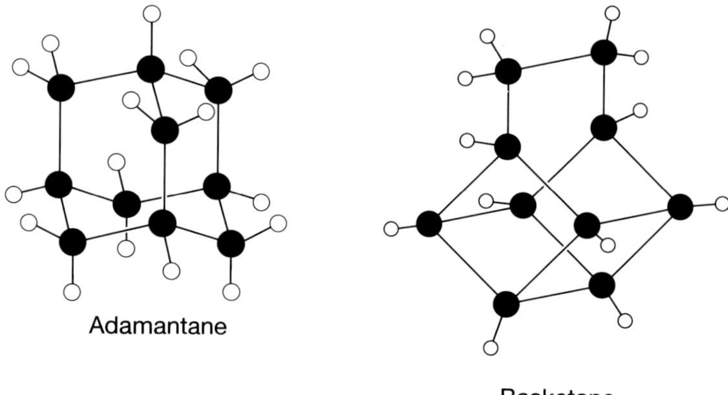


**Figure 1.19** *Adamantane and basketane contain several carbon rings joined along their edges.*

The molecule adamantane, shown in Figure 1.19, can be regarded as a fragment cut out of the carbon network of diamond (Figure 1.2*b*). This remarkable molecule was discovered in 1933 by the Czechoslovakian chemists S. Landa and V. Machacek, who isolated it from petroleum. The name is derived from the Greek *adamas,* meaning diamond.

The void inside the adamantane framework makes it a kind of molecular cage. A more obvious hydrocarbon receptacle is the aptly named basketane (Figure 1.19). Basketane is very closely related to a structure that has long fascinated chemists, the perfect hydrocarbon cube. This molecule, called cubane (what else?), was first made by Philip Eaton and coworkers at the University of Chicago in 1964. It is just one member of a whole family of prism-shaped hydrocarbons called the prismanes (Figure 1.20). The pentagonal prism competes with the simpler two-ring molecule mentioned earlier for

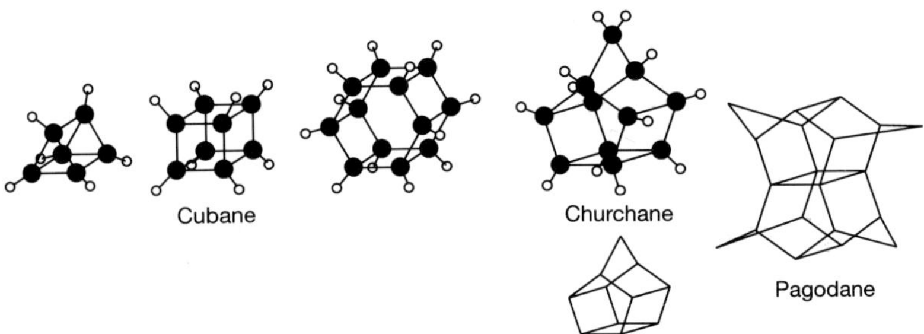


**Figure 1.20** *The prismanes are hydrocarbon polyhedra; one of the most well studied is cubane. More ornate variants are churchane and pagodane. For the latter two I show here the carbon framework in "stick" form.*

the name of housane; Gerald Kent of Rider College in New Jersey was moved to add a steeple to this molecular domicile to produce churchane, while Horst Prinzbach and colleagues of the University of Freiburg proposed that the ornate edifice which they built in 1983 from two churchane-like units should be christened pagodane.

As you might imagine from all of this, synthetic chemists cannot resist a challenge. One such, however, presented so daunting a prospect that it remained unconquered by several decades of assault, thereby earning a reputation as "the Mount Everest of polycyclic chemistry." This refractory item is a dodecahedron built from carbon – dodecahedrane (Figure 1.21). Two groups in the 1970s came within spitting distance of cracking the problem, but neither could find a way to make the final links. Philip Eaton's team in Chicago built half of the molecule in 1977, a bowl-like structure of six pentagons. All this needed was a cyclopentane roof linked to the five corners of the rim; with this in mind, Eaton called the bowl molecule peristylane, from the Greek *peristelon*, a group

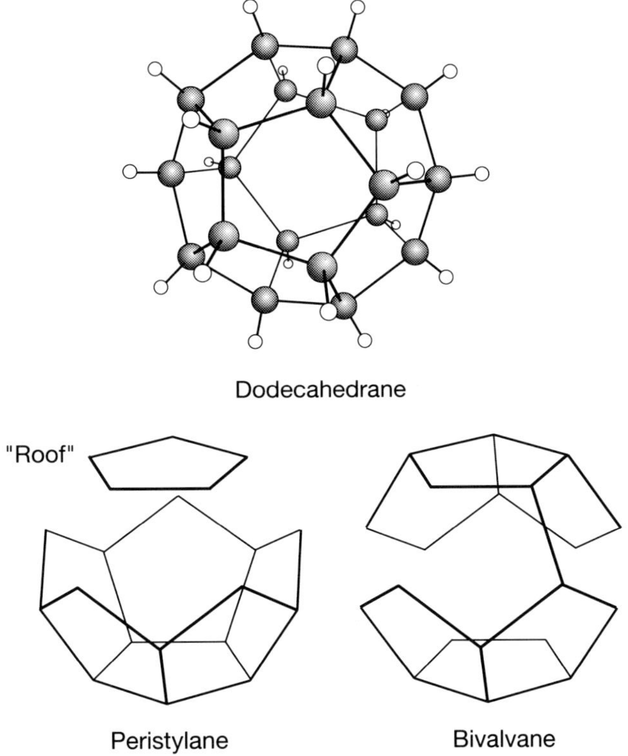


**Figure 1.21** *The dodecahedrane molecule is one of the most spectacular hydrocarbons synthesized to date. It consists of a perfect dodecahedron of carbon atoms, each capped with hydrogen. Philip Eaton's group couldn't quite manage to put the roof onto peristylane to create the molecule in 1977, but Leo Paquette had more success in 1981 by sealing shut the jaws of bivalvane.*

of pillars designed to support a roof. But they couldn't fix the roof in place. At Ohio State University, Leo Paquette took a different approach: he created two smaller, three-pentagon fragments of the dodecahedron, which were linked via corners (Figure 1.21). This produced a molecule shaped like a clam or a bivalve, suggesting the name bivalvane.

It was not until 1981, however, that Paquette and his coworkers managed to get the clam to close. Their first version of dodecahedrane had the slightly untidy feature of two methyl ($CH_3$) groups sticking out of it; this compound was a solid with the highest melting point known for any hydrocarbon (greater than 450 degrees Celsius). But by 1982 Paquette's group had tidied up their act to produce the genuine article – perfect dodecahedrane.

## Chemistry comes round

### *It came from outer space*

The construction of dodecahedrane represents a pinnacle of achievement in chemical synthesis. It was the product of a long and intricate series of steps in a procedure that utilized the knowledge accumulated over decades of work on carbon compounds. But this elegant molecule has now been eclipsed utterly by one more remarkable still, yet which has been created in a variety of ways so crude that they scarcely warrant description as syntheses at all.

While, moreover, dodecahedrane is strictly a laboratory curiosity whose value resides in a symbolic demonstration of the prowess of modern chemistry, the newest and most stunning of exotic carbon structures promises many practical applications and has already acquired a dedicated field of research all of its own. The study of this molecule is conducted not only by chemists but by physicists, astrophysicists, materials scientists, engineers and biologists. Entire conferences have been devoted to it, newspaper articles and television programs have expounded its virtues. In molecular terms, it is undoubtedly a superstar.

This molecule is a new, hitherto unknown form of pure carbon. As such, its discovery engenders in chemists not a little humility, since they have tended to assume that the natural states of the pure chemical elements were rounded up long ago. Even hoary old chemistry texts that dwell on how to prepare "sal ammoniac" are on sure enough ground when they come to describe the yellow crystals of natural sulfur, the dull gray powder of pure silicon or the pungent green gas that is chlorine. And as for carbon – its natural states diamond and graphite have been known since pretty much the beginnings of civilization. As Joseph Conrad said in 1914, "every schoolboy knows [that there is] a close chemical relation between coal [for which, crudely, read graphite] and diamonds." Yet even today, when schoolgirls too may be familiar with the structure of DNA, it seems that we did not know everything about pure carbon. How is it that a third form of this element passed undetected for so long?

Its discovery involves a tale scarcely less exotic than the molecule itself. The story starts in earnest (although not, as we shall see, in spirit) in the 1970s, when chemists Harry Kroto and David Walton from the University of Sussex were pondering on the nature of certain molecules detected in interstellar space. As Chapter 3 explains, it is possible to "see" molecules across thousands of light years of space by detecting the radiation that they absorb or emit, in particular the emission at radio frequencies. Chapter 3 discusses in more detail how molecules interact with light, infrared, radio waves and other forms of electromagnetic radiation.

The Sussex group were studying long, chain-like molecules called polyynes (the double "y" should be pronounced as in "polly-ines"). These molecules are mostly carbon – they consist of chains in which carbon atoms are linked by alternating single and triple bonds, giving each atom its complement of four bonds without the need for additional atoms such as hydrogen. In polyynes themselves the two ends of the chains are capped by hydrogen atoms, but the chemists were also investigating cyanopolyynes, in which one end terminates in a triple bond to nitrogen, forming a cyano (CN) group (Figure 1.22). A student of Kroto and Walton, Anthony Alexander, succeeded in 1974 in preparing the five-carbon cyanopolyyne $HC_5N$, and he measured how it absorbed microwave radiation.

Kroto became excited by the possibility that these molecules might be produced in space, and a collaboration with Canadian astronomers Takeshi Oka, Lorne Avery, Norm Broten and John MacLeod led to the detection of the "fingerprint" of $HC_5N$ in radio waves from a cloud of molecules near the center of our galaxy. Tremendously excited by this discovery, they went on to detect the seven-carbon molecule $HC_7N$ soon after.

Why should these peculiar molecules, which had taken considerable effort to prepare in the laboratory, turn up in outer space? Kroto believed that such molecules might be produced in the atmospheres around certain members of a class of old stars called red giants – stars that are nearing the ends of their lives and have begun to swell into huge bodies that turn red as their energy supplies dwindle. Some red giant stars contain large quantities of carbon atoms in their outer atmosphere, and these atoms can, when sufficiently far from the heat of the star's interior, combine with each other and with other atoms present (such as hydrogen, nitrogen and oxygen) to form molecules both familiar – such as formaldehyde, methane and methanol – and exotic. It seemed likely that the polyynes, which are mostly carbon, could be created in this carbon-rich environment.

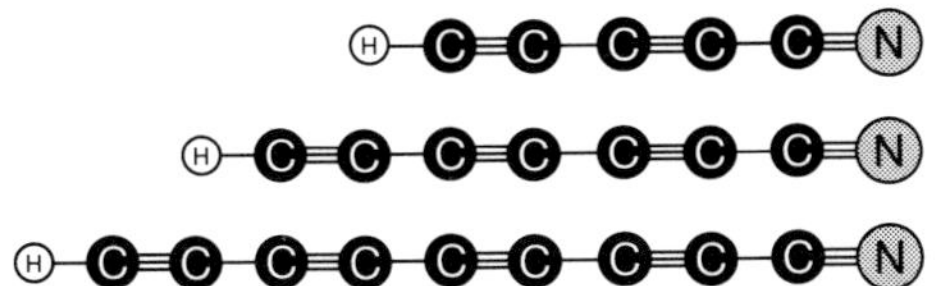


**Figure 1.22** *Cyanopolyynes with five, seven and nine carbon atoms in their chains, identified by Harry Kroto and colleagues in the atmospheres of carbon stars.*

When Kroto and the Canadian astronomers discovered the radio signature of a cyanopolyyne with no less than nine carbon atoms ($HC_9N$), Kroto could not but suspect that far longer carbon chains, perhaps with as many as thirty or so carbon atoms, might eventually be found in interstellar molecular clouds. A relative of the 32-carbon polyyne $HC_{32}H$ had in fact been synthesized by David Walton and his students in 1972.

In 1984, Kroto's musings along these lines took a new turn. He was invited by Robert Curl to visit the laboratory of Curl's colleague Richard Smalley at Rice University in Houston, Texas. Smalley had developed an experimental technique for generating small clusters of atoms by using a laser beam to vaporize solid targets. The atoms in the vapor would recombine into small clusters of a hundred or so, which, it was thought, might exhibit interesting properties that were somewhere in between those of molecules and those of bulk solids. This means of vaporizing solids, called laser ablation, was capable of producing temperatures of tens of thousands of degrees within the small region of the sample on which the lasers were focused. To identify the products the researchers used a technique called time-of-flight mass spectrometry, in which the clusters, which are formed as positively charged ions, are accelerated in an electric field and passed down a long tube. Because heavier clusters are accelerated less than light ones, the time taken for a cluster to travel down the tube to a detector at the far end provides a measure of its mass. One thus obtains a "mass spectrum" of the products – a breakdown of the relative abundances of clusters containing different numbers of atoms.

Smalley was at that time concerned primarily with studying clusters of semiconducting materials such as silicon and gallium arsenide, partly because they are such important materials for microelectronics technology. But Kroto realized that the high temperatures involved in the laser ablation technique might enable one to mimic the kind of chemistry that goes on in the atmosphere of a carbon-rich red giant star: all that one need do was replace the silicon target with carbon – that is, with graphite. Perhaps here was a way to generate long-chain polyyne-type molecules in the laboratory.

To Kroto this was an exciting idea, but the Rice group were more concerned with their silicon work, and so studies on graphite were postponed until a convenient time presented itself. That same year, however, Donald Cox, Andrew Kaldor and colleagues from the Exxon company in Annandale, New Jersey, embarked upon this very experiment and reported the mass spectra produced by laser ablation of graphite. For smaller clusters the Exxon team saw a series of mass peaks spaced twelve atomic mass units (that is, the mass of a carbon atom) apart, indicating that the carbon atoms accumulated one at a time (Figure 1.23). But for carbon clusters larger than forty atoms or so the peaks were no longer separated by twelve mass units but by 24, corresponding to clusters with only even numbers of carbon atoms. The Exxon team could offer no real explanation for this even-number preference of large carbon clusters, because the mass spectrometric technique provided no information about the *structure* of the clusters.

It was not until the end of August of 1985 that Smalley and Curl were able to initiate the collaboration with Kroto on carbon clusters. This endeavor was also to involve research students James Heath, Sean O'Brien and Yuan Liu. Smalley and his colleagues

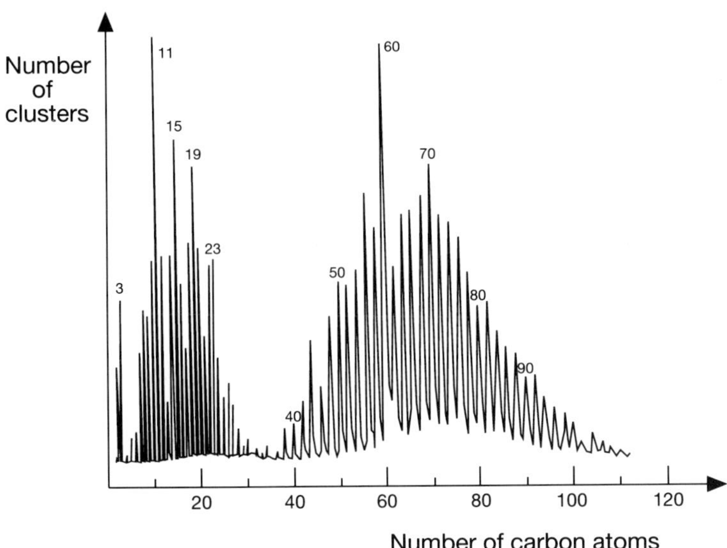


**Figure 1.23** *The mass spectrum of carbon clusters produced by researchers at Exxon in 1984. For clusters with more than forty atoms, there is a clear preference for those with even numbers of atoms.*

fully expected to spend "no more than a week or so" working on graphite – for all the intrigue of the astrophysical implications that Kroto foresaw, it was hard to regard the work with anything like the urgency that adhered to the studies of semiconductor clusters.

Kroto and the Rice team first studied the molecules produced when the carbon clusters produced by vaporizing graphite were allowed to react with gases such as hydrogen, oxygen and ammonia, which might be expected in carbon-star atmospheres. They found the cyanopolyynes that Kroto had expected, along with other chain-like molecules. The mass spectra from graphite alone, meanwhile, were very similar to those that the Exxon group had described: here again was the rather mysterious preference for even-numbered clusters above about forty atoms. After several experimental runs, however, something became apparent in these spectra of which the Exxon report had made no mention: the peak corresponding to a sixty-atom cluster was sometimes appreciably – as much as three times – larger than those to either side. It seemed that formation of the sixty-atom cluster, denoted $C_{60}$, was at times more favorable than the formation of the other even-numbered clusters (Figure 1.24). In fact, the Exxon group had noticed this result too, but because they had no explanation for it they had made no mention of it in their paper.

On the afternoon of Friday September 6th, Kroto and the Rice team decided that they should try to identify the experimental conditions under which the sixty-carbon peak was most prominent. Heath and O'Brien were more than happy to sacrifice the weekend for the possibility of uncovering a new result, and offered to spend the two

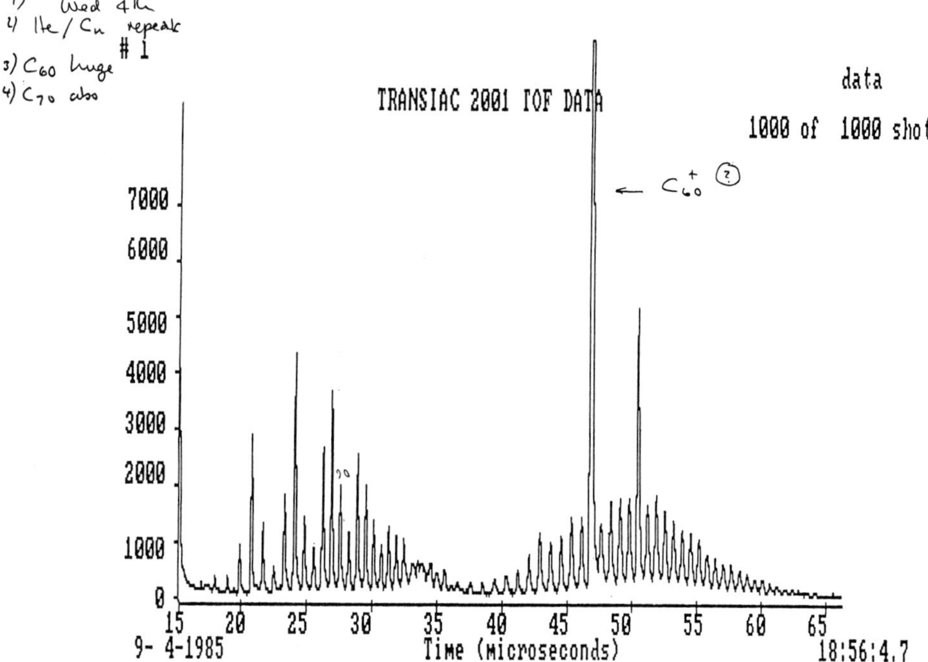


**Figure 1.24** *In the mass spectrum of laser-vaporized carbon obtained by Harry Kroto and Richard Smalley's group at Rice University, the peak corresponding to the sixty-atom carbon cluster ($C_{60}$) stands out loud and clear. In this plot – the raw data from one of the early runs – Harry Kroto picked out the most prominent peak with a tentative "$C_{60}^+$?," and the researchers noted in the corner that the peak presumably due to $C_{70}$ is also clearly evident to the right. (Picture courtesy of Harry Kroto, University of Sussex.)*

days trying to adjust the experimental conditions so as to identify how to enhance the formation of $C_{60}$. On the morning of the following Monday they revealed the fruits of these efforts: mass spectra in which the $C_{60}$ peak was like a mountain surrounded by little hillocks. At the same time the seventy-atom cluster ($C_{70}$) was also prominent – it seemed to recur as an inseparable companion to the huge $C_{60}$ signal.

The researchers set about trying to figure out how it was that $C_{60}$ came to be so much more stable than the other clusters. There must be, they decided, something about its structure that would account for this stability. For these large clusters, a chain structure was most unlikely. An alternative candidate was a graphite-like structure in which the carbon atoms were bonded into small flat sheets stacked one atop the other. Kroto suggested that a symmetrical arrangement of $C_6$, $C_{24}$, $C_{24}$ and $C_6$ sheets would explain the magic number 60 (Figure 1.25). But all such structures leave the carbons at the edge of the sheets with just three bonds apiece, rather than the requisite four, giving them an unsatisfied "dangling" bond; this would be likely to make the cluster very reactive.

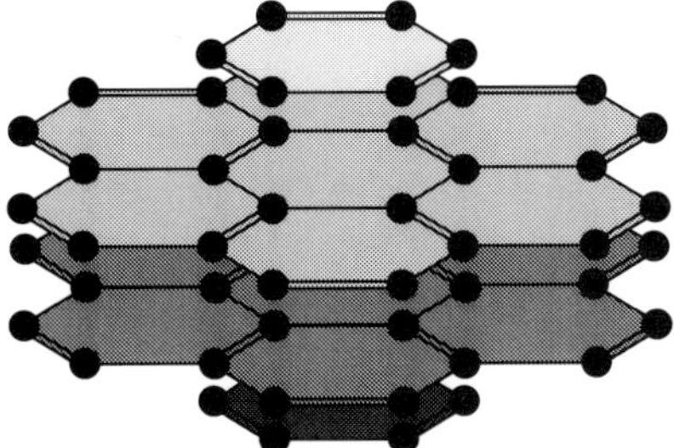

**Figure 1.25** *Attempts to construct a $C_{60}$ cluster from graphite-like layers of carbon hexagons leave "dangling bonds" at the edges, which would make the clusters highly reactive. This version, postulated by Harry Kroto, was nevertheless somewhat appealing as it suggested a possible interpretation of the preference for 60 carbon atoms: a quadruple-decker sandwich of 6, 24, 24 and 6 carbon atoms. But Kroto and his colleagues were to dream up a more startling alternative for the sixty-atom structure.*

One way of getting around the problem of dangling bonds would be to have the sheets curl up on themselves to form closed shells. Although this was a promising idea, it wasn't clear how a graphite-like sheet of carbon atoms, which is perfectly flat, could be induced to bend. For Kroto, however, these curved hexagonal sheets awoke memories. In 1967 he had visited the Montreal Expo exhibition and become captivated by the US Pavilion, designed by the American architect Richard Buckminster Fuller. It was a geodesic dome built from flat polygons (Figure 1.26). Buckminster Fuller was something of a maverick amongst architects, and these domes were a characteristic of his work. Kroto recalled that the Montreal dome had been fashioned from hexagonal units. Could $C_{60}$ be a miniature version of Buckminster Fuller's bizarre designs? But while it was easy to arrange hexagons into a flat sheet, it was not obvious to the researchers how to construct a closed dome.

**Figure 1.26** *Richard Buckminster Fuller's geodesic dome at the Montreal Expo of 1967. The dome is made from triangulated polygons joined at their edges. (Photograph by Robin Whyman, kindly supplied by Harry Kroto.)*

A mathematician would have given them the answer immediately: to construct a dome from hexagons is simply not possible. This had been proved in the eighteenth century by the Swiss mathematician Leonhard Euler, and the fact was no doubt known to Buckminster Fuller. Kroto did recall, however, that five-sided shapes – pentagons – might also have featured in the domes. He remembered also a cardboard kit that he had once purchased and assembled for his children, a sphere-like map of the heavens which he called a "stardome." This too had been built from hexagons and pentagons. But none of the researchers was quite sure what the rules were that governed such objects. As Kroto was due to return to England on the Tuesday, there was little time left to solve the puzzle. In the library at Rice, Smalley located a book on Buckminster Fuller's work, *The Dymaxion World of Buckminster Fuller* by Robert W. Marks, which he took home on the Monday evening to ponder.

It is remarkable to find that, despite what we like to think of as the sophistication of modern science, some of the most significant insights can still be obtained by sitting down with a can of beer and fiddling with cardboard or ball-and-stick models. Yet this is how the structure of $C_{60}$ was deduced. That night James Heath collected sixty "Juicy Fruit" gum balls and attempted to stick them together with toothpicks in an effort to model the sixty-atom cluster. Hours later he and his wife had little to show for their efforts but sticky fingers and an empirical awareness of Euler's dictum that a closed shell cannot be made purely from hexagons. Meanwhile Smalley, having abandoned efforts at cracking the problem on his home computer, took to fiddling with cardboard hexagons and sticky tape. His attempts to create from these a curved structure came to nothing, but as midnight came and went he remembered Kroto's comment about pentagons. Once he added these to the makeshift construction kit, everything began to fall into place. Five hexagons arrayed around a pentagon with their adjacent edges touching automatically curl upwards into a bowl (Figure 1.27). On adding further hexagons and pentagons, Smalley was able to produce a hemisphere. The rest was easy: constructing another hemisphere on top produced a ball-like polyhedron containing twelve pentagons and twenty hexagons. Counting corners (where the atoms would sit), Smalley found to his delight that he had built a sixty-atom cluster.

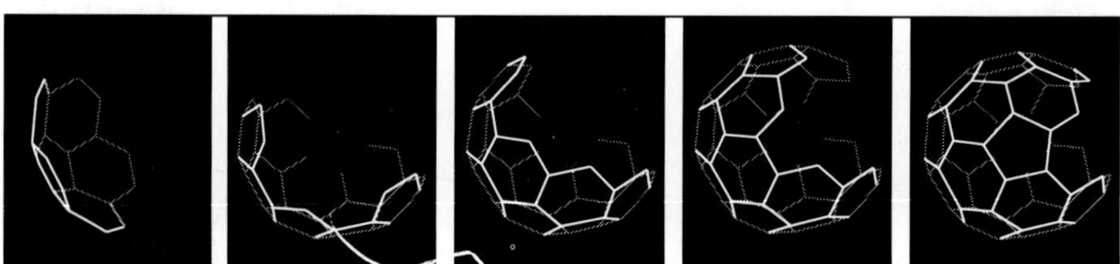

**Figure 1.27** *Building a closed cage from hexagons and pentagons. The pentagons are essential to get the sheets to curl up and the cage to close.*

*The carbon soccer ball*

The dome structure (Figure 1.28*a*) had everything. It was beautifully symmetrical, it was robust, it explained the "magic" number of 60, and it had the pleasing feature that every corner (that is, every atom) was equivalent. How could it be anything other than correct? This was Smalley's view when he called together his colleagues the following morning and showed them the model. They were struck at once by the aesthetic appeal of the solution, but Bob Curl cautioned that the model also had to make sense in chemical terms. Each carbon atom had to have four bonds, and as each was attached to three other carbons, one of those three bonds had to be a double bond. Could one arrange double bonds on the object in such a way as to satisfy this requirement for every atom? Using sticky labels to represent double bonds, Curl and Kroto quickly established that one could (Figure 1.28*b*).

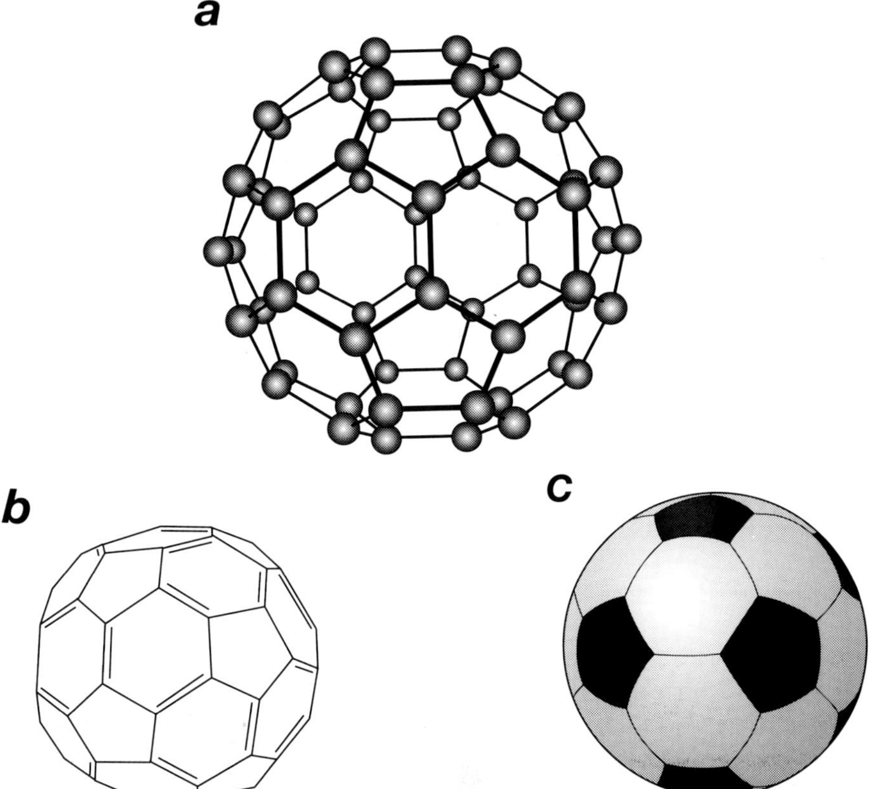


**Figure 1.28** *The structure of the sixty-carbon-atom cluster, called buckminsterfullerene (a), and the pattern of single and double bonds which allows all of the atoms to form four bonds (b). Each carbon atom in the cage is equivalent. The pattern of hexagons and pentagons in this highly symmetrical shape is the same as that in a soccer ball (c).*

A structure so symmetrical and harmonious as this must surely be familiar to mathematicians, reasoned Smalley. So he telephoned the chairman of Rice's mathematics department, William Veech, and described his model. Did it have a name? Veech called him back shortly, but his answer was hardly a technical one: what Smalley had described to him, said Veech, was "a soccer ball." The leather patches on a soccer ball are stitched together in precisely the same pattern (Figure 1.28*c*).

The soccer ball structure does have a technical name, however: it is a truncated icosahedron. This object is just one member of an infinitely large family of closed shells made up from pentagons and hexagons. It was Euler again who showed that 12 pentagons can be assembled with any number of hexagons (except one) into closed structures. The truncated icosahedron is a particularly symmetrical member of the family; others are generally less so. The researchers realized that this family provided an explanation for the even-number bias in the large carbon clusters generated by laser ablation of graphite. Each time an additional hexagon is added to make the next successive member of the family, it turns out that two new corners appear in the resulting structure, corresponding to the addition of two more carbon atoms.

Kroto delayed his return from Houston in order to write up these striking results and the "soccer ball" conjecture in a short paper, which was sent to the journal *Nature*. Putting the findings on paper forced the issue of what to call this remarkable new sixty-carbon molecule. Although several suggestions were put forward in the paper, such as "spherene" and "soccerene" (the "-ene" ending signifying the presence of double bonds in the structure, as in benzene), Kroto's favored proposal gave a nod to his source of inspiration for the cage structure: buckminsterfullerene. This one stuck, apparently capturing the imagination of the chemistry community (who often reduce it, however, to the less dignified "bucky ball").

The team subsequently turned their attention to the 70-atom peak in the mass spectrum. They suggested that this corresponded to another highly symmetric and therefore particularly stable cage-like cluster. A molecule with the formula $C_{70}$ can be created from $C_{60}$ by inserting an extra ring of hexagons around the equator of the sphere, producing an elongated shell more like a rugby ball (Figure 1.29). Other particularly symmetrical shapes can be derived for $C_{32}$, $C_{50}$ and $C_{84}$, and there seems to be no limit, in principle, to the size of the closed shells that can be postulated. The whole family of closed-shell carbon clusters has now become known by the generic name of "fullerenes."

Soon after the *Nature* paper was published, it began to emerge that $C_{60}$ had a prior history. The formation of closed-cage molecules from graphite-like sheets had in fact been proposed way back in 1966 by David Jones, a chemist from the University of Newcastle in England. Jones has for decades been writing a speculative science column under the pseudonym of Daedalus, in which he describes the usually far-fetched but ingeniously plausible inventions that take shape in Daedalus's imagination. No sooner are these ideas hatched, suggests Jones, than they are seized upon and developed into marketable products by Daedalus's company, Dreadco. Jones' insight and inventiveness draws much admiration from his fellow scientists, who take delight when Daedalus's

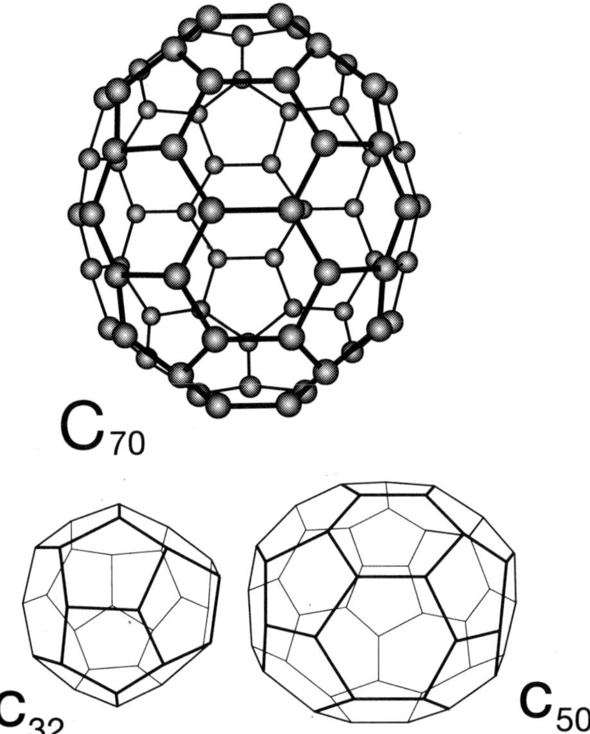


**Figure 1.29** *The structure of the fullerene $C_{70}$ has been deduced experimentally – it corresponds to a rugby-ball shape. Closed-cage structures for the smaller clusters, such as $C_{32}$ and $C_{50}$, were postulated by the Sussex/Rice researchers. Like $C_{60}$ and $C_{70}$, these small fullerenes contain twelve pentagonal rings.*

speculations are borne out by subsequent research. Buckminsterfullerene, a rudimentary form of which featured in Daedalus's column in *New Scientist* in 1966, provides by no means the only example of this foresight.

While Jones had considered the idea of "curved graphite" more broadly, some had foreseen buckminsterfullerene itself. The Japanese chemist E. Osawa had suggested the possibility of the spherical $C_{60}$ molecule in 1970 and had speculated about its properties. Russian scientists had conducted a theoretical study of the then-hypothetical molecule in 1973, and at the University of California at Los Angeles, Orville Chapman had made some attempts to build the molecule using techniques of organic synthesis like those that later enabled Leo Paquette to make dodecahedrane. But Chapman had never published this work, while that of the Japanese and Soviet scientists was buried in obscure journals and had passed unremarked.

In 1986 Kroto and the Rice researchers conjectured that, if the fullerenes really were hollow carbon cages bearing a resemblance to graphite, might they not be formed in the soot produced from combustion of carbon-rich substance? Soot is, after all, nothing more than disordered fragments of graphite-like sheets, and fullerenes appeared to represent the perfect way of avoiding dangling bonds at the sheet edges. But combustion chemists seemed unwilling, on the whole, to take up the challenge of looking for $C_{60}$ in soot. When, in 1987, Klaus Homann and colleagues at the Institute of Physical Chemistry in Darmstadt performed a mass spectrometric analysis of the ions formed in sooty flames and found that the most abundant carbon ion with more than ten atoms was that containing sixty, the observation was explored no further, even when Kroto pointed out the possible connection with the experiments at Rice. In those days, either you were a believer or you weren't! Homann's result was recalled in 1991, however, when researchers at the Massachusetts Institute of Technology reported that by carefully controlling the combustion rate and mixture of gases in flames of natural gas ignited in air, they could produce $C_{60}$ and $C_{70}$ in respectable quantities. The implication is that mankind may have been making $C_{60}$ unwittingly for millenia.

## *Fullerene frenzy*

After 1985, buckminsterfullerene acquired something approaching cult status. Everyone had heard of the *Nature* paper but most regarded it as a rather quaint curiosity. Kroto, however, became convinced that the molecule might provide the key to a whole new facet of carbon chemistry. Maybe here was a way to wrap up toxic or radioactive metal atoms in individualized carbon coatings. Perhaps $C_{60}$ would make an excellent lubricant, since the carbon balls might roll over each other like ball bearings. The speculations were many and colorful; but they remained speculations all the same, because the stuff had been produced only in minute quantities, mixed up with all the other products of laser ablation. And because no one had succeeded in isolating significant amounts of the pure material, the experiments that could confirm the hypothetical soccer ball structure remained beyond reach. In 1990, all that changed.

Like Harry Kroto, the physicists Donald Huffman at the University of Arizona in Tucson and Wolfgang Krätschmer at the Max Planck Institute for Nuclear Physics in Heidelberg had in the early 1980s become interested in the possibility that novel carbon molecules might be formed in stellar atmospheres and interstellar space. The two physicists collaborated in 1982 on experiments in which they vaporized graphite by electrical heating and measured the properties of the black soot that was produced. Huffman and Krätschmer studied how this soot absorbed ultraviolet light so as to compare the absorption spectra with those measured by astronomers. They found that their soot behaved much like that produced in ordinary combustion except for some features in the ultraviolet spectrum that were not observed for normal soot. At that time, they concluded that the extra features were due to impurities that had crept into the vaporization chamber, perhaps the oil used in the vacuum pump. It was not until three years after the 1985 *Nature* paper had been published that Huffman realized that perhaps what he and Krätschmer had seen back in 1982 was the signature of $C_{60}$.

Huffman's idea was received with some skepticism by Krätschmer, but the two agreed that the idea was worth testing further. When Krätschmer's group at Heidelberg measured the mass spectrum of the carbon soot produced by heating graphite with an electrical arc discharge, they saw immediately the prominent $C_{60}$ peak. After some trial-and-error adjustments, the researchers were able to produce $C_{60}$ in relatively large quantities – a few milligrams – with their simple arc-discharge technique. These amounts would certainly be sufficient for the experiments that could establish the structure beyond doubt, but the obstacle that remained was the need to extract pure $C_{60}$ from the rest of the detritus. Early in 1990, Krätschmer, Huffman and their respective students Kostantino Fostiropoulos and Lowell Lamb tried heating the soot so that part of it sublimed; when cooled, the vapor condensed back to a solid. Part of this solid dissolved in liquid benzene to give a deep red solution, and evaporation of the benzene solvent left reddish brown crystals (Plate 1). Mass spectrometric analysis revealed that they contained 90 per cent $C_{60}$, the remainder being $C_{70}$. Here at last was the opportunity to put the soccer ball structure to the test. By bouncing X-rays off the crystals (a technique described in Chapter 4), the researchers were able to deduce that the crystals comprised stacks of spherical molecules with their centers about a nanometer (one thousand-millionth of a meter) apart – just what was expected for regular arrays of $C_{60}$ balls. In August 1990, Krätschmer and Huffman described their new method for isolating $C_{60}$ and their evidence for the soccer ball structure in a paper in *Nature*. To Kroto the news of this success came both as a delight and as a bitter blow. It showed that he and the Rice group had guessed right after all in 1985; but Krätschmer and Huffman's breakthrough had beaten his own efforts by a whisker. Kroto had experimented with a similar arc-discharge technique in 1986, but had been severely hampered by a lack of financial backing. He had caught wind of the advances that Krätschmer and Huffman were making in 1989 when they presented a preliminary report of their work at a conference, and had resurrected the arc-discharge apparatus. But he and his colleagues at Sussex faced the same problem of extracting the $C_{60}$ from the soot. By August of 1990 Kroto's colleague Jonathon Hare hit on the benzene separation method independently, and obtained the red solution. But by then the race was already run, since Krätschmer and Huffman had managed the crucial last step of getting crystals out of the solution. When *Nature* asked Kroto to act as a referee of the paper sent by Krätschmer and Huffman he realized that his group had been beaten to the finishing post.

All the same, the Sussex team took advantage of the fact that, with their red solution already to hand, they were at least ahead of other rival groups at that stage. By the end of August they had performed the test that clinched the case for the predicted structure of the molecule – a test that was missing from the paper of Krätschmer and Huffman, although their conclusions could scarcely be doubted for all that. The experiment involved the use of a technique called nuclear magnetic resonance (NMR) spectroscopy, which showed that every one of the sixty carbon atoms is equivalent – just as is predicted for the soccer ball structure. These NMR experiments also confirmed the rugby ball structure of $C_{70}$.

The jury was in: $C_{60}$ is a molecular soccer ball. Within months of publication of Krätschmer and Huffman's recipe for mass-producing $C_{60}$, everyone was playing the game. Soon the molecules were to be seen lined up in neat rows under a new kind of microscope called the scanning tunneling microscope (Plate 2). Organic chemists started to explore how $C_{60}$ behaves in chemical reactions. Most theoretical calculations predicted that it would be a relatively stable, unreactive molecule, like benzene. But it turns out to be not so difficult to open up the double bonds in the carbon cage: hydrogen and fluorine, for example, can be attached to the carbon atoms. The balls have been attached to the backbone of long, chain-like polymer molecules, like a string of lucky charms. In an electrochemical cell, $C_{60}$ was found to take up additional electrons to form negative ions such as $C_{60}^{-}$ and $C_{60}^{2-}$, implying that it should form "salts" with metals, just as if it were an unusually large atom resembling, say, chlorine.

This led researchers at AT&T Bell Laboratories in New Jersey to react $C_{60}$ with the alkali metals lithium, sodium, potassium, rubidium and cesium. While ionic salts were indeed formed, these compounds proved to be far stranger than the researchers had anticipated, as we will see in Chapter 6. It will suffice to say here that these experiments showed $C_{60}$ to be not only the most interesting molecule to have come the way of chemists for many years, but also to have some astonishing revelations in store for physicists too. Compounds of $C_{60}$ and metals now provide one of the main focuses of $C_{60}$ research.

A particularly intriguing variety of these compounds appear to contain metal atoms *inside* the cage. These so-called endohedral ("inside the polyhedron") structures are made by forming fullerenes from a composite of graphite and a metal compound. The first were synthesized by Jim Heath almost immediately after he, Kroto and the Rice team discovered the molecule itself in 1985, by applying the laser ablation method to rods of graphite mixed with lanthanum oxide. They found that individual lanthanum atoms became intimately associated with the $C_{60}$ cage; the obvious corollary was that they were trapped inside. As many as four metal atoms have now been trapped inside a fullerene shell.

## *The ball keeps rolling*

$C_{60}$ has proved so easy to make that countless scientists have been unable to resist the temptation to dabble with it in the hope of uncovering some new and unexpected property. For those who do not want to go to the lengths of setting up their own little $C_{60}$ factory, the molecule and its relatives ($C_{70}$, $C_{84}$ and so on) are now available in gram quantities from several commercial companies in the United States. It is not exactly cheap at present (about forty times the price of gold), but some expect $C_{60}$ to cost no more than aluminum in a few years time.

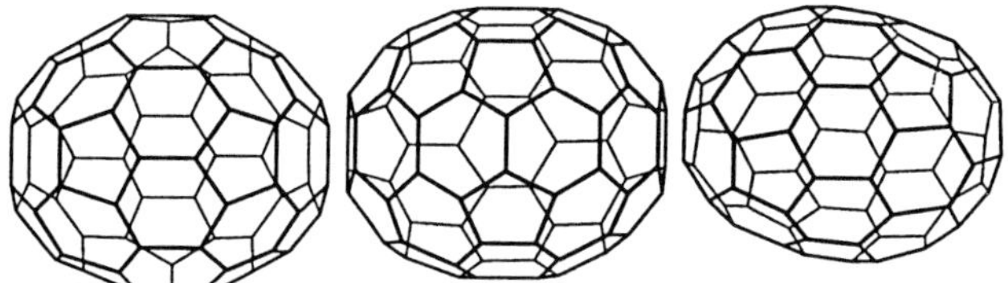

**Figure 1.30** *Some of the larger fullerenes, such as $C_{76}$, $C_{78}$, $C_{82}$ and $C_{84}$, have been isolated and their structures deduced. Although in principle a huge number of isomers can be postulated for each, the requirement that no two pentagons be adjacent reduces the range of possibilities to just a few. Here I show some of the structures deduced for the $C_{78}$ molecule.*

For those in search of vistas still more exotic, the properties of many of the larger fullerenes are still largely unexplored. All, it seems, must contain twelve pentagonal carbon rings as predicted from Euler's work. Moreover, while there are in principle a huge number of possible arrangements of the five- and six-membered carbon rings, the situation is greatly simplified by the fact that the five-membered pentagonal rings avoid being placed side by side – this would give rise to an unstable bonding pattern, which would rapidly rearrange. For $C_{60}$, for instance, there are 1,812 ways of arranging the pentagonal and hexagonal rings, but only one in which no pentagons are adjacent.

Molecules for which the same constituent atoms are arranged differently in space are called *isomers*; we will encounter several examples in the later chapters. The rule that all pentagons must be isolated from one another reduces the number of observed

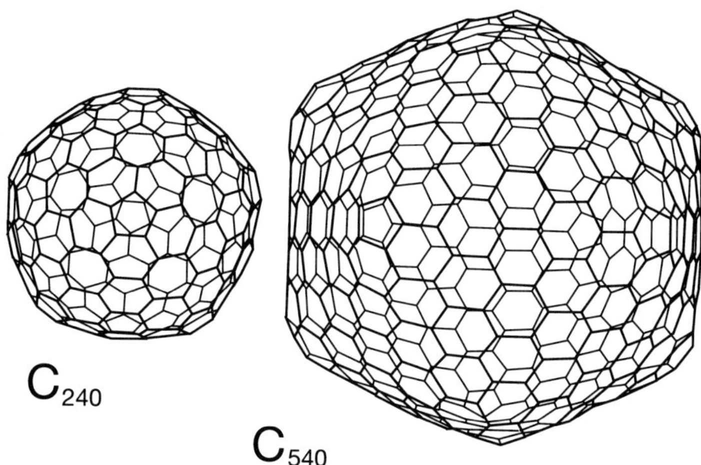


**Figure 1.31** *Harry Kroto and his student Ken McKay found that highly symmetrical structures can be built for the giant fullerenes, such as $C_{240}$ and $C_{540}$. Again they contain just twelve pentagons, which allow the cages to close. As the cages get bigger, the "corners," where the pentagons sit, get sharper. These giant fullerenes have not yet been isolated and purified in sufficient quantities to allow detailed structural studies, but it seems likely that they will exist in several isomeric forms.*

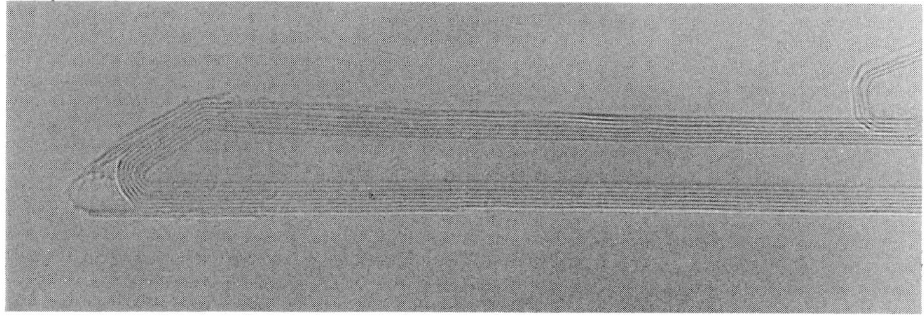

**Figure 1.32** *Tubular fullerenes? These hollow carbon tubes, consisting of concentric graphite-like sheets, were discovered by Sumio Iijima in 1991. They are capped at the end by polyhedral or conical shells. The tubes, shown here in cross-section under the electron microscope, are typically one to fifty millionths of a millimeter in width. It is possible that the smallest tubes have hemispheres of* $C_{60}$ *at the end. (Photograph courtesy of Sumio Iijima, NEC Corporation, Tsukuba.)*

isomers of the larger fullerenes to a manageable level. $C_{76}$ has just two, it seems, and $C_{78}$ perhaps eight or so (Figure 1.30). The giant fullerenes $C_{120}$, $C_{240}$ and $C_{540}$ are expected to have some particularly symmetrical isomers (Figure 1.31) but these molecules haven't yet been isolated in sufficient quantities to put this to the test.

Larger still are the fullerene-related structures discovered in 1991 by Sumio Iijima at the NEC Corporation in Tsukuba, Japan. He found that under certain conditions, the electrical-arc method of fullerene formation produces instead fine carbon fibers growing on one of the electrodes. Examination of these fibers under the microscope showed that they are hollow tubes comprised of graphite-like sheets curled into cylinders (Figure 1.32). Each tube contains several cylinders nested inside each other like Russian dolls. The tubes are capped at the end by cones or faceted hemispheres, which presumably contain pentagonal rings to allow the sheets to curl up. These graphite-like tubes, some just a nanometer in diameter (close to that of $C_{60}$) and up to a thousand times as long, are predicted to have some interesting properties. They should be the strongest carbon fibers known, and may also conduct electricity. At the end of 1992 Iijima and his colleague Pulickel Ajayan succeeded in breaking open the caps at the tube ends, whereupon the tubes sucked up liquid lead as if they were drinking straws. These carbon "nanotubes," and the related hollow, concentric carbon particles that have also been created in the carbon-arc method (Figure 1.33), are now developing into an entire subdiscipline of fullerene research.

It is now clear that graphite-like sheets provide an almost limitless capacity for forming carbon structures – the sheets can be curled up and folded like paper. This field of research has exploded so rapidly since the breakthrough of fullerene mass production in 1990 that it is virtually impossible to predict what tomorrow might bring. Richard Smalley has suggested, however, that of one thing we can be certain: "Buckminster Fuller would have loved it."

a

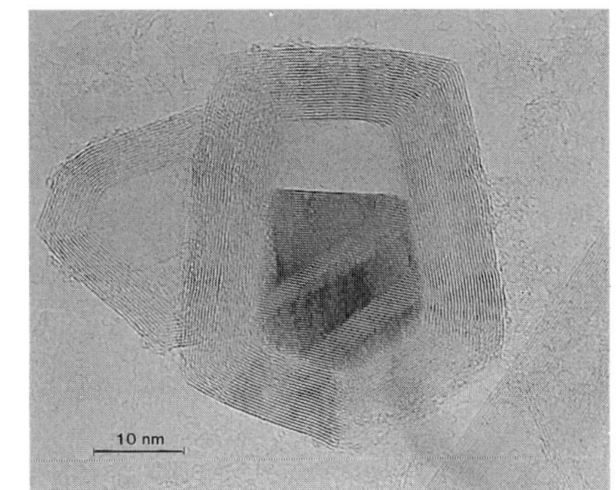


b

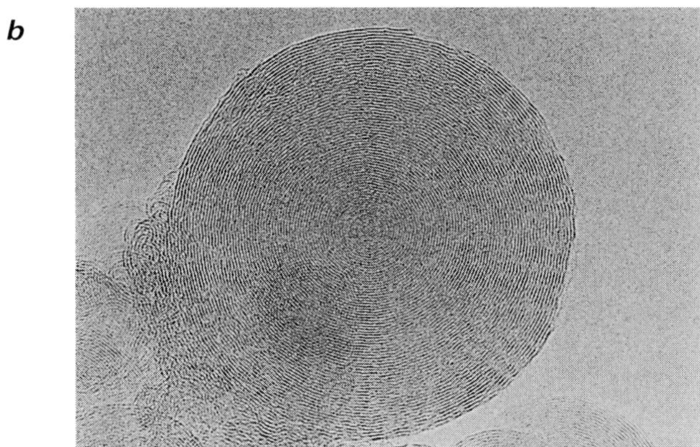

**Figure 1.33** *The known range of shell-like structures formed from graphite-like sheets of carbon becomes ever more diverse. Of particular interest are hollow, nested polyhedral particles (a), which are directly analogous to Iijima's tubes and can encapsulate metal crystals (the particle here has lanthanum carbide inside, evident as darker material), and compact, concentric graphitic "onions" (b), first studied in detail by Daniel Ugarte while at the Ecole Polytechnique Fédérale de Lausanne, Switzerland. (Photograph (a) courtesy of Y. Saito, Mie University, Japan; (b) courtesy of Daniel Ugarte.)*

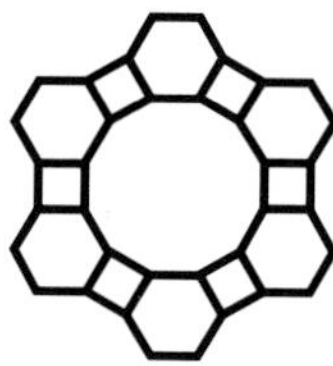

# 2

# Bringing Down the Barriers

## Getting chemical reactions to go

*However laborious and difficult this task may be, whatever impediments and obstacles may lie in the way of its accomplishment, this transmutation does not go counter to nature ...*

Paracelsus

As a unifying theme for the diverse and perhaps disparate topics discussed in this book, we might alight on the concept of transformation. First and foremost, chemistry is about transformation: a change from one substance into another, or a change in the physical state of a substance. Generally speaking, what we recognize as a *chemical* (as opposed to a physical) transformation – a chemical reaction – involves the breaking and making of chemical bonds; that is to say, atoms are traded, exchanged, plucked out of one location and placed back in another.

Very few of the materials that we encounter and utilize today are "raw"; most are the product of some kind of chemical reaction. Plastics, those revolutionary fabrics of the modern world, are not substances that can be dug out of the ground: they must be synthesized from crude oil, which contains all the requisite atomic building blocks but assembled in the wrong manner. There is far more to paper than thin slices of wood; its whiteness, toughness and flexibility are the result of a considerable amount of chemical processing. Most metals must be extracted chemically from their ores, and are often then mixed into alloys such as stainless steel, brass and pewter. Our homes are full of compounds produced in extremely sophisticated chemical reactions: drugs, food additives and the constituents of cosmetics, which generally bear little resemblance to the substances that went into their making. To many people, the very word "chemicals" denotes these highly synthetic substances ("this food contains too many chemicals"), as if to suggest that all chemicals are man-made. But the growing desire for a return to the use of natural products is not so much a flight from chemistry *per se* as a tacit acknowledgement that perhaps the chemistry of our own bodies is better attuned to that of nature, whose synthetic prowess (Berthelot notwithstanding) far exceeds our own.

The problem facing both the chemical industries and nature herself is how to rearrange often crude, simple starting materials into more useful compounds. This is generally much more than a matter of selecting starting materials that contain all the right atoms and mixing them up in a big pot – that is most likely to lead to a mixture of all manner of useless substances, amongst which that one desired may appear in tiny yields. Or perhaps nothing will happen at all – one will be left with a pot of well-stirred reagents, like an unbaked cake mix. What is needed for systematic chemical syntheses is a set of rules that will identify the likely outcome of a chemical process – what will be the major products, and under what conditions will the reaction occur?

It is relatively rare to find a useful chemical reaction that will proceed simply by adding together the reagents. But heat alone, or even just stirring or shaking, is sometimes sufficient to get things going. Light and electricity will do the job in other cases. Yet there are many important industrial processes for which these things are not enough, but for which a rather more mysterious agent of transformation is used: a substance that is not itself a reagent (since it is not consumed in the process) but which nevertheless initiates the reaction, which would not take place in its absence.

Substances of this sort, called catalysts, seem to have an almost magical ability: without becoming altered by losing or gaining atoms, they somehow enable a reaction to proceed. Some catalysts are particles of a bulk solid such as a metal or a metal oxide: others are single molecules that share a solvent with the reagents. Some are there just to initiate a reaction, but often they also determine the mixture of products obtained. In the United States, about 43 per cent of the total production of the chemical industries relies on the action of catalysts. The role of catalysts in the chemistry of living organisms, meanwhile, cannot be overemphasized. Scarcely a reaction in the body would be possible if it were not for the gentle assistance of nature's own catalysts, called enzymes.

While they may be in many ways the alchemical *lapis philosophorum* of modern chemistry, there is nothing mystical about the way that catalysts work. This chapter explains how a catalyst does its job, and provides a few illustrations of the exquisite control of chemical reactions that they can afford. It is important to appreciate, however, that catalysts cannot work miracles. They can help a reaction to occur, but only if the reaction is "feasible" in the first place. This is to say that we must make a distinction between whether a reaction *can* occur in principle and whether it *will* in practice. Catalysis can influence only the latter. To understand how it does so, we need first to see what it is that determines the feasibility of a reaction. Given a set of starting materials and a desired outcome in which they have combined into some useful substance, how can we know whether the transformation is possible?

## The driving force of chemistry

### *The end product*

Many chemical reactions appear to be strictly one-way processes. We begin with a set of compounds (the reactants) which, when mixed together under suitable conditions, undergo a transformation involving the rearrangement of atoms to yield a new set of compounds (the products). Chemists represent this process of reaction in a simple, schematic form according to which an arrow shows the transformation of reactants on the left-hand side to products on the right:

$$\text{reactants} \rightarrow \text{products}$$

In many cases it is necessary, as noted already, to supply some encouragement to get the reaction going, perhaps by shaking the reactants together or by heating them. (In either case, what we are really doing is supplying energy.) But once the reaction has taken place, there appears to be no going back. The products can be left standing together for as long as we like without ever showing an inclination to rearrange their constituents back into the reactants. Consider, for example, burning a lump of wood in air. The reaction here is between the rather complex organic molecules in the wood fibers, mainly the carbohydrate called cellulose, and the oxygen molecules in air. The products will be chiefly carbon dioxide gas and water, along with nitrogen oxides from the nitrogen-containing molecules in the wood, and also a fair quantity of pure carbon in the form of soot and charcoal. The reaction has to be triggered by igniting the wood, but once underway it should continue of its own accord. Yet even if we carry out the burning in a closed box (containing sufficient oxygen) so that none of the products escapes, we will never at a later date open the box to find that the oxide gases and soot have miraculously recombined into the original block of wood and oxygen gas. In other words, the reaction

$$\text{wood} + \text{oxygen} \rightarrow \text{carbon and nitrogen oxides} + \text{water} + \text{soot}$$

is feasible enough, but the reverse process is not.

The irreversibility of this example may come as no surprise: it is hard to conceive of the various oxide gases suddenly rearranging themselves into the complicated molecular structure of cellulose. There are plenty of simple reactions, however, for which the preferred direction is not obvious at all. Will a piece of zinc metal react with a solution of sulfuric acid into which it is dipped? Yes – but a piece of silver will not. Why, by the same token, does iron corrode in air to form the red iron hydroxide that we know as rust, while gold retains its pristine luster? Why does hydrogen gas combine explosively with oxygen gas to form water, rather than water explode into its constituents hydrogen and oxygen? The world is full of such mysteries.

That these are not *really* mysteries any longer is due to the efforts of the nineteenth century scientists such as the American Willard Gibbs, the Englishman James Prescott

Joule and the German physicist and physiologist Hermann von Helmholtz, whose work helped to establish the rules governing the direction of transformations in the natural world. Their studies were directed not primarily at chemical matters, however, but towards the rather more general question of how heat is transferred within and between different systems – a discipline known as thermodynamics (literally, the motion of heat). Thermodynamics is the fundamental science of transformations: it provides a scientific framework to describe all processes of change in the world, from the formation of black holes to the metabolic pathways of the body, from the way in which weather patterns are dictated by the supply of heat from the Sun to the consequences of an expanding Universe. Common to all these phenomena is the question: why do they proceed in the manner and direction that they do, and not in some other? Why does a drop of ink in water always disperse, but the uniformly tinted solution never unmix to reform the original droplet? Why doesn't water ever run uphill? Why, indeed, does time itself appear to flow only in one direction? Clearly, thermodynamicists have found themselves in the position of having to explain some very profound questions.

There is, however, a universal answer, which is embodied in the so-called Second Law of thermodynamics. (What, you might ask, is the First Law? It is that principle which is commonly known as the "conservation of energy" – energy is never destroyed, but only transformed from one kind to another.) The Second Law states that all realizable transformations are accompanied by an increase in the total amount of *entropy* in the Universe. (Strictly speaking, it says that the entropy cannot decrease; there is a class of transformations – those that can be reversed exactly – for which the entropy content of the Universe can remain unchanged.)

These days, entropy is a term not uncommon in everyday parlance, but it has acquired a certain air of mystery. There is, however, nothing very mysterious about it at all. It can be regarded as a measure of disorder – a pile of bricks, for instance, has more entropy than a house. Similarly, a liquid has a greater entropy than a crystal, since the former the molecules tumble about in disarray while in the latter they are stacked in an orderly, regular pattern. The Second Law is therefore saying that the Universe is bound to become ever more disorderly. This too can appear to be a very recondite and mysterious statement, but in fact it is saying nothing more than that things tend to happen in the most probable way: there is simply a greater probability that things will become disordered than the reverse. The Second Law is therefore actually a statistical law, which does not prohibit absolutely the possibility of a change that induces an increase in entropy, but says only that such a change is overwhelmingly unlikely when we are considering huge numbers of molecules.

## *Uphill or downhill?*

Although the Second Law of thermodynamics provides a universal arrow for specifying the direction in which change, chemical or otherwise, will occur, it is not actually of very much practical use to chemists. The problem is that the Second Law considers only the entropy of the entire Universe, which, as you might imagine, is not an easy thing to measure. In order to predict which way a chemical reaction will go, we need to know not just how the entropy of the reactants differs from that of the products, but also how the heat given off (or consumed) changes the entropy of the surroundings. How heat produced in a reaction changes the surroundings is hard to establish in detail – it will depend on the nature of the surroundings themselves. But fortunately we do not need to worry about these details – the entropic effect of heat dished out to the surroundings depends just on *how much* of this heat there is. If the loss or gain of heat by the chemical system is accompanied by a change in volume (if a gas is given off, for example), this also has an effect on the entropy of the surroundings. When there is a volume change of this sort, the chemical system is said to do work on the surroundings (this work can be harnessed, for example, by allowing the change in volume to drive a piston), and this work must also be taken into account in determining the total entropy change.

We can therefore determine the direction of a chemical change as specified by the Second Law on the basis of just the change in entropy of the reactants, the amount of heat consumed or evolved, and the work done on the surroundings. All of these can in principle be measured. Willard Gibbs expressed the directionality criterion in terms of a quantity called the Gibbs free energy, which quantifies the net effect of these various contributions on the total change in entropy during the transformation. The Gibbs free energy represents the balance between the change in entropy of the system and the change in entropy of the surroundings; the latter is represented by a quantity called the enthalpy, which is the sum of the heat change (due largely to the making and breaking of chemical bonds) and the work done (due to a change in volume).

A chemical reaction is feasible if there is an overall increase in entropy of the system and its surroundings (the latter being an effective representation of the rest of the Universe); this means that, for example, if the products have less entropy than the reactants, this decrease must be more than balanced by an increase in entropy of the surroundings due to the heat given out or the work done via volume changes. This translates into the rule that the Gibbs free energy must decrease. (Strictly speaking, this is true only when the temperature and pressure of the system are held constant. Under different conditions, other kinds of free energy must be considered instead of that defined by Gibbs.) The change in Gibbs free energy therefore defines the "downhill" direction for the reaction. In the same way that a ball perched atop a hill will run down it, thereby reducing its potential energy (the value of which depends on the ball's height above the ground), a chemical reaction will tend to proceed in that direction in which it loses free energy (Figure 2.1).

## The kinetic hurdle

### *The possible and the actual*

The "decreasing-free-energy" criterion seems to suggest that neither this book nor you the reader should exist. That is to say, free energy would be decreased significantly if

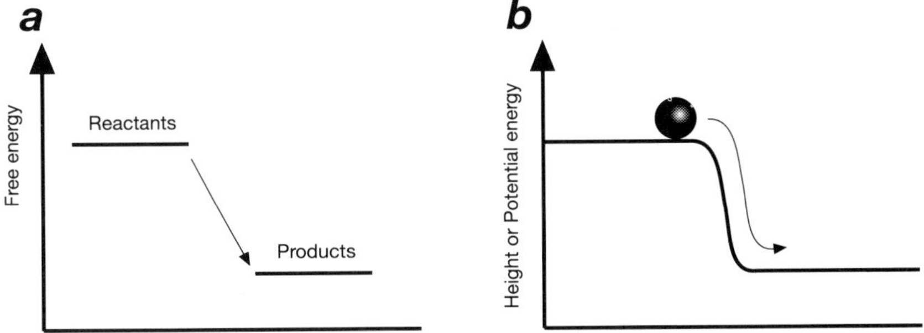


**Figure 2.1** *The "downhill' direction of a chemical reaction is determined (under conditions in which the temperature and pressure are held constant) by the Gibbs free energy: the reaction can proceed in that direction for which the Gibbs free energy of the end products is less than that of the starting materials (a). Similarly, a ball can lower its potential energy spontaneously by rolling down a hill to a lower level (b); but it will not spontaneously roll uphill.*

both were to burst into flames. Indeed, not only does combustion of organic matter give out a lot of heat, increasing the entropy of the surroundings, but converting the orderly molecular structures in our bodies or in the fibers of paper into disorderly molecules of gaseous carbon dioxide and water is accompanied by a vast increase in entropy of the products. So all is geared towards a considerable boost for the total entropy of the Universe, which is to say, a large decrease in Gibbs free energy. Yet unless the book or (heaven forbid) the reader is thrown onto a bonfire, both will continue to exist in our oxygen-rich atmosphere for a considerable time to come. So what is wrong with Gibbs's criterion for determining whether a chemical transformation is up- or downhill?

Happily, there is nothing wrong with it. But it is a criterion for deciding only whether the reaction is feasible *in principle* (which it clearly is). It says nothing about what will happen *in practice*. The vast majority of chemical reactions that are downhill processes turn out to be hindered by a barrier that prevents them from occurring, at least at any significant rate. What determines the feasibility of the reaction is the thermodynamics—considerations of enthalpy, entropy and free energy. But what hinders the reaction from proceeding is the so-called "kinetics" of the transformation.

To understand kinetic hindrance, we need to consider what is happening during a reaction at the molecular level. The way in which atoms are linked together in the product molecules is invariably different from their arrangement in the reactants; this is the very essence of chemical change. The transformation therefore requires that bonds be broken and/or formed. Regardless of what the relative intrinsic energies of the reactant and product molecules might be, the breaking of bonds requires energy. In other words, the initial step of a reaction is generally an uphill process: energy must be supplied to snap the molecules apart before their atoms can recombine (with a concomitant release of energy) into new configurations. Even if a reaction can potentially release a large quantity of free energy, it can proceed only if energy is first supplied in order to

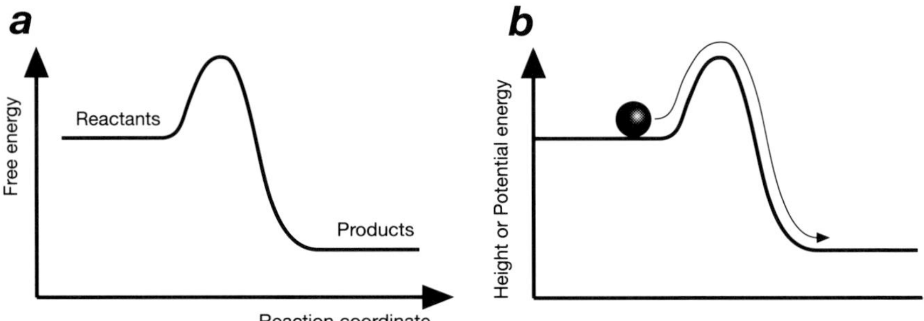


**Figure 2.2** *There is generally a free energy "barrier" that must be overcome in order for a reaction to proceed. Initially the free energy of the reactants increases as bonds are weakened, and only after the barrier is surmounted does the free energy decrease (a). So, like a ball that is prevented from rolling off a hill by a hump (b), the reaction must be given a "push" to get it over the free energy barrier.*

get things underway. From our initial analogy of a ball at the crest of a hill, we now have a picture in which the ball is on a plateau off which it is prevented from rolling by a hump at the edge; to get the ball down into the valley, we must give it a kick to get it over the hump (Figure 2.2). Chemists generally depict the thermodynamic aspects of a reaction using a diagram such as that in Figure 2.2*a*. Horizontal "motion" over the free energy barrier and down into the valley corresponds to the progress of the reaction. The horizontal axis of the graph (called the reaction coordinate) represents the degree to which the bonding between the constituent atoms has changed, or in other words the "extent" of reaction.

Although some reactions do indeed begin with the reactant molecules falling apart into fragments which then recombine in new ways, many others exist in which the breaking of old bonds and the formation of new ones happen at more or less the same time. This is often accompanied by a gradual rearrangement of the other, nonparticipant atoms in the reactant or product molecules. The transformation from reactants to products is then a smooth process involving unstable intermediate atomic configurations. Take, for instance, the reaction between methyl bromide ($CH_3Br$) and a hydroxide ion ($OH^-$) in which the methyl groups swaps its bromine for the hydroxyl group, forming methanol:

$$CH_3Br + OH^- \rightarrow CH_3OH + Br^-$$

This represents the fate of methyl bromide in a solution of sodium hydroxide, since the products have a lower free energy than the reactants. In ousting the bromine atom, the hydroxide ion takes a devious approach: it sneaks up from behind and pushes the bromine out from the opposite side. At some stage during the reaction, the molecule finds itself in a rather uncomfortable "halfway" position where the central carbon atom has a bromine atom on one side and a hydroxyl group on the opposite side (Figure

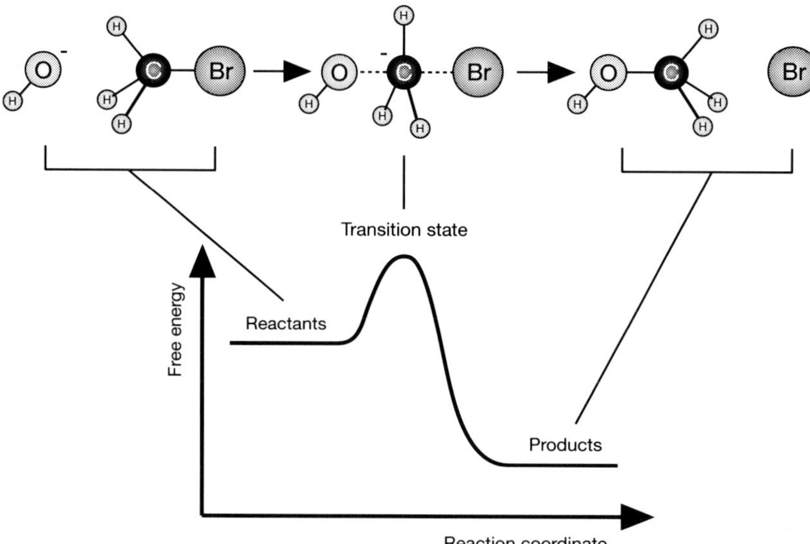


**Figure 2.3** *When methyl bromide is hydrolyzed by a hydroxide ion, the reactants must go through a high free energy state in which the carbon atom is surrounded, uncomfortably, by five others. This corresponds to the transition state.*

2.3). Not surprisingly, this strange creation has a high free energy – it is crowded, and the bonds from the carbon to the bromine and hydroxyl are weak. This transient species lies at the top of the energy barrier that separates the reactants from the products, and is called the transition state. A small push in one direction will send the transition state down towards formation of the products, while a push in the other will send it back towards the reactants.

## *The gentle way to faster reactions*

How quickly a reaction will take place depends on how many of the reactant molecules come together with enough energy to surmount the free energy barrier. By heating the system we increase the number of high-energy molecules and thereby increase the rate.

Even for a thermodynamically favorable reaction we can never get rid of all the reactant molecules, however, since a few of the product molecules will always have enough energy to jump *back* over the free energy barrier. The system will eventually settle down into a state in which the average number of molecules on each side does not change: this corresponds to the state of so-called thermodynamic equilibrium. The lower the free energy of the products below that of the reactants, the harder it is for the product molecules to get back over the barrier and so the greater is the proportion of product present at equilibrium. By heating a chemical system we increase the rates of both the forward and the backward reactions, and thus we hasten the progress towards thermodynamic equilibrium.

But heating is a rather crude way of increasing a reaction's rate, and may bring with it complications. For a start, it makes the reaction more expensive to carry out, as we have to supply it with energy. And one may find that either the reactants or the product are unstable at high temperatures – raising the temperature may accelerate other, undesirable processes such as the break-up of the very molecules that one wants to form. The proportion of molecules that have sufficient energy to overcome the barrier to reaction depends not only on the temperature, however, but also on the height of the barrier. If we could lower the barrier, more molecules would be able to react at a given temperature, and equilibrium will accordingly be reached faster.

The height of the barrier is determined by the free energy of the transition state, which is itself dependent on the molecular structure of this ephemeral entity. The function of a *catalyst* is to decrease the free energy of the transition state, thus making it a less unstable species. More precisely, the catalyst must interact with the reactants so as to form a modified transition state at a lower cost in free energy (Figure 2.4). A true catalyst must emerge from this interaction unscathed: once over the free energy hill, the transition state must evolve into the products without permanently changing the state of the catalyst, leaving it free to do its job again on other reactant molecules. If the catalyst were somehow altered or consumed in the process, we would have to renew it constantly; indeed, it would then not really be a catalyst at all, but merely another reactant.

Broadly speaking, catalysts come in two classes. To illustrate the difference between the two, imagine the reactant molecules as a timid couple who, despite being made for each other, would be too shy ever to unite as a pair if left to their own devices. The first way to catalyze the union is to add into the equation a matchmaker who will bring the two together, introduce them and kick off the date by starting a conversation to which they can both contribute. In chemical terms, the catalytic matchmaker is a single

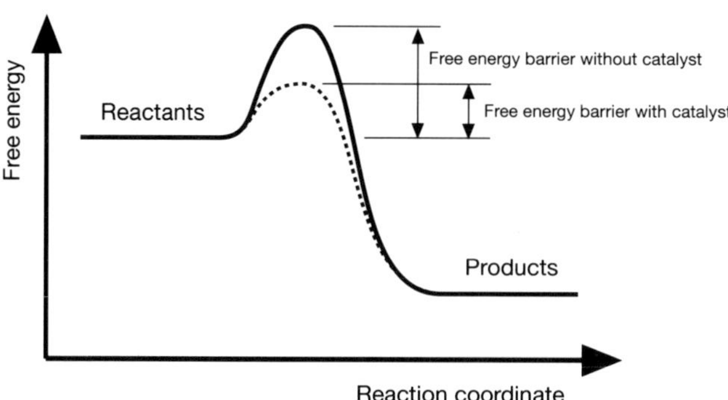


**Figure 2.4** *The function of a catalyst is to lower the free energy barrier to reaction; in other words, to lower the free energy of the transition state.*

molecule, just like the reactants, and will be in the same physical state as the reactants – all, for example, may be in the gas phase or in solution. This type of catalysis is said to be homogeneous.

Alternatively we could arrange for our pair to meet in an environment conducive to their becoming relaxed and friendly; as they are, let us say, both music lovers, we arrange for them a night at the opera, in which situation the relationship should flourish. Catalysts that provide favorable environments for the reaction generally have to be in a different state to the reactants – they are usually solids, while the reactants are liquids, gases or solutes. The environment in which the reactants get together is often the surface of the catalyst, on which the chance of a reaction is enhanced; other catalysts in this class contain molecular-scale cages or tunnels in which the reactants are trapped and held together in intimate proximity. This type of catalysis is called heterogeneous. Some researchers are now turning their attention to catalysts that lie at the boundary between homogeneous and heterogeneous systems: clusters of atoms that are bigger than the reactant molecules, but only a little. It is hoped that this new field of catalysis, still very much in its infancy, might combine some of the advantages of both approaches.

Homogeneous catalysts are generally molecules that have been carefully selected or designed to interact with the reactants in a very specific way, often to ensure that just one of several possible products is formed. The successful design of such catalysts requires a detailed knowledge of the way in which the reactants interact both with each other and with the catalyst, and it is only in recent decades that this sort of information has been easily accessible; in earlier times, homogeneous catalysis was very much a case of trial and error. Homogeneous catalysts are potentially capable of more delicate tasks than most heterogeneous varieties, and there is no better illustration of this delicacy than that provided by enzymes – homogeneous catalysts designed by nature. Although enzyme catalysis is still understood only in a rudimentary fashion, it is clear that it has much to teach us about designing synthetic homogeneous catalysts.

Heterogeneous catalysis is an older, more traditional approach. While these catalysts are commonly cruder and less selective than their homogeneous rivals, they have been and remain the mainstay of much of industrial chemistry, and sometimes play an important role in natural processes too. In recent years, moreover, new heterogeneous catalysts have been developed which demonstrate a degree of selectivity to rival that of homogeneous systems.

## Staying on the surface

Metal surfaces are the archetypal heterogeneous catalysts, in particular those of the transition metals such as nickel, palladium and platinum. These materials are able to induce a wide range of reactions between gases that would otherwise hardly proceed at all (Table 2.1). In the presence of platinum metal, carbon monoxide and oxygen combine to form carbon dioxide, a process central to the operation of catalytic

**Table 2.1** Some important metal-catalyzed industrial reactions

| Metal catalyst | Reaction |
|---|---|
| nickel | hydrogen + unsaturated vegetable oils → saturated vegetable oils |
| iron | nitrogen + hydrogen → ammonia |
| silver | ethylene + oxygen → ethylene oxide |
| platinum/rhodium | ammonia + oxygen → nitric acid |
| iridium/rhodium | carbon monoxide + oxygen → carbon dioxide |

converters. Nickel can induce all kinds of unsaturated hydrocarbons to react with hydrogen to form saturated compounds. This is an important process in the food industry since it converts unsaturated vegetable oils to their saturated counterparts. Ammonia, a crucial component of fertilizers and explosives, is manufactured in vast quantities by the Haber process, the reaction of nitrogen and hydrogen over an iron catalyst. Nitric acid is produced from ammonia and oxygen, courtesy of a catalytic mixture of platinum and rhodium metals; the production of polyethylene from ethylene uses catalysts based on chromium and titanium; and the petrochemical industry, which manufactures from crude oil all manner of hydrocarbon compounds for plastics, fuels and other uses, relies heavily on platinum and other metal catalysts.

In all of these examples, reactions that are hindered by often formidable free energy barriers are greatly accelerated in the presence of the metal surface. The general principles of the catalytic mechanism are the same in every case, relying on the anomalous reactivity of metal atoms at the surface of the material. These exposed atoms have a strong capacity for forming bonds, and will readily bind gas molecules that impinge on the surface – a process known as adsorption.

The strength of the interactions between the surface atoms of the metal and the adsorbed gas molecules (called the adsorbate) can vary considerably, depending on the chemical nature of the two. Sometimes full-blown chemical bonds are formed, and the gas molecule is therefore held in place quite tightly and rigidly. This is called chemisorption, short for "chemical adsorption." Alternatively the bond can be a fairly weak one, strong enough to hold the adsorbate close to the surface but not, perhaps, to prevent it from wandering from one surface atom to the next. This is physisorption ("physical adsorption").

When it undergoes adsorption onto a surface, a gas molecule cannot remain unmoved by the experience. A chemisorbed molecule can form a new bond only by rearranging its existing bonds; and even for physisorption a new bond of sorts is formed, necessarily weakening those within the adsorbed molecule. When ethylene becomes attached to platinum, for example, the double bond between the two carbon atoms is broken open and both carbon atoms form bonds to platinum (Figure 2.5*a*). The weakening of bonds in the adsorbate is often so pronounced that the molecule falls apart, leaving atoms or molecular fragments to wander over the surface. On iron, for example, carbon

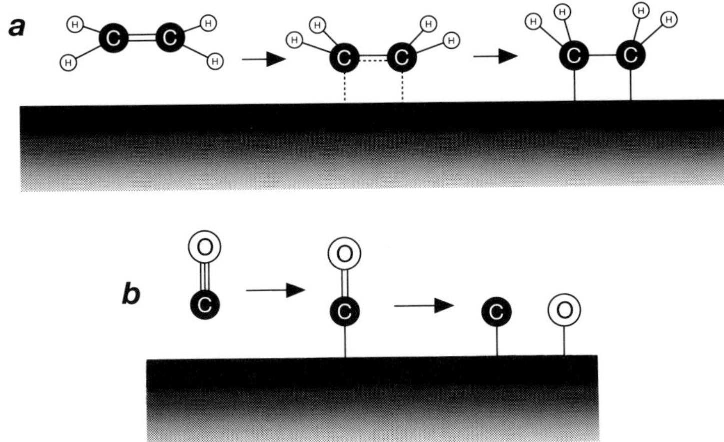


**Figure 2.5** *When an ethylene molecule is chemisorbed onto the surface of platinum, its double bond is broken so that the carbon atoms can form bonds to the surface (a). The chemisorption of carbon monoxide on iron, meanwhile, can result in the complete fragmentation of the molecule into surface-bound atoms (b).*

monoxide is split into its constituent atoms of carbon and oxygen (Figure 2.5*b*). On some metal surfaces, such as platinum, hydrocarbons like ethane ($C_2H_6$) may be broken down entirely into individual, surface-bound carbon and hydrogen atoms.

What this means from the point of view of catalysis is that the surface assists in the disassembly of reactant molecules, the fragments of which can then reassemble on the surface to form the products. To break these bonds in the gaseous phase requires a considerable input of energy, but the metal may allow the process to occur under much more moderate conditions. Not only does the catalytic surface weaken or break bonds in the reactants, but it also provides a meeting place where the reactants can gather in higher concentrations than in the gas, increasing their chance of encountering one another. Both the bond-weakening effect and the enhancement in concentration of species primed for reaction contribute to the efficiency of surface catalysts.

The exact nature of the products resulting from a given set of reactants can often be controlled by varying the conditions under which the reaction takes place, such as the temperature or the relative proportions of the reactants. Equally, the composition of the product may be sensitive to the character of the catalyst itself: different metals may yield different products (or mixtures in different proportions) from the same starting materials. For example, carbon monoxide and hydrogen react on nickel to form largely methane and water, whereas on copper or palladium the main product is methanol. The outcome of a reaction may also depend on the size of the catalyst particles, the chemical nature of the material on which they are supported (particles of catalytic metals are often supported on silica ($SiO_2$) or alumina ($Al_2O_3$) surfaces), or the way in which the catalyst has been prepared. All of these processes tend, however, to produce a mixture of products – various (usually unwanted) by-products are formed as well as the one desired. We should scarcely be surprised at this: if we go to the lengths of pulling apart the reactants into their constituent atoms, it is only to be expected that not all will reassemble in the manner desired. Indeed, there are so many ways in which atoms of carbon, hydrogen and oxygen could assemble, even into simple molecules, that we should perhaps be grateful and a little surprised that we can get any worthwhile quantities of methanol at all from the reaction of carbon monoxide with hydrogen on copper (Table 2.1). Most simple catalytic surfaces, and heterogeneous catalysts in general, suffer from this problem of nonselectivity.

## Molecular sieves

### *The selective networks*

Nonselectivity in heterogeneous catalytic reactions can be a particularly serious headache for the petrochemical industry. The various hydrocarbon components of crude oil can be separated effectively by the process of fractional distillation; but catalysts are needed to convert these fractions into more useful chemicals. Chapter 1 hinted at the bewildering variety of ways in which carbon and hydrogen atoms can be joined together – even rather light hydrocarbon molecules may have a vast array of isomers. Trying to prepare a specific compound via the catalytic reaction of a hydrocarbon is therefore a daunting task when one is faced with the lack of selectivity characteristic of simple heterogeneous catalysts.

There now exists, however, an ever-expanding class of materials that catalyze this kind of reaction with a degree of selectivity that would be unthinkable on traditional metal surfaces. They are called zeolites. The first zeolite catalysts were naturally occurring minerals; the name means "boiling stone," since natural zeolites contain a considerable amount of water which can be removed by heating. A great many novel zeolitic materials have now been synthesized artificially, however.

The natural zeolites, and many of the artificial varieties, are aluminosilicates – compounds whose principal constituents are aluminum, silicon and oxygen. The fundamental building blocks of aluminosilicates are two units shaped like tetrahedra: one in which a silicon atom is surrounded by four oxygens ($SiO_4$), and the other in which aluminum replaces silicon at the center of the tetrahedron ($AlO_4$) (Figure 2.6*a*). Each of the $AlO_4$ units bears a negative charge, while the $SiO_4$ units are uncharged. In zeolites these tetrahedra are joined together via the oxygen atoms at their corners to form continuous frameworks. Because of the charges on the $AlO_4$ units, the aluminosilicate framework as a whole bears a negative charge, and positive metal ions (most commonly sodium) sit in the gaps of the framework to balance the overall charge. The zeolite frameworks comprise cage-like units, such as the sodalite cage (Figure 2.6*b*),

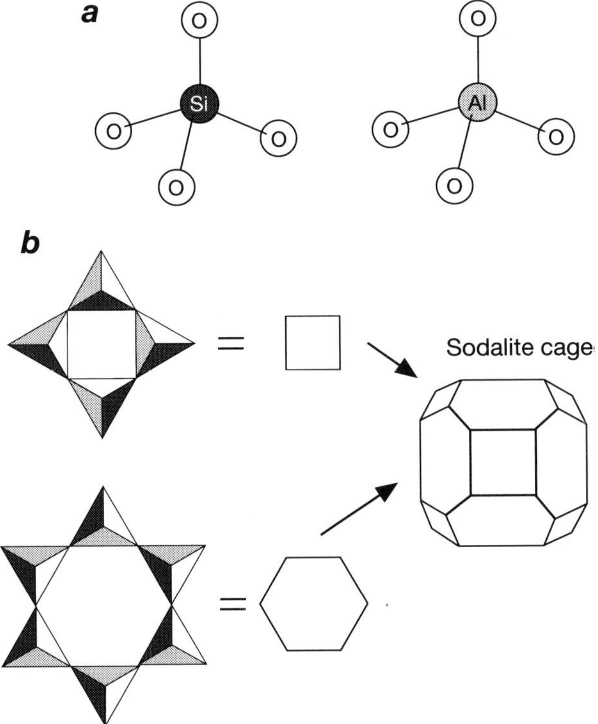


**Figure 2.6** *The building blocks of zeolites: tetrahedral units of $SiO_4$ and $AlO_4$ (a). The former are electrically neutral, while the latter bear a negative charge. These tetrahedra are linked into four- and six-membered rings (b), which are themselves the elements of larger units such as the sodalite cage.*

linked via a series of necks and channels (Figure 2.7). To the eye, the crystals appear to be dense solids, but on the atomic scale they contain a labyrinthine network of pores.

Solids such as pumice and sandstone have much larger pores which can be seen with the naked eye or through an optical microscope; pores of this size are called macropores, and they tend to make a substance weak and liable to collapse. The pores of zeolites, however, are a fundamental feature of the crystal's atomic structure, and measure only a few atomic diameters across, leaving the materials structurally robust. Pores of this size are called micropores. Because of all the open space encapsulated within the zeolite's cavities and channels, the total surface area of the solid is vastly greater than that visible to the eye on the crystal faces. This visible surface area is in fact utterly negligible in comparison with the internal area on the walls of the pores. For example, a zeolite crystal weighing just a gram typically has a total surface area of nearly 1,000 square meters.

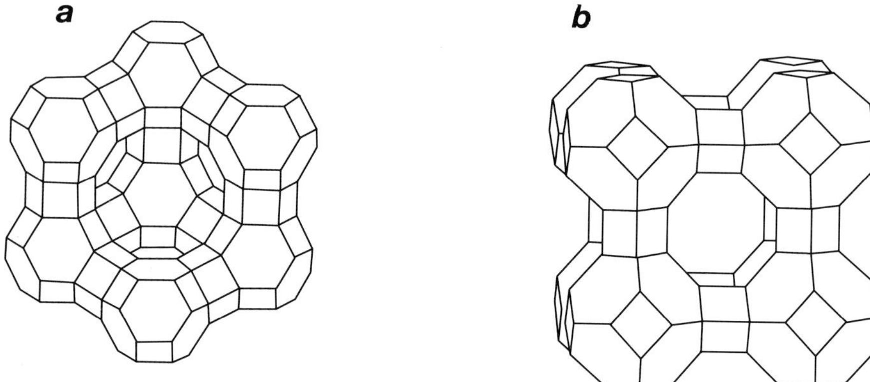


**Figure 2.7** *The structures of Y-type (a) and A-type (b) zeolites. In both of these materials, large "supercages" are accessed by narrower channels.*

When, in the 1950s, zeolites first became used for catalysis, researchers had to make do with those zeolitic materials found in nature, all of which were aluminosilicates. Today, however, the most commonly used commercial zeolites are man-made, such as the zeolite ZSM-5 developed at Mobil Oil Corporation around 1970. Some of these synthetic compounds have frameworks made from atoms other than just silicon, aluminum and oxygen; in particular, phosphorus can replace silicon to give porous aluminophosphates. In 1988 a group of researchers at Virginia Polytechnic Institute created an aluminophosphate with unprecedentedly large channels, accessed via rings of no less than eighteen atoms (Plate 3). This material, known as VPI-5, has demonstrated great catalytic utility by virtue of its wide pores. Other synthetic zeolites have been created that incorporate elements such as gallium, boron and beryllium. This substitution of one kind of atom for another allows fine tuning of the zeolite's catalytic properties.

## *Catalysis in zeolites*

In order to act as heterogeneous catalysts, the zeolites must of course provide catalytically active surfaces. Aluminosilicates are oxide compounds – in chemical terms they are very different from pure metals. Many oxides prove to be versatile catalysts in their own right. Metal oxides such as those of vanadium and molybdenum are used widely for catalyzing the oxidation of hydrocarbons (that is, converting them into oxygen-containing organic compounds by inducing a reaction with oxygen gas). Important industrial chemicals such as acetone and formaldehyde are prepared in this way. But simple aluminum and silicon oxides have different talents – rather than promoting oxidation reactions, they are adept at rearranging hydrocarbons into new forms. Specifically, they are used for hydrocarbon "cracking" (the breaking down of large hydrocarbons into smaller ones), polymerization (the joining of small, unsaturated hydrocarbons into larger chains) and isomerization (the rearrangement of a molecule's atoms into a new structure). These, then, are the processes on which zeolites can exercise their selective capabilities.

There are several mechanisms by which microporous zeolitic materials may achieve product-selective catalysis. The simplest of these relies on pre-selection of the reactants themselves: the size and shape of the channels is such that only certain molecules can get inside. Micropores have a width comparable to the size of many simple molecules, and as a result zeolites have the ability to act as "molecular sieves." Small molecules such as hydrogen, nitrogen and methane can pass through the pores with ease, but larger ones, such as the heavier hydrocarbons, are too big. The pore openings of the so-called X-type zeolites, for example, have a width slightly larger than that of a benzene molecule, so this molecule can enter the channels – but benzene-based compounds with bulky groups hanging off the carbon ring cannot do so. The pores of A-type zeolites, meanwhile, are smaller and impassable to benzene itself. As a consequence of this molecular sieving ability, zeolites can absorb certain gases into their internal warren of tunnels and cages (and in vast quantities, given the amount of internal free space) while excluding others entirely. This provides a route to selective catalysis.

Selective absorption in zeolites may be influenced not only by molecular size but also by shape. The saturated hydrocarbons obtained from crude oil generally come in the form of long, thin, chain-like molecules. Looked at end-on, these are not much bigger than the methane molecule, and so despite the fact that they may contain a considerable number of atoms, these linear hydrocarbons can wriggle like snakes down the pores of zeolites. Other isomers, meanwhile, have branched structures or appendages sticking out from the chain, which block their entry into the pores (Figure 2.8). This imparts to the zeolite a selectivity for absorbing different isomers of a single compound.

Selective absorption of reactants is exploited in the process of "selectoforming," in which the octane rating of petroleum-based fuels is boosted by removing the low-octane

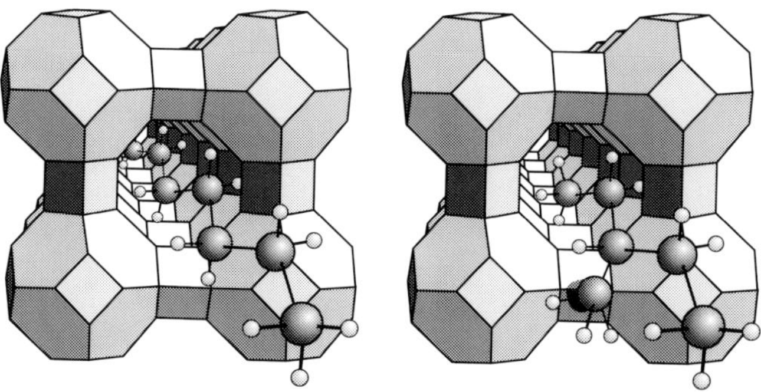

**Figure 2.8** *Straight-chain hydrocarbon molecules can wriggle down the pores of A-type zeolites, but branched chains are too bulky and cannot get inside.*

straight-chain hydrocarbons from petrol while leaving behind those that are branched or that have benzene rings attached. (The octane rating increases as the bulkiness of the hydrocarbons increases.) A small-pore zeolite is used, into which only the straight-chain hydrocarbons can penetrate. Here they are "cracked" (that is, broken down into shorter chains which can be boiled off) while other, bulkier hydrocarbons are left intact. The zeolite ZSM-5 allows a further improvement on this process because its medium-sized pores are big enough also to admit into the catalytic interior hydrocarbons with a single methyl ($CH_3$) branching side group.

Another type of selectivity in molecular sieves results from the limited space available to form products from the molecules that do get inside. Some of the products

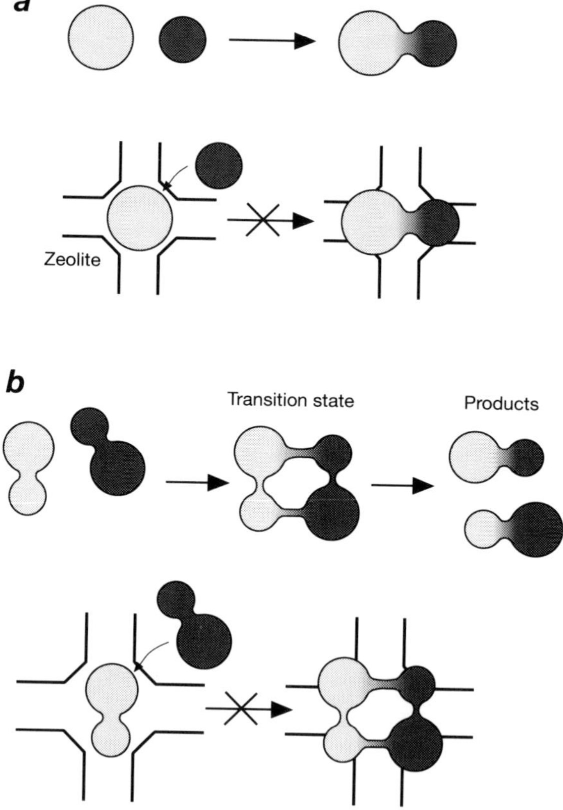


**Figure 2.9** *How catalysis within zeolites is selective. When reactants come together inside the channels or supercages, there may be insufficient room for reactions that would occur on the outside. Potential products that are too big to fit inside the supercages will not be formed (a); and if the transition state of a reaction is too big, that reaction too is prohibited (b).*

that would be formed if the reactants were mixed outside the channels may be too large to be put together inside (Figure 2.9*a*). Alternatively, some of the products that are formed within a zeolite's cages may be too large to escape through the narrower channels, so that the mixture of products that comes out of the zeolite may not reflect the mixture that was actually formed within it. The problem with this trapping of products too large to escape has the drawback, however, that it tends to clog up the pores.

Similar to, but subtlely different from, product selectivity is transition-state selectivity. We saw earlier how reactions pass through a transition state in going from reactants to products. Because it commonly represents a coming-together of reactant molecules (which may then part again in a different form to give the products), the transition state is generally a bigger, bulkier beast than either the reactants or the products. So the various pathways a reaction will take inside a molecular sieve may be restricted by the amount of space available to form the corresponding transition states (Figure 2.9*b*).

## *Zeolites as molecular containers*

Despite their crucial function in many catalytic processes for the petrochemical industry, some researchers feel that zeolites have become typecast into a limited number of chemical roles that does scarce justice to their potential. What zeolites provide, they suggest, is not simply a crude sieve for sifting out one hydrocarbon from another; rather, the channels and cages represent a kind of molecular-scale scaffolding that can be exploited for a vast range of new and exciting uses.

One such is the "ship-in-a-bottle" catalyst developed by Norman Herron and colleagues at E. I. du Pont de Nemours & Company in Delaware. Herron's idea was to capture inside the zeolite supercages a catalytic molecule that is too large to escape through the narrow exits from the cavity. The molecule would then be permanently encased within the crystal like a spider in its lair, waiting to perform its catalytic function on small molecules admitted through the passageways. The result would be a reuseable "composite" catalyst which could be easily separated from the products.

In order to encapsulate the catalytic molecule, it must be assembled within the supercage from smaller components fed through the channels, just as a ship must be constructed through the neck of the bottle that encloses it. For their prototype system, the du Pont team chose to trap within zeolite Y a compound called cobalt salen, which consists of the organic molecule salen bound in a "complex" with a cobalt ion (Figure 2.10). First the researchers infused the supercages of the zeolite with bare cobalt ions; then they fed down the channels the salen molecules. These molecules are flexible and able to pass down the channels; but the assembled metal complex is too big to escape from the supercage.

The du Pont researchers went on to develop this ship-in-a-bottle approach to create a catalytic system capable of mimicking the natural enzyme cytochrome P450. This enzyme is extremely adept at adding oxygen atoms to organic molecules under the

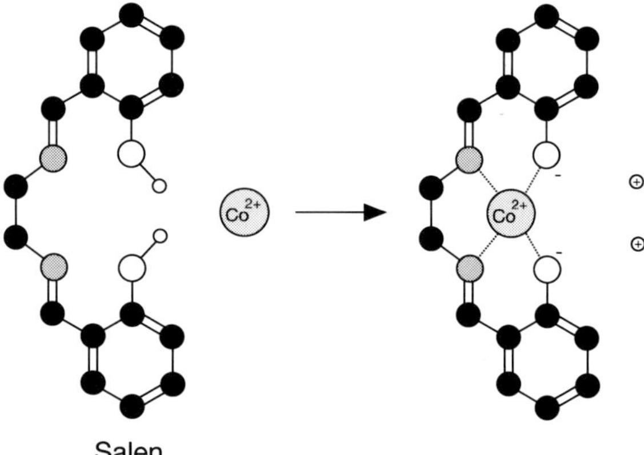


**Figure 2.10** *Norman Herron and colleagues assembled the catalytic cobalt salen complex inside zeolite Y by passing its components—cobalt ions and the organic molecule salen—down the channels. When they came together inside a supercage, the resulting complex was too big to escape.*

very mild conditions that prevail in the body. In cytochrome P450 the active catalytic site is a complex consisting of an iron ion sitting inside the ring of a porphyrin group (Figure 2.11*a*). Herron and colleagues assembled within zeolite Y an iron-containing complex called iron phthalocyanine (Figure 2.11*b*), which they expected to be a good mimic of the natural cytochrome, by passing the iron and phthalocyanine components down the channels separately and letting the complex form in the supercages (Figure 2.11*c*). When floating around freely in water, these complexes have a tendency to react with each other rather than doing their intended job of oxidizing other organic molecules. But trapped within the zeolite supercages, the iron phthalocyanine molecules are presented from encountering one another, and so they retain their catalytic ability. When loaded with oxygen, the trapped complexes did indeed show a capacity for oxidizing simple hydrocarbons.

In these examples, the zeolite cages were being used as a kind of "molecular bottle" in which to carry out chemical reactions. A related idea is to exploit the shape and size of passages and cavities in the pore network as "molds" for growing materials. Crystals grown inside a zeolite supercage, for instance, should all end up with more or less the same size and shape, containing perhaps only a few dozen atoms each. Researchers are interested in tiny crystals (or "clusters") of this sort for a variety of reasons, amongst which is the possibility that they might themselves show useful catalytic activity. Of particular interest are clusters of semiconducting materials, which are expected to have interesting light-absorbing properties as a result of their finite size. Clusters of this sort, often called "quantum dots," might provide elements for use as transistors or switches in optical computers (which use light rather than electricity to carry and process information). The du Pont researchers, amongst others, have created clusters of the

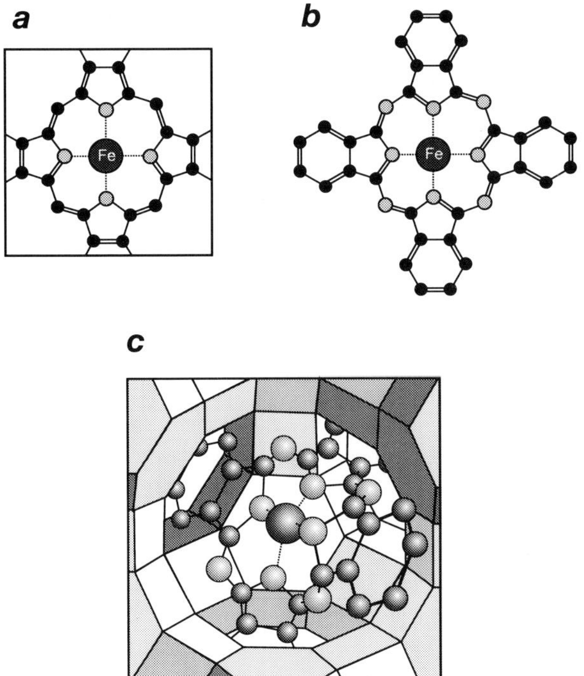


**Figure 2.11** *Iron phthalocyanine (b) mimics the catalytically active porphyrin unit of the natural enzyme cytochrome P450 (a). When assembled inside the supercages of zeolite Y, the iron phthalocyanine complex becomes a trapped "ship-in-a-bottle" catalyst (c). (Picture (c) after Norman Herron, E. I. du Pont de Nemours Company, Delaware.)*

semiconductor cadmium sulfide within the cavities of zeolites – the clusters have almost uniform sizes determined by the amount of space available.

The possibility of attaching molecules to the walls of the pores extends the use of zeolites as molecular molds to that of molecular scaffolding: the network can act as a kind of template on which assemblies of molecules can be created. This kind of template assembly is somewhat reminiscent of certain biological processes, particularly that in which DNA molecules make copies of themselves (Chapter 5).

## Catalysis the natural way

### *Enzymes: nature's engineers*

The way in which zeolites act as highly selective catalysts bears some striking resemblances to the functioning of nature's catalysts, the enzymes. These are huge protein molecules, often containing many thousands of atoms, which are structured in a way that makes them very sensitive to the shapes of the molecules on which they perform their catalytic function. Like zeolites, they have cavities into which the target molecules are admitted, and within which chemical processes take place at a catalytic surface that often contains embedded metal ions as the "active" components.

But enzymes are in most respects much more complex and dexterous examples of molecular-scale engineering than zeolites. While the latter derive their selectivity from the rather crude determinants of pore size and cavity shape, enzymes are designed by nature to receive their targets as a lock receives a key – usually only a single kind of target molecule (called the enzyme's substrate) will have the right shape to fit within the enzyme's catalytic site. Thus most enzymes serve only a single, well-defined catalytic function, in contrast to the propensity of synthetic catalysts to induce general *classes* of reaction. This tremendous specificity of enzymes is ultimately the result of the kind of "molecular recognition" processes that we will encounter in Chapter 5.

Just about all of the chemical processes that take place within a living organism rely on enzyme catalysis. There are thought to be something like 7,000 different sorts of enzyme that occur naturally; this may sound like a lot, but when you consider that there is estimated to be between 3 and 30 million different species of living organisms, it becomes clear that certain types of enzyme must do the same task in very different types of organism. Many have been found to be shared by organisms ranging from fungi and bacteria to fish and humans. The implication is that, once evolution hits on a good way of carrying out a biochemical process, it tends to stick with it.

All enzymes are essentially protein molecules – polymers in which the basic structural units are amino acids. Twenty varieties of amino acid have been identified in proteins. They possess both an "acid" group, the carboxyl group COOH, and a "basic" group, the amino group $NH_2$ (Figure 2.12). These two groups can react together so as to become linked by a chemical bond, called a peptide bond. Chains of amino acids linked by peptide bonds are generically termed polypeptides; proteins are giant polypeptides that occur naturally. Many of the fibrous tissues of the body are made from nonenzyme proteins, such as keratin (which forms skin, hair and nails), collagen (tendons) and myosin (muscle). Enzymes adopt their highly specific shapes (Plate 4) by an extremely well orchestrated process of folding of the polypeptide chain, a process that is still far from understood.

Enzymes typically measure up to a tenth of a micrometer across, and many are assemblies of more than one molecular unit, held together by noncovalent bonds. It is therefore unclear whether they should be thought of as homogeneous or heterogeneous catalysts – they are certainly bigger than the single molecules that chemists generally recognize as belonging to the former class, but are much smaller than the "bulk" particles of metals or other inorganic solids used for heterogeneous catalysis. The truth is that they display aspects of both classes.

The tasks that enzymes must perform are often very difficult, if not presently impossible, for chemists to achieve artificially. In general, an enzyme's function will be to cut away a part of a molecule or to stitch its component parts together. Some, for

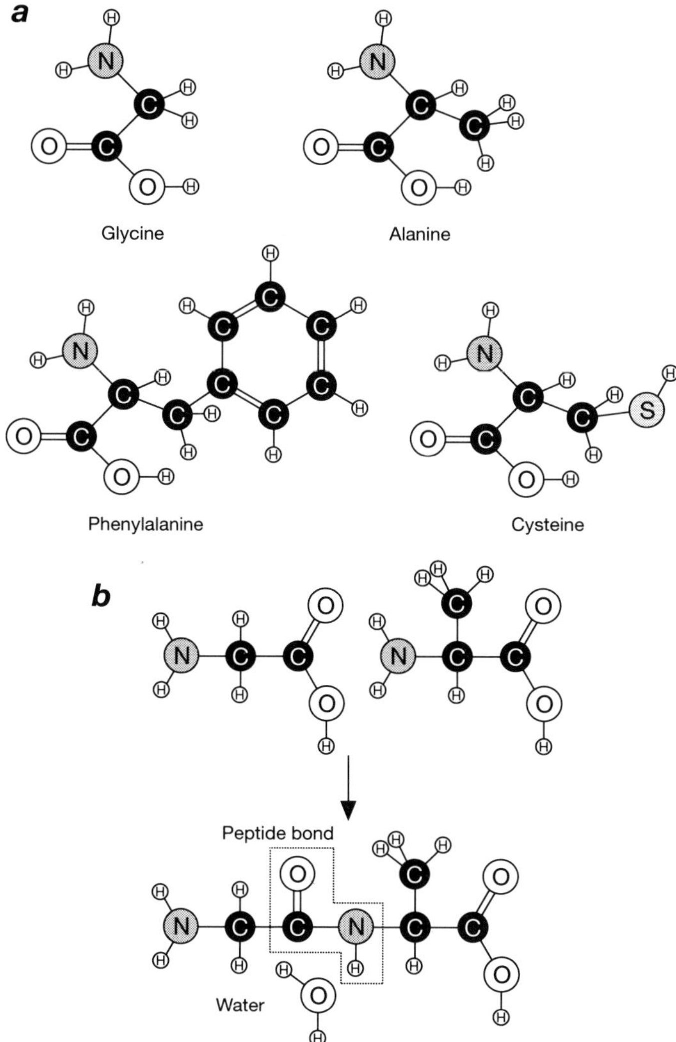


**Figure 2.12** *Some representative amino acids, the components of protein molecules: glycine, alanine, phenylalanine and cysteine (a). In water, the carboxyl groups tend to shed a hydrogen ion, forming the negatively charged carboxylate (—$CO_2^-$) group, and the amino groups are protonated to yield the —$NH_3^+$ unit. Molecular ions like these, bearing both positive and negative charges, are called zwitterions. In proteins, amino acids are joined together via peptide bonds (b).*

instance, break down large carbohydrate molecules into the sugar glucose and then further into carbon dioxide and water – the reaction sequence of our principal metabolic pathway. Breaking down large molecules into smaller ones is something that is not so difficult to achieve by nonbiological catalysis, but the trick that enzymes perform is to do this in a controlled manner that allows the energy released by bond breaking to be stored as chemical energy rather than being wasted as heat. Other enzymes help to put together protein molecules from their constituent amino acids, or assemble the gene-carrying molecule DNA – tasks that lie well beyond the present capabilities of organic chemists.

### *Enzymes for industry*

Enzymes represent perhaps the industrial chemist's ideal: a catalyst that can perform a very specific task efficiently and under extremely mild conditions. They are therefore now finding ways to put nature's catalysts to work to facilitate reactions in the chemicals industry.

There are two ways in which enzymes can be used in this way. The first is to deploy them within their natural environment: within living cells. Microorganisms such as bacteria are used as "living factories" to convert raw materials enzymatically into the desired product. The advantage of this approach is that the enzyme is provided with everything that it needs to do the job. Many enzymes must interact with additional molecules called cofactors, or with certain metal ions, in order to be able to function properly, and these will be present already within living cells. But the outcome of such a process can be unreliable – while the action of any specific enzyme is highly selective, a microorganism may send the reactants down a variety of reaction pathways to yield a mixture that includes unwanted by-products.

The second alternative is to separate the enzyme from the cells that produce it, and to use the pure form for catalysis. Microbes may be bred to provide a source of some of the enzymes used in this way, while others may be extracted from the cells of more advanced animals or of plants. The isolated enzyme can be used to carry out extremely delicate chemical syntheses, but it must be supplied with all the cofactors that it requires, which themselves must be isolated and purified. In addition, some way must be found to immobilize the enzyme molecules in the reaction system, so that they do not get flushed away with the product. This method of immobilization must not affect the enzyme in such a way as to render it inactive, by distorting its shape for example.

The first of these two methods – the "whole-cell" approach – is at least as old as ale. Brewers have for centuries exploited a certain kind of yeast (the organism *Saccharomyces cerevisiae*) to ferment sugar – that is, to convert saccharides into alcohols. Yeast cells employ several types of enzyme for this task. But anyone who has tried to brew their own beer will recognize another of the hazards of the whole-cell approach: the microorganisms can be inadvertantly poisoned and killed off unless the conditions are kept just right.

Baker's yeast proves useful for a number of industrial processes, particularly ones involving the addition of hydrogen atoms. For example, one of the crucial and most delicate steps in the synthesis of the compound coriolin, which is used in cancer treatment, involves the conversion of a carbonyl group (C=O), attached to a five-membered carbon ring in the precursor molecule, into an alcohol group (CHOH) (Figure 2.13). This involves the addition of two hydrogen atoms, one onto the carbon

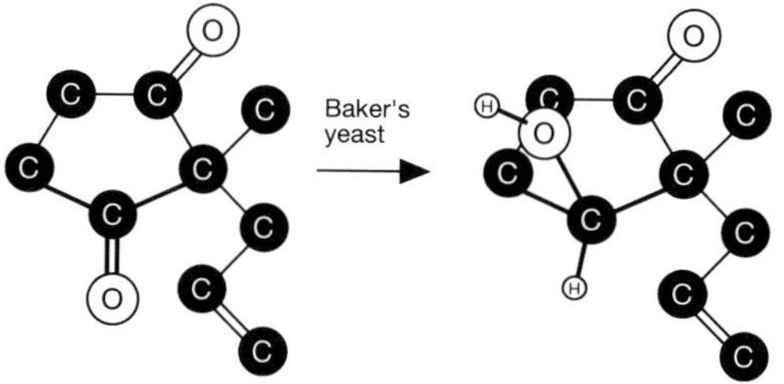


**Figure 2.13** *A crucial step in the synthesis of the drug coriolin involves the addition of hydrogen to a carbonyl (C=O) group. A natural enzyme in baker's yeast ensures that the C—H bond is formed on the correct side (here the lower face) of the five-atom ring. (Hydrogen atoms attached to carbons are not shown.)*

ring and the other onto the oxygen of the carbonyl group. Getting the former to add consistently onto the correct side of the ring would be extremely difficult with standard techniques of organic synthetic chemistry; but for the enzymes in baker's yeast the task is routine.

This reaction illustrates one of the most valuable aspects of enzyme-catalyzed chemistry: it can be *enantiospecific*. The majority of natural biomolecules contain atomic groupings that can exist in two mirror-image forms, like a left and a right hand. These groupings, called "chiral centers," are generally carbon atoms to which four different substituents are attached (Figure 2.14). Molecules possessing a single chiral center are ascribed a "handedness" as a result of their ability to rotate the plane of polarized light either clockwise or anticlockwise. As the two mirror-images of such chiral molecules are identical but for their handedness, they are isomers of one another: these isomers are called enantiomers. In the body, chiral molecules are generally found in only one enantiomeric form – all natural amino acids (bar glycine) are chiral, and are found

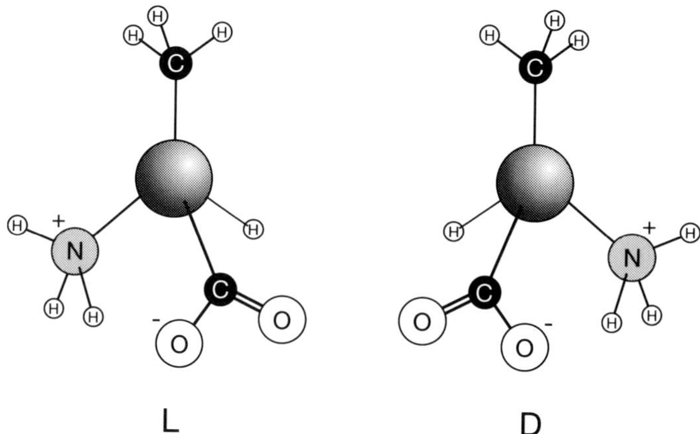


**Figure 2.14** *There are two distinct arrangements of four different substituents around a carbon atom (such as those in alanine, shown here in zwitterionic form), giving molecules that are mirror-images of one another. Such molecules are said to be chiral, and the two forms are called enantiomers. They cannot be interconverted without breaking or grossly distorting the bonding arrangement. Chiral molecules can rotate the plane of polarized light, a property called optical activity. The enantiomers are denoted* L *or* D *depending on whether the sense of rotation is to the left or the right.*

only in the "left-handed" form (denoted L), while all sugar molecules are right-handed (the *dextro* or D form).

As natural chiral molecules are the result of enzyme-mediated biochemical processes, it is clear that enzymes must have an extraordinarily efficient capacity for distinguishing between different enantiomers; in other words, they show a high degree of enantiospecificity. In an enzyme-catalyzed reaction involving a chiral molecule, the enzyme will usually enhance the rate of reaction only for one of the enantiomers; and in a reaction that produces chiral molecules from nonchiral starting materials, enzymes will generate only one of the two possible enantiomers of the product molecule.

Making chiral molecules from nonchiral ones poses a great challenge to organic chemists, because unless handedness can somehow be incorporated into the synthetic process, the result will be an equal ("racemic") mixture of both enantiomers. In general, this means that the catalyst must itself be chiral. Although chemists have now developed a variety of purely artificial catalysts of this sort, their successful synthesis involves a certain amount of trial-and-error and their application generally yields some of the "wrong" enantiomer mixed in with the desired product. As illustrated by the tragedies that resulted from the use of the chiral drug thalidomide, the physiological consequences of this can be profound. The ability of enzymes to churn out the right enantiomer with essentially absolute fidelity is therefore a tremendous boon to industrial and pharmaceutical chemists.

Aspartic acid – a precursor to the artificial sweetener Aspartame (better known as Nutrasweet) and a valuable starting material for many pharmaceutical products – is produced in purely left-handed form when ammonia and fumaric acid are fed to colonies of the microbe *Escherichia coli*, which is commonly found in the gut. These microorganisms contain the enzyme aspartase, which joins ammonia and fumaric acid (both nonchiral) in such a way as to form left-handed aspartic acid only (Figure 2.15).

Some enzymes, called isomerases, can convert one enantiomer to the other. One such, called glucose isomerase, is used in the so-called "corn syrup process" to transform the sugar glucose into its enantiomeric partner fructose. Fructose tastes sweeter than

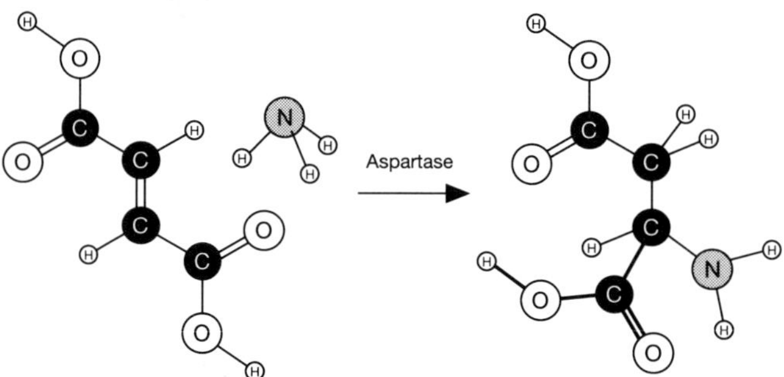


**Figure 2.15** *An enzyme in the bacterium* Escherichia coli *links ammonia and fumaric acid to give just the* L *form of aspartic acid.*

glucose, and is therefore widely used as a sweetening agent in soft drinks and confectionery, because it gives a much sweeter flavor for the same amount of sugar. But glucose is the more readily available of the two sugars since it can be manufactured in great quantities from starch, a carbohydrate obtained from corn. Glucose isomerases are used to boost the fructose content of commercial corn syrup to between 40 and 90 per cent. These enzymes are used either in "whole cell" processes or in pure form, extracted from living cells and immobilized in polymers or on surfaces.

A great many more of the reactions of value to the chemical industry that are currently catalyzed by "inorganic" means appear to have the potential to become enzyme-mediated processes in the future. The synthesis of methanol, for instance, which is currently carried out by reacting hydrogen and carbon monoxide at temperatures of around 280 degrees Celsius in the presence of a copper/zinc oxide catalyst, can be effected by the action of the enzyme methane monooxygenase on methane and oxygen at room temperature. Glucose and oxygen can be converted into a whole range of useful compounds, including methanol, butanol and acetic acid, by the enzymes in several different microbes. The use of enzymes in these cases has not yet been made sufficiently efficient to create commercially viable processes. There remain, however, some industrial syntheses for which enzymes are most unlikely ever to prove useful, because the products are highly toxic or otherwise harmful – the synthesis of nitric and sulfuric acids, of immense importance industrially, are two such cases.

*Designer enzymes*

These uses of enzymes in industry are very well and good; but they rely on our being able to find an enzyme in nature that does just the job we want it to. How much better it would be if we could design or own artificial enzymes for catalyzing whatever process we wish!

An approach pioneered by Richard Lerner at the Scripps Research Institute in La Jolla, California, and Peter Schultz of the University of California at Berkeley is now allowing us to do precisely that. What these researchers have discovered is a way to persuade nature to make protein molecules to order that fit any given substrate.

Organisms are remarkably adept at churning out proteins: the immune system generates a whole family called immunoglobulins, otherwise known as antibodies, which can identify and bind to foreign molecules in cells and thus tag them for destruction. It is necessary for the immune system to "raise" antibodies against all manner of rogue organic flotsam: in each case, the antibody has a binding cavity that fits the target molecule (called the antigen) closely.

The key to generating antibody proteins that can catalyze a reaction, and thus show enzyme-like behavior, rather than just binding a given molecule, lies with the way in which enzymes work. I have suggested above that the enzyme's catalytic region fits the substrate like a lock and key. This is a good approximation to the truth, but it is not quite accurate. In fact, the enzyme is structured so as to fit the *transition state* that the substrate will adopt when it is transformed in the enzyme-induced reaction. In this way, the enzyme stabilizes the transition state and lowers the free energy barrier to reaction. The transition state generally looks rather similar to the pristine substrate.

Lerner and Schultz reasoned that, if one could raise antibodies against the transition state to a specific reaction that we might wish to carry out, the protein should stabilize the entity and thus catalyze the reaction. But the transition state is an ephemeral species; so instead the researchers used a stable molecule that was thought closely to resemble the transition state in shape and structure. The immune systems of living cells exposed to these transition-state analogs are induced to generate the desired antibody.

These and other researchers have now demonstrated many instances in which "catalytic antibodies" isolated from living cells do indeed show the anticipated ability to enhance reaction rates. This discovery promises to open up a whole new field of selective catalysis.

### *Sense and sensors*

Like enzyme molecules, we ourselves have an extraordinary ability to distinguish between many thousands of different organic and inorganic compounds. While of course some of these can be identified by their appearance or texture, we achieve our extensive chemical discrimination primarily by means of the sense of smell. Smell is controlled by a class of proteins in the nasal membrane, which possess an enzyme-like facility for distinguishing subtle differences between organic molecules. But to function as a sensing device, the olfactory apparatus must be able to do more than simply recognize and distinguish substrates: it must have a means of converting the act of recognition into a neural response – a signal. It is this combination of recognition capability and consequent triggering of a signal in some form of processing "hardware" that researchers are now seeking to mimic in devices designed to detect specific biomolecules. These devices, called biosensors, are tools for chemical analysis in which the specificity of enzyme-induced chemical reactions is exploited to induce an electrical signal that indicates the concentration of a particular chemical species (the "analyte").

Biosensors promise to be particularly valuable for medical applications, since they can be used to monitor concentrations of important biochemicals in the blood. A biosensor that produces a changing electrical signal in response to levels of glucose in the bloodstream of diabetics, for example, could be used to control the release of insulin so as to maintain the concentration of blood sugar at a steady, safe level.

Much of the earliest work on biosensors was directed towards the development of a glucose sensor. A primitive device of this sort was designed during the 1950s by Leland Clark of the Children's Hospital Research Foundation in Cincinnati. It was based on a sensor for measuring the concentration of oxygen in the bloodstream, a factor that is of critical importance for patients undergoing surgery. Clark's oxygen sensor was an electrochemical device consisting of a platinum electrode wrapped in a plastic membrane, through which gases such as oxygen could diffuse. The electrode was hooked up to an electrical circuit, and the current flowing through the circuit depended on the surface voltage of the electrode. Oxygen diffusing through the membrane and onto the electrode altered the electrode's voltage and thereby caused a measurable change in current, the magnitude of which indicated the amount of oxygen adsorbed and thus its concentration in the liquid around the electrode (Figure 2.16*a*).

The chemical interactions at the platinum surface are unspecific and ill-suited to adapting the sensor for monitoring concentrations of a biomolecule such as glucose. To achieve this goal, Clark coated the plastic-wrapped electrode with a gel in which was trapped an enzyme called glucose oxidase. As the name implies, this enzyme assists the oxidation (reaction with oxygen) of glucose molecules. As glucose molecules were oxidized within the enzyme-containing gel, the oxygen around the sensor was consumed, and this was registered by a change in the electrode voltage (Figure 2.16*b*). But now the oxygen concentration, and thus the sensor's signal, was determined by the amount of glucose present.

Clark's device was clearly a somewhat *ad hoc* adaptation of the oxygen sensor, and its size (about 1 centimeter in diameter) precluded its use for monitoring levels of blood sugar directly within the body. But it incorporated the elements that feature in most biosensors today: a catalytic molecule, usually a natural enzyme, immobilized in some way inside the device, which interacts with the analyte in such a way as to produce a signal.

Not all biosensors rely on the generation of electrical signals, however. Some devices have been developed that emit light in response to the analyte. Biosensors of this sort generally use very thin optical fibers to guide the light to and from the sensing region.

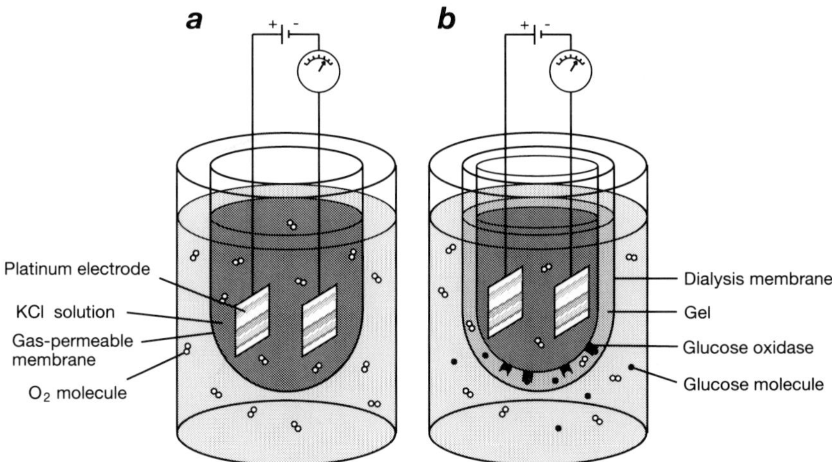


**Figure 2.16** *The oxygen sensor developed by Leland Clark (a) contained metal electrodes surrounded by a membrane permeable to dissolved oxygen gas. The electrode potential depends on the oxygen concentration in the surrounding solution. By coating the membrane in a gel containing the enzyme glucose oxidase, and immobilizing the enzyme using a dialysis membrane permeable to glucose, Clark turned the device into a glucose sensor (b). The oxygen concentration around the electrodes decreases when oxygen is used up in glucose oxidation catalyzed by the enzyme.*

Optical fibers are plastic cables that can "conduct" light much as a copper wire conducts electricity. The light rays remain trapped within the fiber as they pass along it, being constantly reflected back from the fiber's sides. These fiber-optic sensors incorporate molecules that fluoresce (emit light when they are illuminated) on interacting with the analyte. Fluorescein, for example, becomes fluorescent when it picks up a hydrogen ion, and so can be used to monitor acidity (which is determined by the concentration of these ions). A sensor of this sort is in itself hardly a biosensor, but the same process can be used to generate an optical signal from an enzyme/analyte reaction that produces a change in acidity. By using bundles of optical fibers to transmit simultaneously several individual light signals, it should become possible to develop compact sensors that can monitor several different analytes at once.

The very small temperature changes due to the heat produced in the reaction between analyte and enzyme can also be used as the basis of a sensing technique, for example by incorporating into the device very sensitive temperature gauges called thermistors. Biosensors of this sort, called "enzyme thermistors," have been developed to monitor glucose and penicillin.

One of the major medical applications envisaged for the biosensors is as a component in an artificial pancreas for diabetics. The technologies necessary for such a device are not yet available, however. What is needed is a biosensor with a long life that can continuously monitor glucose concentrations in the blood and which can be hooked up to an insulin-releasing device, the whole assembly being biocompatible and small enough to be implanted in the body. The principles behind a device of this sort are clear enough, but the engineering problems remain to be solved. Once the latter have been overcome, the medical benefits will be tremendous.

Biosensors should benefit public health in other ways too. They can be used to monitor levels of toxic or otherwise hazardous substances in the environment or in food. Indeed, biosensors are used already for assessing the quality of meat and fish by measuring the levels of the harmful organic compounds that accumulate as the foods age. There is considerable military interest in the use of biosensors to detect nerve gas or biological weapons. In the pharmaceutical industry they will be employed to monitor the composition of fermentation vessels in which drugs are produced. A particularly intriguing development involves immobilizing the enzymes in an environment that simulates the conditions in which they occur naturally, by embedding the molecules in artificial membranes like those that constitute cell walls (discussed in Chapter 7). It is conceivable that the biosensors of the future will be less like microelectronic devices and more like the biological sensors, such as those found in the olfactory system, that have helped to inspire their creation.

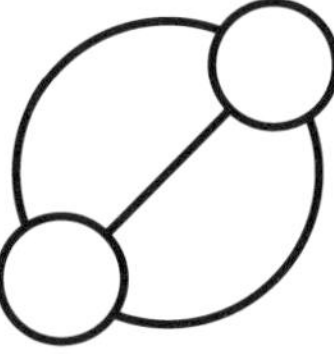

# 3

# Caught in the Act

## Watching atoms dance

*It's a wild dance floor there at the molecular level.*

Roald Hoffmann

How can we see into the microworld of molecular chemistry? This has surely been a burning question ever since – indeed, before – the atomistic view of matter came to be accepted. It is all well and good to draw ball-and-stick pictures of molecules depicting the beautiful carbon cages of the fullerenes or the weird and wonderful architectures of hydrocarbons. Yet what chemists crave is to see these things for real rather than as plastic models on the benchtop or computer images on TV screens. And more – they would like to see this world in action, with molecules whizzing through space and chemistry unfolding before their eyes. They are now finding ways to do so.

The most literal approach to investigating molecular-scale phenomena is to try to build a microscope powerful enough to show them to us. Microscopes have revealed a tremendous amount about the world that is too small for the naked eye to see: we can watch fertilized ova divide and develop into embryos, or blood corpuscles course through capillaries and veins. But these objects are typically a few thousandths of a millimeter in size, whereas molecules are very much smaller. Even today, the most powerful optical (light-based) microscopes can see only down to scales of about a few hundred millionths of a millimeter, while simple molecules such as methane and ammonia are about a thousand times smaller. Microscopes need not rely on light alone, however. Electron microscopes, which use beams of electrons instead of light, and the recently developed scanning probe microscopes, which provide images by electronic or mechanical means, can now take pictures of single molecules (see Plate 2), in some cases enabling us to see individual atoms. However, it is often necessary to know more or less what the molecule looks like beforehand in order to make sense of these images.

A second way to look into the atomic microcosm is described in Chapter 4. It involves bouncing beams of X-rays off crystals, in which the atoms or molecules are packed together in regular arrays. The pattern formed by the reflected beams can be turned into a map of the atoms in the crystal. This technique, called X-ray diffraction, is immensely useful for determining atomic arrangements with high precision, but on the whole it is revealing only for those compounds that can form crystals; the method gives far less detailed information about the molecular structure of liquids, gases and noncrystalline solids. Moreover, the structure of some molecules (particularly biomolecules) in crystals may differ significantly from that in the "active" natural state.

Both microscopic and diffraction techniques provide mainly *static* information – the molecules must be frozen in place in order for us to see them. These techniques are therefore of very limited value for finding out anything about the way that molecules move (that is, about molecular *dynamics*). But molecules at room temperature are far from static – they spin and vibrate much faster than the eye could ever hope to follow. Furthermore, the process of chemical change, of transformation, is inevitably a dynamic one, in which atoms and molecules encounter each other, interact and form new unions. These motions take place on timescales appropriate to the tiny size of the molecules: a second in time is, to a molecule, much as the age of the Earth is to us. So how can we see these ephemeral events in the microcosmos?

The technique called spectroscopy, one of the oldest approaches to the investigation of molecular shape and behavior, is equally a tool for studying structure *and* dynamics. At first sight, spectroscopy might appear to be an awfully crude way of finding out about the way that molecules behave – in its simplest implementation it involves nothing more than shining light through a sample and measuring the amount that is absorbed as one changes the light's color. What one obtains is therefore not a picture of molecules at all, but simply a wiggly graph indicating the variation in light absorption as one scans through the spectrum. But this "absorption spectrum" may contain a great deal of information about the chemical make-up and motions of the molecules. Spectroscopy today represents perhaps the chemist's primary investigative tool, and it has acquired a sophistication that allows it to reveal things no other technique can disclose. In particular, it provides a means of studying the extraordinarily fast processes that are the norm in the atomic world, allowing us to make "molecular movies" which catch chemical processes in the act.

The basis of spectroscopy is the interaction between molecules and light. We are perhaps used to thinking of light as a kind of "passive" signal that allows us to see objects but does not disturb them. The truth is, however, that we are able to view objects at all only because light *does* perturb matter. The central process of spectroscopy – the absorption of light – requires that molecules take up some of the energy that the light carries. This absorption of energy can sometimes alter significantly the physical and chemical properties of the absorbing molecules, even to the extent of splitting them apart. For this reason, spectroscopy – the use of light to study molecules – is intimately linked with photochemistry, in which light induces or influences chemical processes.

The importance of photochemical processes in the natural world would be hard to overestimate. Chemical reactions induced by sunlight, for instance, drive the biochemistry of plant growth: this is the basis of photosynthesis, which provides us with the air we breathe. Photochemistry is also of critical importance to many of the chemical reactions that take place in the atmosphere, a topic discussed in Chapter 10. And to the practising chemist, photochemistry is becoming a potential tool for exquisite control of chemical reactions – we will see how photochemical "scalpels" may one day be used to perform the most delicate surgery on molecules.

## And there was light

### *What's in a color?*

Color is one of the best understood characteristics of objects in the everyday world, and also one of the least. That is to say, we have a very good idea of why things possess the colors they do, and we now also have a reasonable (though far from complete) understanding of how the body's visual system recognizes these colors. But colors have an aesthetic value too, comparable to that conveyed by music or literature – it is far from a matter of indifference to us that Rembrandt chose to utilize deep, rich golds, reds and browns, that Cezanne felt his skies might benefit from greens and pinks. At present, our emotional response to color remains mysterious.

In its simplest manifestation, spectroscopy is a kind of recognition of chemical composition according to color. In this sense it is scarcely different from the means by which alchemists monitored the progress of their transformations from base metals towards gold, which were thought to require a very specific sequence of color changes. Spectroscopists rely, however, not on direct visual impressions but on devices called spectrometers which provide a much more accurate and sensitive measure of color. We can easily distinguish silver from gold by eye, but it is far less easy to see the difference between silver and tin in the same way. Likewise, charcoal and lead sulfide look much the same shade of black to the naked eye – but a spectrometer is able to see the subtle differences in their "blackness." And spectroscopy provides a means to identify substances that are apparently colorless: it can tell apart oxygen gas from nitrogen gas, while to our eyes they are both invisible. How does the absorption of light provide this kind of discrimination?

Historical theories of the nature of light have tended to consider two possibilities: that light rays are comprised of tiny particles, or of waves travelling through some kind of medium, just as sound waves propagate through air. Isaac Newton favored a corpuscular (particle) theory of light, whereas his contemporary Christiaan Huygens developed a description based on waves traveling through an all-pervading medium called the ether. Both theories gave reasons for why light seemed to travel in straight lines, and both, too, seemed able to explain the rules of optics known at that time.

The dichotomy of wave and corpuscular models persisted into the early nineteenth century, when Thomas Young began to find ways in which they could be tested. In 1669, the Dane Erasmus Bartholin observed that light rays traveling through certain crystals, such as Iceland spar, have different properties when they pass in one direction than in another. This phenomenon, known as double refraction, can produce beautiful colorations. Young suggested that it can be explained on the basis that light consists of waves of a transverse nature, which means that they are like up-and-down undulations in a plane. An implication of this idea is that light rays can be *polarized*: that is, the plane of undulation can have a preferred orientation. The Frenchman Augustin Jean Fresnel demonstrated between 1818 and 1821 that Young's suggestion accounted for the observed behavior of double-refractive crystals.

Just as sound waves require a medium through which to travel, so it was thought that light waves must move through the invisible ether. By the late nineteenth century, the picture of light as transverse waves in an ether was widely accepted, but hardly a dent had been made on the question of what the ether really was. In the earlier half of that century the work of Michael Faraday had been instrumental in securing the long-standing belief that electricity and magnetism were related: an electrical current was known to be capable of exerting a magnetic force, and changes in the strength of a magnetic field were known to induce electrical currents. In 1845 Faraday demonstrated that the plane of polarized light could be rotated by passing the light through a magnetic field, suggesting that light too was related to magnetism and electricity. The Scotsman James Clerk Maxwell took the bold and brilliant step of proposing that light was itself a combined electrical and magnetic – that is, an electromagnetic – disturbance in the ether. In Maxwell's theory, light waves consist of *electromagnetic radiation* traveling through the ether. The wave has two components: an electric field whose amplitude increases and decreases in an oscillatory way in one plane, and a magnetic field which oscillates in step with, but in a plane at right-angles to, the electric field (Figure 3.1). A light ray therefore serves to convey electromagnetic energy through the ether. The frequency of the undulations (which is related in a simple way to their wavelength) determines the ray's color: the electromagnetic waves of red light, for instance, undulate at about one hundred million million times per second, while blue light has a frequency of oscillation about four times greater. In 1887 the German physicist Heinrich Rudolph Hertz first proved that these electromagnetic waves exist.

While Maxwell's description of light as electromagnetic waves will serve us well for understanding certain aspects of the interaction between light and matter, twentieth-century science has modified it in ways that are essential for a more complete insight. An experiment conducted in 1887 by Albert Michelson and Edward Morley led to the inescapable conclusion that the ether, the medium through which light waves propagate, was nothing but a fictitious construct. Light, it seems, does not need a medium through which to travel; the electric and magnetic fields are self-supporting, being able to propagate through completely empty space. Modern physics presents this idea from a

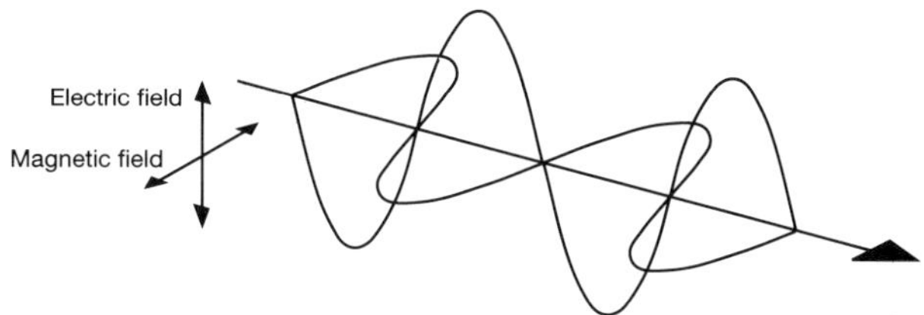


**Figure 3.1** *According to the electromagnetic theory of James Clerk Maxwell, light rays consist of oscillating electric and magnetic fields. The plane of the electric field is at right-angles to that of the magnetic field. The former is commonly called the plane of polarization of the light ray. Light from sources such as electric bulbs contains rays with their polarization planes oriented at random, but by passing the light through a polarizing filter, a beam can be obtained in which all the rays have the same polarization.*

slightly different perspective, according to which "empty" space is no longer regarded as empty at all – it is permeated by an electromagnetic "field" which can be set oscillating when excited by an energy source, like a violin string waiting to be plucked. These vibrations are then what we call light.

The early-twentieth-century revolution of quantum theory, meanwhile, insisted that regarding light as waves gave only half the picture: light could sometimes behave instead as if it were comprised of discrete particles. In 1905 Albert Einstein rationalized the photoelectric effect (p. 21) in terms of this particulate view of light. Einstein proposed that light consists of photons – little packets of electromagnetic energy characterized by the frequencies of Maxwell's oscillating electromagnetic fields. Photons are a most unusual breed of particle – they possess energy but no mass. The energy is equal to the photon frequency multiplied by a constant value, called Planck's constant (after Max Planck). A photon of red light therefore has a lower energy than a photon of blue light.

Photon frequencies are by no means limited to those of visible light. Our eyes are sensitive to only a small part of the complete span of the electromagnetic spectrum, which continues to frequencies both higher and lower than the visible range (Figure 3.2). Electromagnetic radiation with frequencies slightly less than (and therefore wavelengths slightly greater than) that of red light is called infrared radiation; although our retinas do not register infrared photons, we experience the energy that they carry in the form of heat. Beyond infrared radiation come microwaves (with wavelengths of a few millimeters) and radio waves (with wavelengths ranging from meters to kilometers). Off the visible scale to the high-frequency end lies ultraviolet radiation, followed by X-rays and then by gamma rays (emitted by some radioactive materials) and cosmic rays, whose origin in deep space is not yet understood. Photons of X-rays and gamma rays, as well as some ultraviolet radiation, contain enough energy to break apart chemical bonds. Intense sources of these types of radiation can therefore be very damaging to matter, particularly to the rather delicate compounds that comprise living tissue.

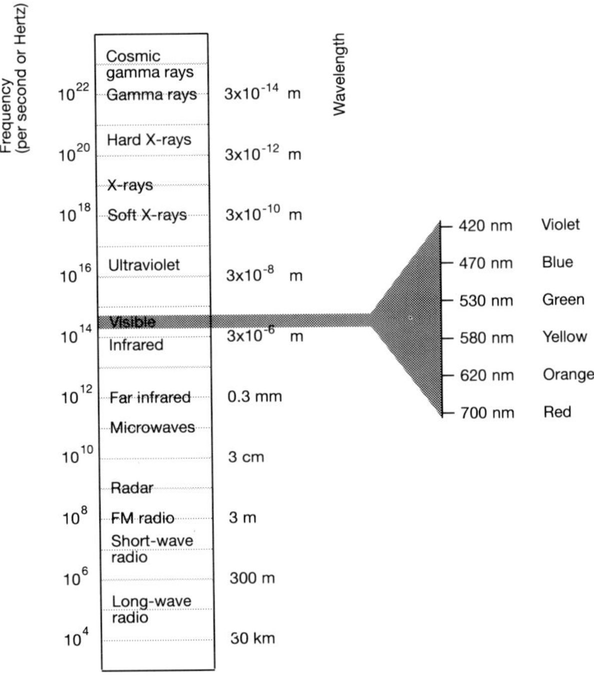


**Figure 3.2** *The spectrum of electromagnetic radiation runs from radio waves at the low-frequency end to gamma rays and cosmic rays at the high-frequency end. Visible light represents a fairly narrow band around the middle of the spectrum. (Here the frequencies of the radiation are represented in exponential notation – multiples of 10 in which the superscript denotes the number of zeros that follow the "1." Thus 1,000 is written as $10^3$. While the frequencies are very large numbers, the wavelengths are very small – here, a negative superscript indicates that the specified number of zeros follows a decimal point. So 0.001 (or one thousandth) is written as $10^{-3}$.)*

## Exciting times

### *How the leaf gets its color*

We saw in Chapter 1 that atoms and molecules are swathed in clouds of electrons, which are particles bearing a negative electrical charge. It is no surprise, therefore, that matter and electromagnetic radiation often interact strongly. Perhaps we should be more surprised that they appear sometimes *not* to interact, as when glass transmits almost all the light that falls on it. But how light gets through any medium – even empty space – is a far from trivial matter, described by an elegant and wonderfully successful theory called quantum electrodynamics. The transparency of glass is not an indication that photons of light pass right through it without interacting with the atoms in the material; the photons are simply not *absorbed*. (Actually, glass does absorb electromagnetic radiation strongly in the infrared region of the spectrum. If our eyes were sensitive to infrared, "clear" window glass would look "colored.")

Thus, although light and matter both contain electric fields, their interaction is not indiscriminate. When matter is irradiated with electromagnetic radiation, it will absorb photons of some frequencies but not others. Those not absorbed will be either transmitted through the substance or reflected. When some of the frequencies absorbed lie within the visible range, under white light the object takes on the color corresponding to those visible frequencies that remain. A leaf, for instance, absorbs red and blue light and therefore only the greenish part of the spectrum gets reflected. A cornflower absorbs the reds and yellows; a carrot the greens and blues. Objects that reflect all visible light are white, while those that absorb it all appear black.

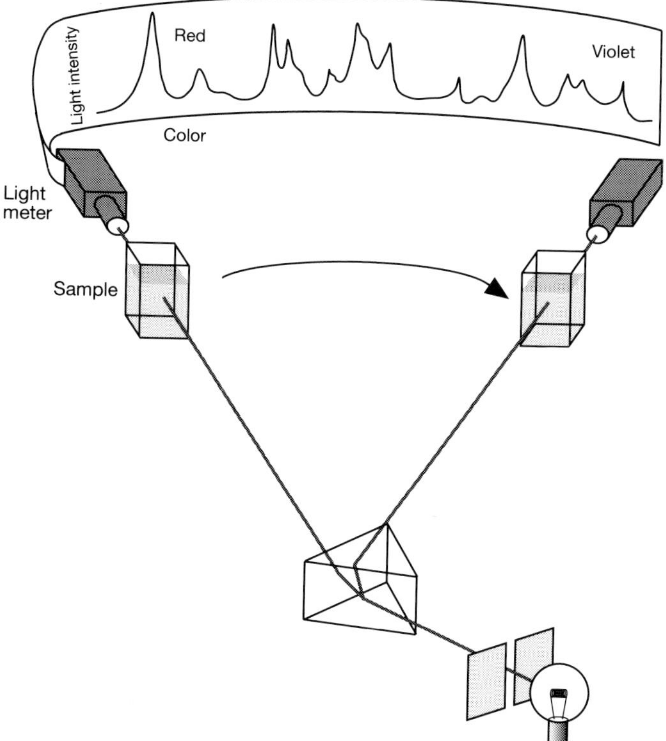


**Figure 3.3** *A typical spectrometer measures the absorption of a "probe" light beam when it is passed through a cell containing the sample. Photodetectors register the change in brightness of the transmitted light as the color (that is, the wavelength) is changed. The result – a plot of the change in amount of absorbed light with changing wavelength – is the absorption spectrum. Peaks in the spectrum (absorption bands) provide characteristic signatures of the molecular motions or the molecule's electronic structure. In this schematic, the sample is moved through the spectrum generated by passing white light through a prism. In a real spectrometer, it is more convenient to rotate the prism than to move the sample and detector.*

Leaves owe their greenness to the pigment chlorophyll *a*, the compound that enables a plant to capture sunlight and use the energy to build the substances it needs for growth. In telling us that a leaf is green, our visual system is providing a crude impression of the absorption spectrum of chlorophyll *a* within the wavelength range of visible light.

To measure this spectrum more accurately, scientists employ a spectrometer. Imagine using a prism to separate white light into its constituent colors, and then moving through this spectrum a cell containing a solution of chlorophyll *a*. When the cell is held in the red part of the spectrum it will absorb most of the light that falls on it, and from the opposite side the cell will appear almost black (similarly, a leaf viewed through a sheet of red cellophane seems dark and almost colorless). In the green part of the spectrum, meanwhile, the cell will let almost all of the light through. So if we set up a light meter on the far side of the cell to measure the intensity of the light transmitted through it, this will give a reading that rises and falls as the cell is moved through the spectrum, depending on the color of the incident light (Figure 3.3). In a real spectrometer it is more convenient to keep the cell fixed and to sweep the color-separated light across it. For chlorophyll *a*, the variation in absorption registered by the light meter – the absorption spectrum – is shown in Figure 3.4*a*.

The exact shape of the absorption spectrum provides a kind of fingerprint of the compound – chlorophyll *a* has the same spectrum regardless of which plant it is taken from. A solution of a nickel salt such as nickel sulfate is also green; but its absorption spectrum (Figure 3.4*b*) is readily distinguishable from that of chlorophyll.

Our eyes can detect absorption only in the visible range, but spectrometers can be designed that incorporate "light" meters sensitive to frequencies outside this range, most commonly in the infrared and ultraviolet. This extends the amount of information that we can extract from an absorption spectrum: compounds that are colorless to the eye, for example, may contain strong absorption bands in the infrared or the ultraviolet.

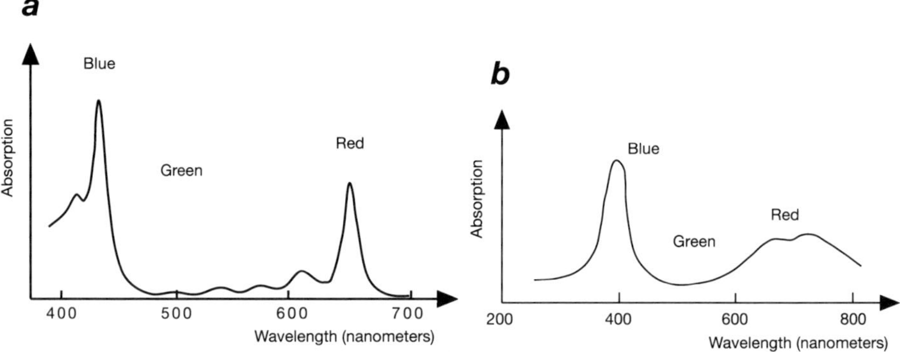


**Figure 3.4** *The chlorophyll* a *molecule absorbs light strongly in the red and blue parts of the spectrum, but allows green light to be transmitted (a). This molecule is responsible for the color of leaves. Nickel salts (such as nickel sulfate) dissolved in water commonly have a green color too, but the absorption spectrum of dissolved nickel salts (b) reveals how their "greenness" differs from that of chlorophyll.*

Water, for instance, absorbs strongly at infrared frequencies, a fact that has important consequences for the role of water vapor in the atmosphere.

## *Shakes and ladders*

When a molecule absorbs light, its energy increases – it gets "hotter," by which I mean that the molecule may shake, jiggle or spin more vigorously. It is then said to have been "excited" by absorption. The reason why molecules absorb photons of light at some frequencies but not at others is that their motions are determined by the rules of quantum mechanics, which place restrictions on which motions are allowed. We saw in Chapter 1 that the electrons in atoms and molecules are not free to adopt any old orbit around the nuclei, with correspondingly arbitrary energies; instead, they are confined to a ladder of specific energy levels, with the energies in between the rungs being forbidden. The electronic energy levels are said to be quantized. In the same way, the energies of the motions that the molecules execute in space may also be quantized. The two-atom oxygen molecule $O_2$, for example, both rotates (tumbles end over end) and vibrates (the bond between the two atoms stretching and contracting) in free space (Figure 3.5). The energies associated with these motions may take only certain values, forming a ladder of energy levels like those for the electron energies. As the energies of rotation and vibration depend respectively on the speed of rotation (the number of revolutions per second, say) and the frequency of vibration (the number of in-and-out

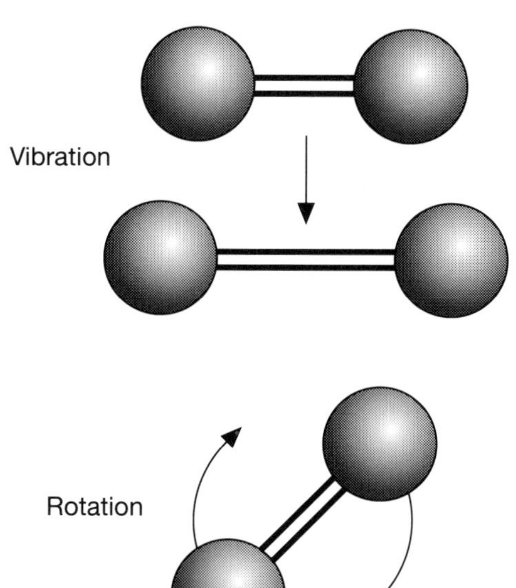


**Figure 3.5** *Oxygen molecules are like miniature dumbbells. In oxygen gas, where their motions are unhindered, they rotate by tumbling end over end, and vibrate through stretching of the bond.*

oscillations per second), the latter two quantities are quantized too. The $O_2$ molecule, therefore, can vibrate only at specific frequencies and spin only at specific speeds.

This quantization of molecular motions is perhaps even more counterintuitive than quantization of electronic energies. We are not, after all, familiar on an everyday basis with the things that electrons and nuclei get up to, and you might be perfectly ready to accept that the internal structure of atoms and molecules cannot be readily likened to a system of billiard-ball particles. But vibrations and rotations are not at all remote from our everyday experience, and the suggestion that they can be quantized runs contrary to this. It is as if one were asserting that a wheel can be spun only at intervals of 10 revolutions per minute: 10 r.p.m., 20 r.p.m., 30 r.p.m. and so on. But we know perfectly well that we can drive a wheel at any arbitrary rotational speed just by supplying the necessary energy. With a wheel quantized in 10 r.p.m. intervals, the speed would stick at, say, 10 r.p.m. even if we were to crank up the driving power, until we had turned it up far enough for the speed to jump abruptly to 20 r.p.m. This does not happen in the real world, of course, and mercifully so. Imagine a poor cyclist trying to ride a "quantized bicycle." She would be restricted to specific speeds, and attempts to increase her speed by trying to push harder on the pedals would be to no avail until the amount of exertion was sufficient to let the bicycle lurch jarringly to the next allowed speed.

The reason why we do not observe quantization of rotation and vibration in our normal lives is not due to any fundamental difference between the atomic and the macroscopic worlds; it is simply a question of scale. The spacing between the rungs on the energy-level ladder of a rotating or vibrating object is, crudely speaking, related inversely to the object's mass. The more massive the object, the more closely spaced are the rungs. For objects as big as a bicycle wheel (or even, say, the tiny flywheel of a watch), the spacings are too small to be measurable, so that to all intents and purposes there are no "forbidden" gaps – all energies are permitted. This is a general feature of quantum mechanics: at the scale of atoms and molecules it predicts some unfamiliar effects, but as the scale increases towards the everyday, these effects become smaller and smaller until they are essentially negligible. The assertion that quantum effects should vanish in this way for large systems is known as the Correspondence Principle.

Amongst all this discussion of quantization of molecular motions, I must now come clean and confess that there is, in addition to rotation and vibration, a third type of motion which I have been sweeping under the carpet. The former two kinds of motions can be executed on the spot as it were. But in gases and (in a more restricted manner) liquids, molecules are free also to move from one point in space to another. (In fact, this also happens to a small extent in solids.) Scientists call this kind of motion translation. Translational motion can also be quantized, but even for molecules it turns out effectively to be "continuous" – that is, the energy of translational motion can be increased smoothly rather than in stepwise fashion. The reason for this is that the spacing between translational energy levels depends not only on the object's mass, but also on the size of the vessel in which it is contained: as the vessel gets larger, the spacing gets smaller.

For molecules in a macroscopic laboratory beaker (or the cell of a spectrometer), the translational energy levels are spaced so closely that this motion is essentially unquantized. This means that translational motion can be largely neglected in the succeeding discussion.

The quantization of molecular rotation and vibration means that molecules can increase their rotational or vibrational energy only by absorbing energy in "lumps" or packets of certain sizes. Electromagnetic radiation provides these packets in the form of photons, whose energy depends on their wavelength. Whether or not the molecule can absorb a photon is determined first and foremost (but not exclusively) by whether the photon energy is equal to the energy gap between the level occupied by the "nonexcited" molecule and one of its higher levels.

As well as increasing its rotational and vibrational energy, a molecule can become excited by increasing its electronic energy – that is, by absorbing energy that boosts electrons into higher-energy orbitals. Molecules therefore possess a hierarchy of three energy-level ladders. On the whole, the rungs of a molecule's rotational energy-level ladder are spaced more closely than those of the vibrational ladder, which in turn are closer than the rungs of the electronic ladder. Photons with energies typically in the microwave range have the potential to excite a molecule rotationally, whereas higher-energy (that is, higher-frequency or shorter-wavelength) photons, generally in the infrared, can induce vibrational excitation; and photons of still higher energy, in the visible or ultraviolet, excite jumps between electronic levels (Figure 3.6). The jump from one energy level to another is called a transition.

The sequence of rotational, vibrational and electronic energy levels is unique to each type of molecule, being dependent on the masses of its constituent atoms, their arrangement in space (in other words, the molecule's shape), the strengths of the bonds and, for electronic energy levels, several more subtle factors. The molecule might absorb a photon, producing a peak in the absorption spectrum, whenever the energy of the photon corresponds to the difference in energy between one level and another. Photons with energies in between these values will not be absorbed and so can pass right through the material.

I say "might absorb," not "will absorb," because there is more to the question of whether or not a molecule will make a transition than the need for the photon to have the right energy. Ultimately, the interaction of a molecule with a photon is electrical in origin: it is an interaction between the electrical component of the photon's oscillating electromagnetic field and the molecule's electron cloud. To absorb (or to emit) a photon of a given frequency, the molecule must itself generate an electric field that oscillates at that frequency. Rotational motion will do this if the molecule possesses some imbalance in its distribution of electronic charge – if, crudely speaking, the molecule has an excess of negative charge at one end and of positive charge at the other. In the hydrogen chloride molecule, for instance, the electron cloud is pulled towards the chlorine atom, creating an excess of negative charge there and a corresponding positively charged region around the hydrogen (Figure 3.7). A molecule with this kind of asymmetric distribution is said to possess an electric dipole moment; it is rather like

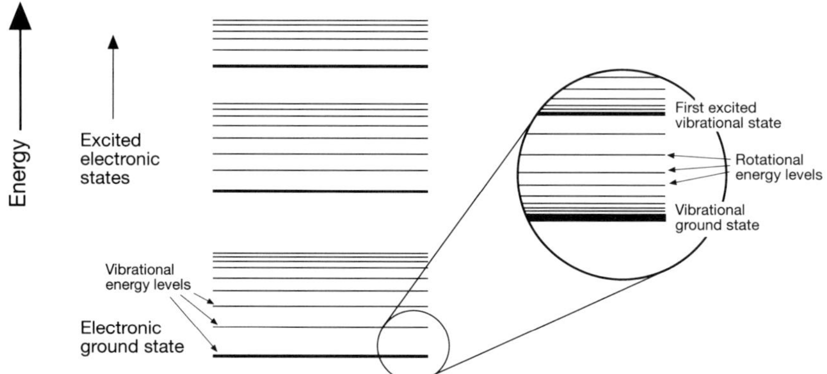


**Figure 3.6** *The energy levels of simple molecules form a hierarchy of three ladders. The rotational levels are spaced most closely together—it takes only a small "quantum" of energy, as contained within a microwave photon, to excite a transition from one level to the next one up. Vibrational energy levels are further apart, and there is a system of rotational levels for each rung on the vibrational ladder. Electronic energy levels are further apart still, and each has its own vibrational ladder. Because the vibrations are not like those of "perfect" springs, the vibrational levels become more closely spaced towards the top of each ladder. The chemical properties and shape of molecules in different electronic states can be very different. The distance between electronic energy levels generally corresponds to the energy in photons of visible or ultraviolet light.*

the electrical equivalent of a magnet, with positive and negative "poles." The molecule's rotation causes the direction of the electric field to alternate, allowing it to interact with the oscillating electric field of a photon and permitting absorption of the photon's energy.

Molecules that have no electric dipole moment, such as $O_2$ and $N_2$, cannot undergo rotational transitions. Carbon dioxide is also prohibited from rotational excitation, even though parts of the molecule possess different electric charges. There is a slight excess of negative charge around the oxygen atoms at either end and a corresponding positive charge around the carbon atom in the middle (Figure 3.8), but the $CO_2$ molecule nevertheless possesses no net electric dipole because the distribution is symmetrical: in effect, there are two electric dipoles pointing in opposite directions, which therefore cancel out.

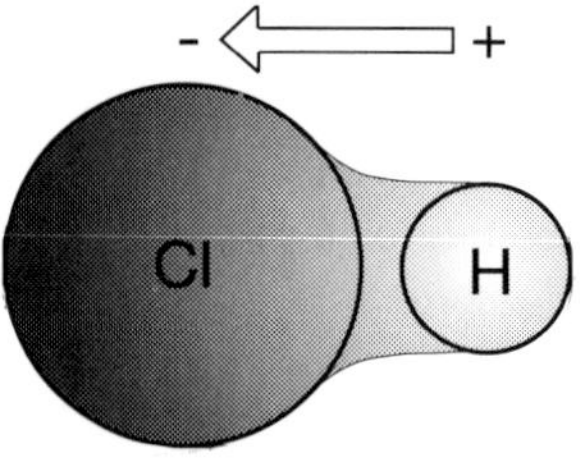


**Figure 3.7** *In the hydrogen chloride molecule, the chlorine atom draws the electron cloud towards it, leaving the hydrogen atom relatively denuded. There is thus an excess of negative charge around the chlorine and of positive charge around the hydrogen, so the molecule has an electric dipole moment pointing along the axis between nuclei (indicated by the white arrow). The highly schematic distribution of charge shown here should not be taken too literally, however.*

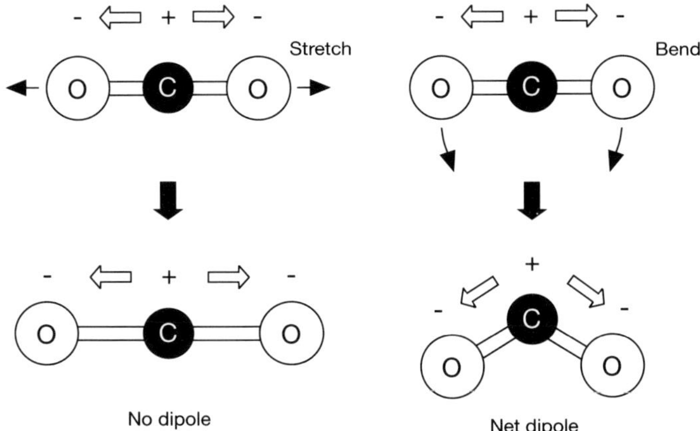


**Figure 3.8** *In carbon dioxide the oxygen atoms have electron "pulling power," but this creates a symmetrical distribution of charge: the two "dipoles" along the C—O axes cancel out, and there is no net electric dipole moment. Stretching vibrations of $CO_2$ along the axis through the nuclei retains the symmetrical distribution of charge, and so does not induce a dipole moment. Bending of the molecule, on the other hand, induces a temporary asymmetry in the distribution of charge, and so introduces a dipole.*

Vibrational transitions are allowed only when the relevant vibrational motions change the molecule's electric dipole moment. The vibration of $CO_2$ in which both bonds between carbon and oxygen stretch equally cannot therefore be excited into more vigorous oscillation by absorption of radiation, because this stretching motion maintains the symmetrical arrangement of charge and so leaves the molecule still without a dipole (Figure 3.8). The *bending* vibration of these bonds, however, is different: it brings both oxygens out of line with the carbon atom to form a V shape, giving the molecule a net negative charge at the two ends of the V and a positive charge at its apex. This vibration bestows a dipole on the molecule, whereupon it can interact with and absorb photons.

Thus transitions between energy levels are "allowed" or 'forbidden" on the basis of so-called selection rules which hinge on whether or not there is a change in the distribution of electric charge in space. For electronic transitions the selection rules are governed by rather more subtle factors; in essence, each allowed transition is characterized by an electronic "transition moment," a little like an electric dipole moment, which points in a specific direction in space and reflects the redistribution of electronic charge between the initial and excited states. An electronic transition can be excited when the transition moment is aligned with the electric field of a photon of appropriate energy. These various selection rules make absorption spectra much simpler and thus easier to interpret than if all "energetically possible" transitions of a molecule were allowed.

Generally speaking, rotational transitions are not terribly informative about the structure of a molecule (in liquids and solids, the freedom of a molecule to rotate is in

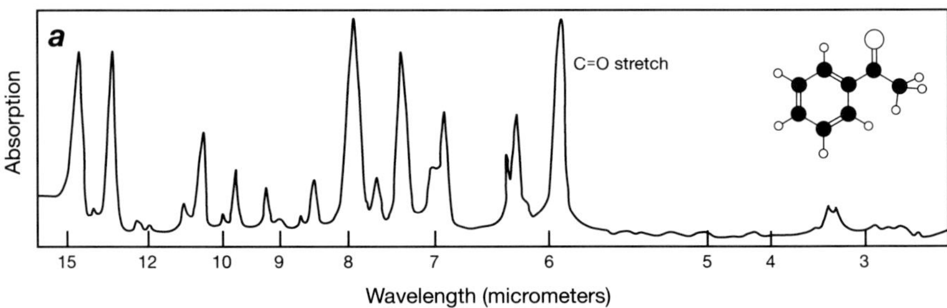


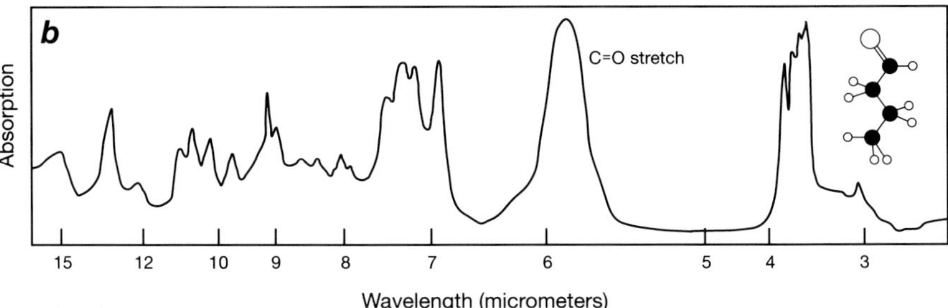


**Figure 3.9** *The absorption band at a wavelength of about six thousandths of a millimeter (six micrometers) in many organic compounds is diagnostic of the presence of the carbonyl group (C=O), which has a stretching vibration at the corresponding frequency. Here the "carbonyl stretch" band can be seen in the spectra of butyraldehyde (a) and acetophenone (b).*

any case often restricted). The most telling part of the absorption spectrum lies between infrared (IR) and ultraviolet (UV) wavelengths. IR spectra provide information about molecular vibrations, while visible and UV spectra tell researchers about the electronic structure of the molecule.

Certain groupings of atoms tend to have characteristic vibrational transitions in whichever molecule they appear. The carbonyl group, an oxygen atom double-bonded to a carbon, executes vibrations with much the same frequencies in a wide range of compounds; a vibrational transition of this group can generally be excited by absorption of an infrared photon with a wavelength of about 5.5 to 6 thousandths of a millimeter. Compounds containing the carbonyl group therefore have a distinctive absorption band in their IR spectra at these wavelengths (Figure 3.9). A small, lightweight hydrogen atom attached to carbon vibrates in and out at a much higher frequency; so it takes a photon with a wavelength of about 3.5 thousandths of a millimeter to induce a transition in this group. These characteristic frequencies are what makes infrared spectroscopy a valuable analytical tool for chemists seeking to identify the structure of an unknown compound: if it has an absorption band at a wavelength of about six thousandths of a millimeter, say, then it probably contains a carbonyl group.

The colors of chemical compounds are determined by their electronic transitions, since these generally lie in the visible range (as well as in the UV). For example, the

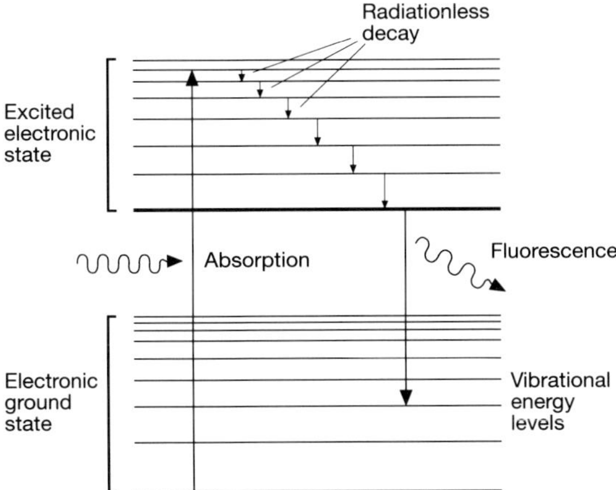


**Figure 3.10** *Fluorescence is the emission of light from a molecule previously excited to a higher energy level by absorption of a photon. When absorption takes the molecule to a high rung of the vibrational ladder in the electronically excited state, it may tumble down the ladder by losing energy through collisions with other molecules (called radiotionless decay), before jumping down to the lower electronic level by emission of a photon. The energy of the emitted photon is then less than that of the absorbed photon, and the wavelength is correspondingly longer.*

green color of nickel salts results from absorption of blue and red photons, which kick an electron on the nickel ion up to higher energy levels. The electronic state of lowest energy is called the electronic ground state. Because at room temperature the energy gap between the ground state and the next rung on the electronic energy ladder (the first excited state) is usually much larger than the amount of thermal energy available to the molecules, almost all of them will generally be in the ground state. Molecules that are boosted by absorption up to an excited electronic state will have a different arrangement of electrons to those in the ground state, and their chemical properties will therefore be different too. On the whole, excited-state molecules tend to be more chemically reactive, which is why electronic excitation can induce reactions that would not otherwise take place; these are called photochemical reactions. Provided that they do not undergo some photochemical reaction, electronically excited molecules will "decay" eventually back to the ground state by emitting a photon – this is the process of fluorescence. If this decay does not immediately follow excitation, the excited molecule may first cascade down the vibrational ladder by losing small amounts of energy through molecular collisions. Then the jump down will be smaller than the jump up, so the energy of the photon emitted (and thus its frequency) will be less than that of the photon absorbed (Figure 3.10). This is the mechanism by which fluorescent materials glow under UV light. Light-sensitive molecules in the material are electronically excited by the UV light (which is invisible to us), but the photon that is emitted when the excited state decays is of lower energy than UV, and therefore in the visible part of the spectrum.

## Chemistry in the wink of an eye

### *Quick shooting*

Traditional spectroscopy tells us something about molecular motions – it indicates, for example, how fast a molecule is vibrating or rotating. The speed at which these motions take place is tremendous: an iodine ($I_2$) molecule tumbles end over end typically ten billion times a second. Spectroscopists are now becoming increasingly interested in trying to follow such ultrafast motions in *real time* – in effect, they are striving to make frame-by-frame movies of atoms in motion. The primary motivation behind these studies is that they may provide insights into the process of chemical transformation: similar movements of atoms accompany chemical reactions, during which a molecule may fall apart or an atom be transferred from one molecule to another. Understanding the dynamic processes involved in reactions is an appealing goal in its own right, since it allows us to test theories of chemical bonding. In addition, however, there are potentially practical benefits: an appreciation of the atomic-scale details of a reaction mechanism may allow one to guide it towards a specific outcome.

To obtain a sharp image of an aircraft propellor as it spins, we must use a camera with a shutter that opens and closes in only a small fraction of the time that it takes for the blades to revolve once. In modern high-speed photography such fast shutter speeds are routine. But an iodine molecule spinning ten billion times a second presents a different kind of challenge altogether. To watch something moving this fast, researchers have to use an instrument that can be regarded as the world's fastest camera. It uses a series of laser beams, which are split, reflected and detected by a complex system of mirrors and shutters (Plate 5), to take pictures at the rate of one thousand million million per second. If each of these frames were replayed at the rate of 25 per second used in cinemas, a movie recording the events occurring in just one second of real time would last about a million years.

Laser light is different from the ordinary light that streams from the Sun or from an electric lamp. First, all the photons in laser light have essentially the same frequency: the light is monochromatic, meaning that it is of a single color. Second, the undulating electromagnetic waves of each photon rise and fall in step. A beam of radiation synchronized in this way is said to be coherent. The coherence of laser radiation greatly reduces the fanning-out that occurs for a normal, incoherent beam of light: laser beams may remain pencil-thin over distance of many kilometers.

Laser light is the result of light emission from excited-state atoms or molecules as they fall back to the ground state. Coherence of the emitted radiation is achieved by arranging for all the excited-state molecules to emit in step with one another. Mirrors at either end of the cavity containing the emitting material trigger a kind of chain reaction, whereby emission from just a few excited molecules will bounce back and forth, quickly stimulating all the others to emit in synchrony. The term "laser" derives from this process: Light Amplification by Stimulated Emission of Radiation.

The lasers used for ultrafast spectroscopy have two further strings to their bow. First, the light that they emit is polarized: not only are all the electromagnetic waves in step, but they are also in the same plane. Second, the laser light comes out not in a continuous beam but in short pulses. Indeed, so short are they that they represent perhaps the briefest man-made "event" yet created: pulsed lasers can be constructed that emit, within a single second, about two thousand million million pulses, each lasting about five femtoseconds (a femtosecond is 0.000000000000001 seconds). The crucial point is that this is many thousands of times shorter than the time taken for a molecule to execute a single rotation or vibration. Thus, by using each femtosecond pulse to take a snapshot of a molecule, one can follow its motion frame by frame.

### *Movie time*

Ahmed Zewail of the California Institute of Technology (Caltech) in Pasadena is one of the pioneers of ultrafast spectroscopy. Zewail and colleagues have used femtosecond laser pulses to follow the rotation and vibration of molecules and to watch chemical reactions as they unfold in a mere million-millionth of a second.

Zewail and colleagues watched the real-time rotation of iodine molecules by looking at the effect of rotation on their electronic spectra. Because of the selection rules that govern electronic transitions, an iodine molecule's potential to become electronically excited depends on its orientation relative to the plane of polarization of the radiation: the molecules' transition moment (which points along the axis between the two atomic nuclei) must be aligned with the plane of the electric field in order for absorption to occur (Figure 3.11). This consideration can be ignored in routine measurements of electronic spectra because a sample will contain many molecules with random orientations, so that at any instant there will always be some favorable alignments between the molecules and the probing light (which, in normal electronic spectroscopy, itself has randomly polarized photons). As the molecules tumble, the average number that are favorably aligned will not change in time. The electronic absorption spectrum in a standard spectroscopic experiment is therefore independent of the molecules' rotational motion.

To "see" the rotation, one must eliminate this randomness in molecular orientation. What is needed is some way to start all the molecules rotating from the same orientation at the same time. Zewail's solution is to pick out at some instant all those molecules that happen to have a certain orientation, and perform electronic spectroscopy on those alone. The Caltech team selected a batch of molecules with identical initial orientations by using a polarized femtosecond laser pulse with a frequency tuned to induce a transition from the ground state to the first electronically excited state. Only those ground-state iodine molecules that happened to have their transition moments aligned with the plane of polarization of this ultra-short "pump" pulse got boosted to the first excited state; the others were unaffected (Figure 3.12). A train of pulses from a second laser, each also lasting only a few femtoseconds, was then used to probe this group of excited molecules as they rotated. The probe pulses excited a second electronic transition to a higher excited state, from which the molecules decayed back to a lower state by emitting light (that is, by fluorescing). The researchers monitored the intensity of fluorescence from the second excited state as time progressed. The intensity depends

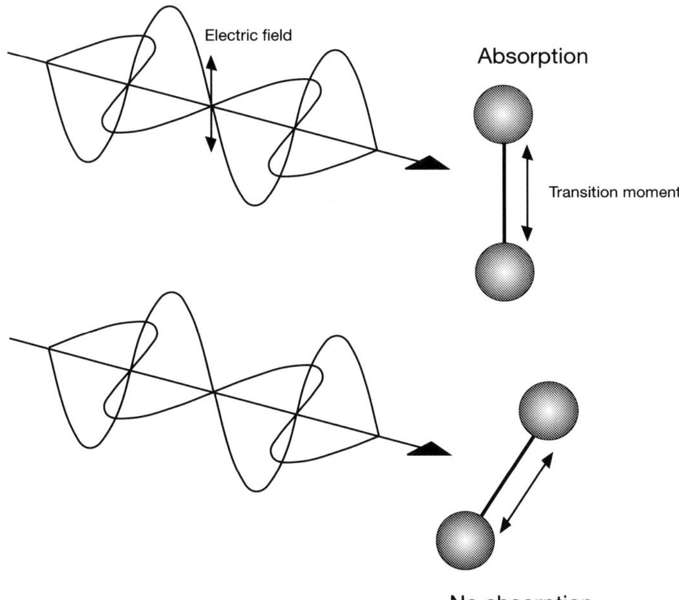


**Figure 3.11** *The iodine molecule will absorb a photon of light only when the plane of polarization is aligned with the molecule's transition moment, which lies along the axis through the two nuclei.*

on how many molecules are excited into the higher state, which in turn depends on the number that have their transition moments aligned with the polarization plane of the probe pulses at each instant. This number rises and falls as the molecules rotate, and so oscillations are seen in the intensity of fluorescence. Although it can hardly be called cinematic, the fluorescence signal is thus a movie of molecular rotation (Figure 3.13).

The wiggles in the time-dependent fluorescence intensity are not, however, all of equal height. The reason for this is that, although all of the excited molecules begin with the same orientation, they do not all have the same speed of rotation. A laser beam can never be perfectly monochromatic – it will always have a narrow *range* of frequencies, and therefore of photon energies. Because rotational energy levels are spaced so closely together, this range in photon energies in the pump pulse will excite ground-state molecules into several excited-state rotational levels, which differ in rotation speed. It turns out that the speeds of rotation in successive levels are related in a simple way. In the time it takes molecules on the lowest rung to rotate once, those on the next will have rotated twice, those on the next three times, and so on. Figure 3.13 shows the result when the molecules are pumped into three rotational levels. Small peaks in the absorption occur when all the fastest molecules have rotated once and come back into alignment, at which point the molecules in the two lower levels are still out of alignment. Another peak comes when the "middle" group are aligned, and a large peak

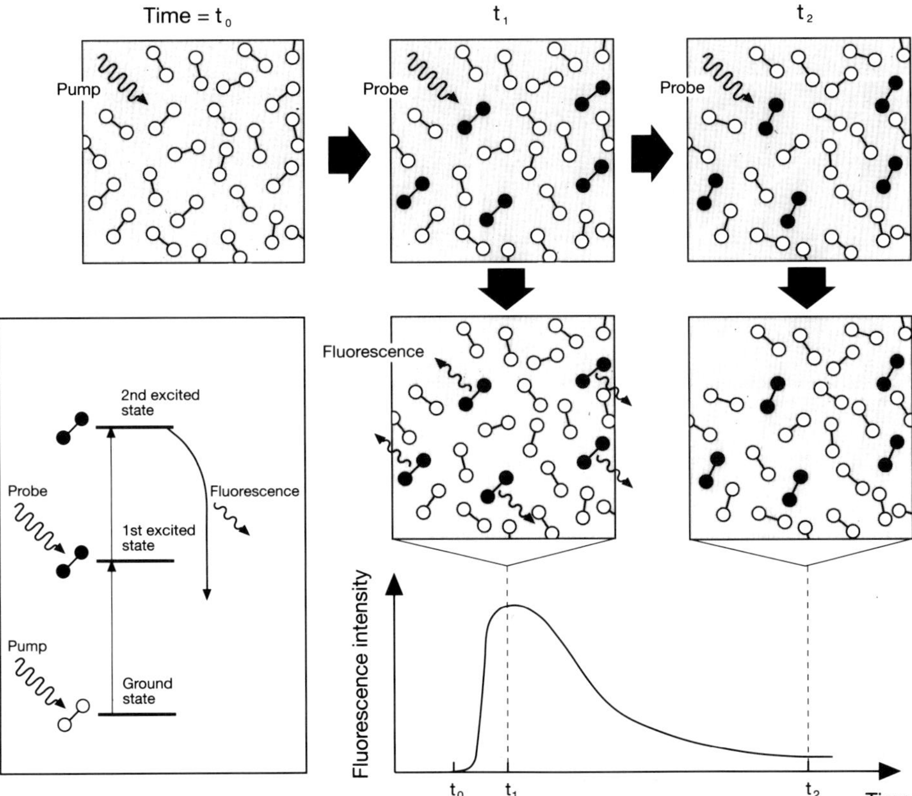


**Figure 3.12** *In Zewail's ultrafast laser "stroboscope" studies of molecular rotation, an extremely short "pump" pulse is used to excite into a higher electronic state all those iodine molecules that happen to be aligned with the polarization plane of the pump pulse at the instant that it is applied (time $t_0$). The excited molecules are shown as black. A sequence of laser "probe" pulses is then used to follow the rotation of these excited molecules. The probe pulses excite the molecules to a still higher state (shown as gray), from which the molecules decay back to the ground state by emitting a photon (that is, by fluorescing). Immediately after the pump pulse (time $t_1$), before the excited molecules have rotated significantly, they are still aligned for absorbing the polarized probe pulse. A few instants later (time $t_2$), many of the molecules have rotated out of alignment with the probe, so they are not excited to the fluorescent state. At still later times, the molecules rotate back into alignment, so the intensity of fluorescence rises again. The sequence of electronic transitions is shown at the bottom left.*

occurs when the slowest molecules have spun round once, since then the other groups are also in alignment, having spun twice and three times.

An iodine molecule vibrates in and out about ten million million times a second, so femtosecond pulses are short enough to take snapshots of this vibrational motion too. The probability that an iodine molecule will absorb a photon and become excited to a higher electronic energy level varies as the atomic separation changes during an

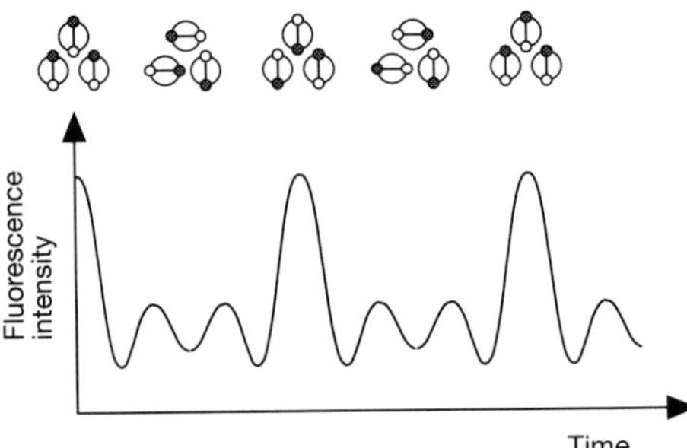


**Figure 3.13** *The ultrafast movie of molecular rotation is a trace of fluorescence intensity as time progresses; this rises and falls as the molecules move in and out of alignment with the probe pulses. Because the molecules do not all rotate at the same speed, some come back into alignment sooner than others, and so the undulations are not all of equal height. Here I show an idealized case for rotation of a group of molecules with three different speeds.*

in-and-out vibration – there is a critical internuclear distance for which the transition probability is greatest. The explanation for this variation in excitation probability, known as the Franck–Condon principle, has its origin in the quantum-mechanical description of the process. If the vibrations of each molecule can be synchronized, it should be possible to see a rise and fall in the number that are electronically excited (via their fluorescent decay) as the atoms in each molecule pass through the critical separation during a vibration.

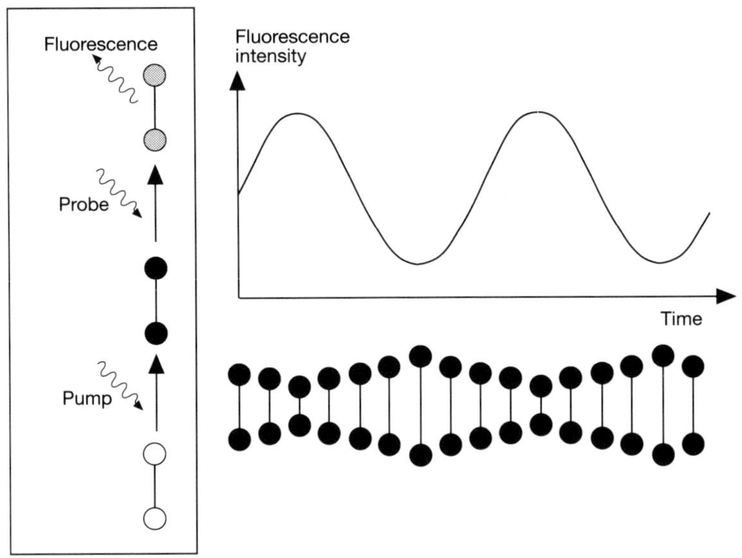


**Figure 3.14** *In Zewail's experiment to follow vibrational motion, the same principles of "pump and probe" are used, but what determines whether or not the excited molecules absorb the probe pulses (and thus fluoresce) is the distance between the two iodine nuclei. The transition to the fluorescent state is most probable when the nuclei are a certain critical distance apart. So the fluorescent signal rises and falls as the molecules stretch in and out through this ideal separation.*

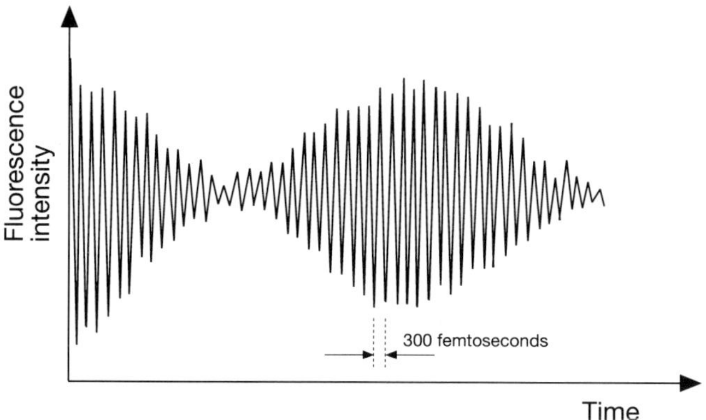


**Figure 3.15** *The result of the vibrational experiment is a fluorescence signal that oscillates in time with the molecular vibrations. A molecule that vibrates like a perfect ("harmonic") spring would give oscillations of constant amplitude, but because the molecular springs are imperfect, there is a slow modulation of this amplitude.*

Zewail and colleagues achieved such synchronization using the same trick as for the case of rotation. A group of iodine molecules was prepared with identical initial separations between the two atoms by pumping all of those with the critical separation at a given instant into the first excited electronic state. The electronic transition from this to a second excited state was then excited with pulses from a probe laser; fluorescence from the second excited state rises and falls in intensity as vibrations take the atoms back and forth through the critical internuclear distance (Figure 3.14). The result (Figure 3.15) is slightly complicated by the fact that the vibration is not quite like that of a simple, ideal spring, but is slightly nonideal or "anharmonic": this imposes a lower-frequency rise and fall on the rapid vibrational oscillations.

Although in both of these experiments the motions are monitored via fluorescence from the second excited state, this is essentially the same as following the change in absorption of the probe pulses. In effect, Zewail and colleagues are recording the electronic absorption spectra of individual molecules (more precisely, of batches of essentially identical ones) with very high resolution in time. It is as though they are watching iodine vapor alternating very rapidly between being colored and being colorless as the molecules all move in synchrony.

## *Quick reactions*

Their early success in probing molecular motions during the 1980s prompted the Caltech researchers to attempt something more ambitious – "filming" a chemical reaction in progress. For their first ultrafast laser movie of a fully fledged reaction, they chose as the subject the photodissociation (that is, the light-induced break-up) of the iodine cyanide (ICN) molecule. This is scarcely the most spectacular of reactions – all that happens is that a molecule absorbs more energy (from an incident photon) than the binding energy between the iodine atom and the cyanide group, and so falls apart. More technically, the ground-state molecule is electronically excited to an *unbound* state. This dissociation process involves just a single molecule (as opposed to, say, the collision of two), and is therefore said to be unimolecular. Of course, the practitioners of femtosecond spectroscopy would dearly love to be able to follow a chemical process of more significance – the action of an enzyme molecule, say. But the goals have initially had to be more modest, and in these early days the systems studied are chosen for their simplicity; after all, one could hardly have expected the Lumière brothers, early pioneers of the motion picture, to make *Star Wars*. Moreover, it is by studying simple, easy-to-understand systems that one can gain insights into the processes that should apply to more complex ones too.

The principles of this experiment, performed by the Caltech team in 1987, were much the same as described above: a femtosecond laser pulse was used to pump the molecules to the excited state, and subsequent probe pulses followed the molecules' fate by exciting a second electronic transition to a state that fluoresces (in this case, the fluorescence came from the cyanide fragment). The energy of the iodine–cyanide pair depends on the distance between them. In the ground (bound) state, there is an energy "well" around the equilibrium separation of the iodine and carbon atoms, in which one can think of the molecules as being trapped, executing small oscillations in the bond length. But in both the first and the second (fluorescent) excited states chosen for this experiment the energy decreases smoothly as the atoms get further apart. Therefore, the excited states dissociate spontaneously because there is no energy barrier to prevent this (Figure 3.16*a*).

Because the variation of energy with interatomic distance is slightly different for the first and the second excited states, their *difference* in energy changes as this distance changes. But the transition from the first to the second excited state can take place only when the photon energy in the probe pulses is equal to the energy difference between the two states. So as the molecules excited by the pump pulse start to fall apart, the CN fragments can absorb the probe pulses, ultimately giving rise to a fluorescent signal, only at a certain distance between the iodine and carbon atoms – that is, at a certain instant during the dissociation process.

Zewail and colleagues exploited this fact to follow the outward motion of the I and CN fragments as the molecules fell apart. By tuning the probe laser to different frequencies, they were able to induce the transition to the fluorescent state at different stages in this process. When they used photon energies corresponding to the energy difference between the two excited states for atoms about 0.3 millionths of a millimeter apart, they saw the fluorescence signal rise initially after the pump pulse was applied, as the fragments reached this separation, and then fall again as they continued outwards to larger separations (Figure 3.16). Once the fragments are about twice this distance apart, the energy curves for both of the excited states are essentially flat – their energy doesn't change as the fragments continue to separate, and so the energy difference is more or less constant. So when they used a probe laser tuned to this energy difference,

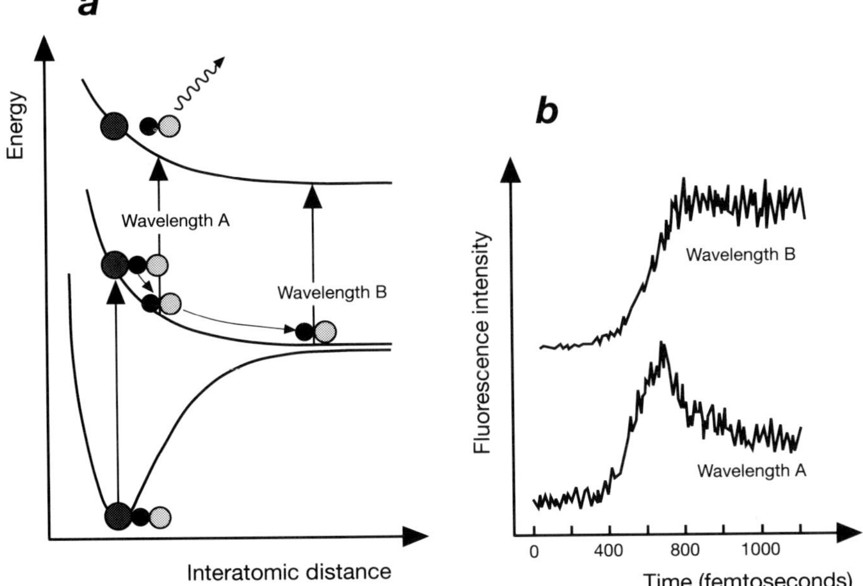


**Figure 3.16** *Snapshots of a chemical reaction: Zewail and colleagues have watched molecules of iodine cyanide fall apart in real time. In the ground electronic state, the I—CN bond vibrates within an "energy well." The Caltech team use a laser pump pulse to excite the ICN molecules to a state in which there is no longer a well, so that the I and CN groups drift apart (a). Ultrashort probe pulses are used to follow this dissociation process. When the photon energy in the probe pulse is tuned to the energy difference between the excited state and a higher, fluorescent state for an I—CN distance of 0.3 millionths of a millimeter (0.3 nanometers; wavelength A), the fluorescence signal rises and then falls as the fragments move apart through this separation (b). For a probe wavelength corresponding to the energy difference at a larger separation of 0.6 nanometers (wavelength B), the fluorescence signal rises later (since it takes longer for the fragments to separate this far) but then stays constant subsequently (b), as the energy difference between the two excited states remains more or less constant at larger separations – the fragments continue to absorb the probe pulses as they drift further and further apart.*

the Caltech team saw the CN fluorescence signal rise more slowly than before (because it took the atoms longer to move this far apart) but then stay constant, since the CN continued to absorb no matter how distant the fragments became (Figure 3.16).

Zewail's group have gone on to observe in real time a rather more complex unimolecular dissociation reaction, that of the sodium iodide molecule (NaI). This molecule is ionic, consisting of a positively charged sodium ion ($Na^+$) and a negative iodide ion ($I^-$) held together by electrostatic attraction. The ground-state energy again varies with interatomic distance in such a way as to form an energy well at small distances, but whereas for dissociation into neutral atoms the energy stays more or less constant for separations beyond those of the well, *charged* fragments continue to attract each other even at large separations. To dissociate the molecule into ions, therefore, energy must be supplied continually to overcome the force of attraction. But if, during

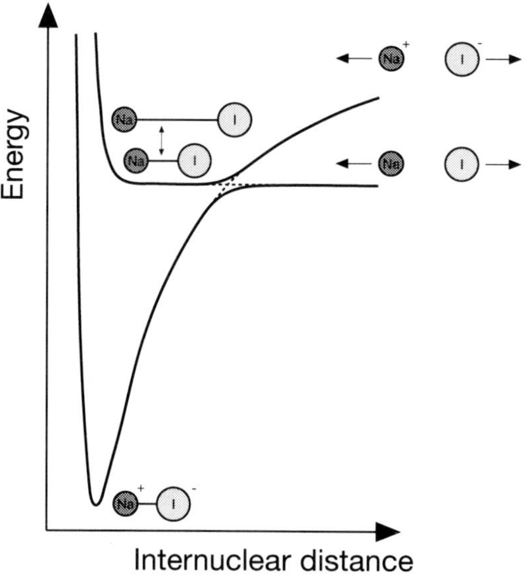


**Figure 3.17** *The sodium iodide molecule can be split apart either into sodium and iodide ions ($Na^+$ and $I^-$) or into two neutral atoms (Na and I). The energy of these two combinations varies in different ways as the separation of the sodium and iodine changes. At a certain separation, the energy curves cross; here, quantum mechanics allows the ionic and covalent forms to be "mixed," so that a crossing from one energy curve to the other is possible. In effect, an energy well is formed from a combination of the covalent curve (at separations smaller than the crossing point) and the ionic curve (at larger separations). A sodium iodide molecule excited onto the covalent curve may then vibrate within this energy well. Every time the atoms reach the outer wall of the well, there is a chance that they will "leak through" onto the covalent curve and continue to separate. In this figure, the dashed lines link the two portions of the "pure" energy curves; curve mixing creates the two hybrid curves shown as solid lines*

dissociation, the sodium ion takes back its electron from the iodide, both atoms become electrically neutral and can drift apart without the need for a further input of energy. The variation of energy with interatomic distance is therefore different for ions and neutral atoms – the ionic curve has a "bound" potential well, while the neutral-atom (or covalent) curve is flat (except at short distances) (Figure 3.17).

One might imagine that dissociation of NaI can be induced by exciting it from the bound, ionic energy curve to the unbound, covalent curve, on which the atoms will just drift apart. But things are not so simple. Where the ionic curve crosses the covalent curve, the energies of the two forms of NaI are equal, and one form can therefore switch to the other with no cost in energy. At this point, the molecule can actually exist in a "mixture" of both states. In effect what happens is that as the atoms on the covalent branch separate they acquire an ionic character at the cross-over point, owing to this mixing, and the electrostatic attraction then pulls them back together again. Mixing of the two states creates a second energy well for the excited state, so that the atoms in the excited molecule simply vibrate back and forth in this well (Figure 3.17). But every time the cross-over point is reached, there is a chance (determined by quantum-mechanical probabilistic criteria) that mixing can produce the covalent state, allowing the neutral atoms to escape the well. The two fragments will then continue their outward motion and dissociation occurs.

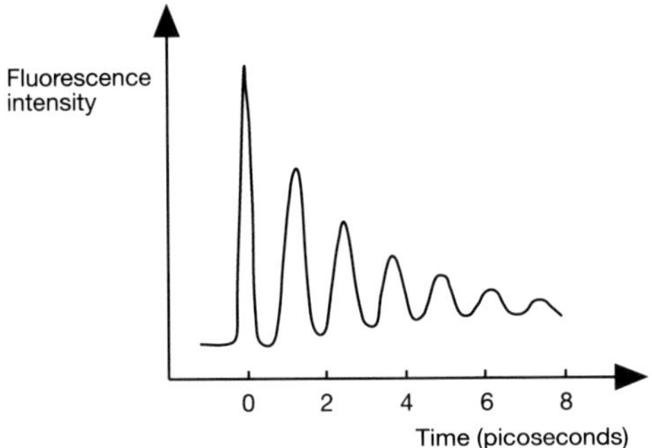


**Figure 3.18** *The light-induced dissociation of sodium iodide has been followed in real time by a pump-and-probe experiment. The vibration within the upper well, and the leaking out of this well onto the covalent curve (leading to dissociation) were followed by exciting the molecules into a higher state in which the sodium fluoresces. As more and more of the molecules "leak out" and dissociate, the amplitude of the fluorescence oscillations decreases. (A picosecond is a million-millionth of a second, or $10^{-12}$ seconds.)*

Zewail's team watched excited-state NaI molecules bouncing back and forth and occasionally leaking through the well and dissociating. They used a pump pulse to induce the transition from the (ionic) ground state to the excited covalent state, and a string of probe pulses to monitor the oscillations of the latter in the upper potential well. The probe was tuned to a frequency that excited free sodium atoms to a fluorescent excited state, just as the CN fragments were excited in their earlier experiment.

The fluorescence signal rose and fell as the vibration of the excited molecule took it in and out of the critical atomic separation for absorbing the probe pulses (once again, this is the effect of the Franck–Condon principle). But the intensity of these oscillations in fluorescence decreased steadily (Figure 3.18) as a fraction of the excited molecules were "lost" via dissociation on each oscillation. Thus both the vibrations of the excited state and the process of unimolecular dissociation became visible in real time.

These are early days for molecular movies, but already Zewail and others have explored more complicated reactions, such as those between two colliding molecules or those taking place within a "cage" of atoms, as they do in solution. Not only do these studies provide insights into the most fundamental processes of chemical transformation, but they give us a more complete picture of the nature of the chemical bond itself. And by developing an understanding of the *timing* of molecular motions, we can begin to exercise a greater degree of control on the atomic scale – to direct the course of reactions.

## The photochemical scalpel

Simple molecules such as ICN and NaI can be dissociated just by heating them up, which pumps the molecules up the vibrational energy ladder until they are shaken apart. While the same is true for more complicated molecules, heat is a highly indiscriminate dissociative agent: it increases the energy in all bonds more or less equally, and so heat-induced dissociation of complex molecules is likely to give a random selection of molecular fragments.

The ability to perform selective dissociation – that is, to break specific bonds within a molecule – would give chemists tremendous power for controlling the course of a chemical reaction. At present, selective bond-breaking is most often carried out by chemical means. In the synthesis of organic compounds, for instance, a common approach is to first shield sensitive parts of the molecule by adding on "protecting groups," before exposing it to the reagent that will break the desired bond. The protecting groups must then be removed afterwards – a long and arduous operation which may be limited in its effectiveness and may decrease significantly the yield of end product. Although selective catalysts of the kind described in Chapter 2 may sometimes be used to enhance the breaking of a specific bond, usually these are developed through painstaking research which often relies to a large degree on hit-or-miss experimentation. By applying the lessons of laser spectroscopy to photochemistry, chemists are now developing new techniques for selective bond-breaking which might ultimately be much cleaner and more efficient than these alternatives.

Laser-induced bond-selective dissociation would represent a kind of "molecular surgery," in which lasers are used to cut away selected areas of a molecule while leaving others intact. This would be done by exploiting the high intensity and color specificity (monochromaticity) of lasers to pump energy into certain bonds and set them shaking fit to bust. We have seen that the vibrations of different bonds in a molecule absorb light only at certain characteristic frequencies. Because the range of frequencies in laser light is so narrow, we might hope to be able to excite vibrations in a single kind of bond within a molecule by irradiating it with laser light of just the right frequency.

But while this is indeed possible in principle, there is a catch in attempting to use it for bond-selective unimolecular dissociation. Just because we deposit a lot of energy into a single bond does not guarantee that it will stay there long enough for the bond to break. Molecules have certain preferred modes of vibration called eigenstates, the nature of which is determined by the shape (or more strictly, the symmetry properties) of the molecule. For example, a vibrational eigenstate of the methane molecule ($CH_4$) is that in which all of the four C—H bonds stretch and contract in concert, preserving the tetrahedral shape of the molecule at all times. When energy is dumped into a specific bond, there will be a tendency for it to get rapidly redistributed so that the molecule executes the preferred eigenstate vibrations. This energy-scrambling effect limits the efficiency with which selected bonds can be broken.

Mindful of this difficulty, researchers have chosen first to set their sights on achieving dissociation simply from a specific vibrational eigenstate. This means preparing the molecule in the chosen eigenstate with enough energy to cause dissociation. Fleming Crim and colleagues at the University of Wisconsin-Madison have used this approach to induce eigenstate-specific dissociation of the hydrogen peroxide molecule, $H_2O_2$. Rather than pumping the molecule directly up to the targeted vibrational eigenstate, however, Crim and colleagues put the energy into the eigenstate via energy transfer within the molecule. They used a laser to excite the O—H stretching vibrations onto the sixth rung of the energy ladder; this energy then became rearranged into a vibrational eigenstate, in which the O—O bond vibrated too. While the energy deposited initially in the O—H bonds was not sufficient to break them, the O—O bond is weaker; when

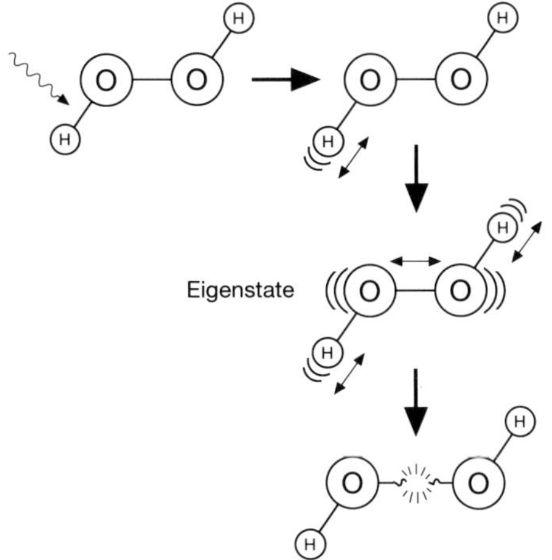


**Figure 3.19** *The O—O bond in hydrogen peroxide can be broken selectively by pumping energy into the O—H bonds. This energy is redistributed rapidly throughout the molecule, placing it in a characteristic vibrational eigenstate. The O—O bond of the excited molecule is then broken in preference to the O—H bonds.*

the molecular vibrations developed into the eigenstate, the molecule therefore fell apart into two OH fragments (Figure 3.19).

Eigenstate-specific dissociation is not, however, the same as bond-specific dissociation. In the former, one ensures that dissociation takes place for a molecule vibrating in a specific manner, but it does not necessarily guarantee that the process will give a single set of products since there may be several different ways in which a given eigenstate can fall apart once it has enough energy to do so. To achieve selective bond-breaking, it is necessary to ensure that all the excitation energy gets channeled into the motions that will cleave the molecule in the intended fashion. This is a more difficult challenge because these motions will not in general correspond to an eigenstate vibration. Bradley Moore and coworkers at the University of California at Berkeley tried to enhance the rate of hydrogen-atom transfer from one carbon atom to another during rearrangement of a cyclic hydrocarbon by using a laser to excite the stretching vibrations of the C—H bond in the hope of breaking it. But they found that the redistribution of energy within the molecules was too quick: pumping the C—H bond had no *specific* effect on the rate of bond-breaking and hydrogen-atom transfer.

There has been more success, however, with very simple molecules. Fleming Crim and coworkers chose to attempt selective bond cleavage in water molecules in which one of the two hydrogen atoms had been replaced by a "heavy" isotope of hydrogen, deuterium (D). Dissociation of HOD can yield either H and OD or D and OH. If the dissociation is induced by heat, a mixture of both pairs of products will be obtained. Because deuterium is heavier than hydrogen, the vibrational frequencies of the O—H and O—D bonds are different, and so in principle it is possible to excite one or the other selectively using laser pulses. The question is whether dissociation can be induced in this way without the excitation energy getting redistributed between both bonds. Crim and colleagues used an ingenious approach to break

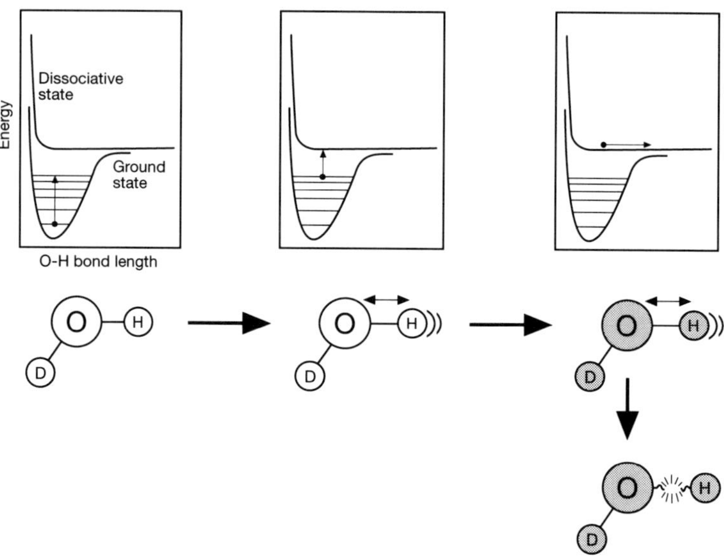


**Figure 3.20** *To break the O—H bond selectively in a deuterated water molecule (HOD), Fleming Crim and colleagues excited the O—H vibration of the ground-state molecules and then pumped these molecules up to a dissociative electronic state, whereupon the "hot" O—H bond broke in preference to the O—D bond.*

the O—H bond selectively. They first excited the O—H vibration up to the sixth rung of the ladder – not enough to break it. By applying a second laser pulse, however, the researchers then boosted the vibrationally excited molecules up to an electronically excited dissociative state, in which the molecule must inevitably fall apart. Because the O—H bond contained more energy than the O—D bond, it was more likely to be the one that broke in this excited state (Figure 3.20); Crim found that O—H bond cleavage was 15 times more common than O—D cleavage. In more recent experiments, Crim's group have been able to use this approach to achieve selective breaking of *either* the O—D *or* the O—H bond, by altering the wavelength of the pump laser accordingly. Richard Zare and colleagues at Stanford University in California have achieved similar selectivity in transferring a hydrogen atom from ammonia ($NH_3$) to "deuterated" ammonia ($ND_3$) – they were able to make this process much more efficient than the reverse transfer of deuterium from $ND_3$ to $NH_3$.

These studies bode well for bond-selective photochemistry, but there is a very long way to go before such techniques can be put to practical use in chemical synthesis. Some researchers suspect that the vibrational motions of larger molecules may turn out to be too complicated for "clean" surgery ever to be possible. Many important industrial processes involve relatively simple compounds, however, and it may not be unrealistic to hope that light will one day provide us with unprecedented control over their outcome.

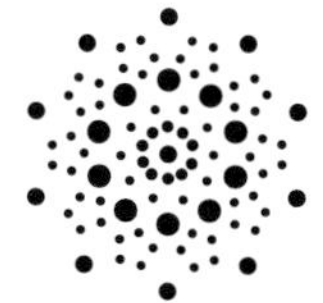

# 4

# Impossible Order

## When atoms meet geometry

*I miss the old days, when nearly every problem in X-ray crystallography was a puzzle that could be solved only by much thinking.*

Linus Pauling

Amongst the many differences between London and New York, most can be appreciated only by experiencing both of those cities at first hand. One of the most striking, however, is apparent from the most cursory of glances at a street plan (Figure 4.1). New York is a city planned and built with foresight: Manhattan in particular has a trellis-work pattern of streets running from the east to the west side, with the great avenues crossing them at right-angles from north to south. The city is thereby cut up into neat, orderly blocks, all much the same size and lying in the same orientation. London, by contrast, displays little sign of regularity. Rather, it sprawls in an apparently random tangle of roads big and small, interweaving in a pattern that has no discernible structure at all. No one ever conceived London on a grand scale; it evolved from muddy tracks between the taverns, churches and hovels of the Middle Ages. A consequence of this difference is that a stranger asking for directions in New York City will receive advice in similar terms regardless of where she happens to be in the city – go four blocks south, say, and then three east. In London there is no concept of a block, and each set of directions has to be specific: go down Fleet Street, across Ludgate Circus and bear right around St Paul's . . .

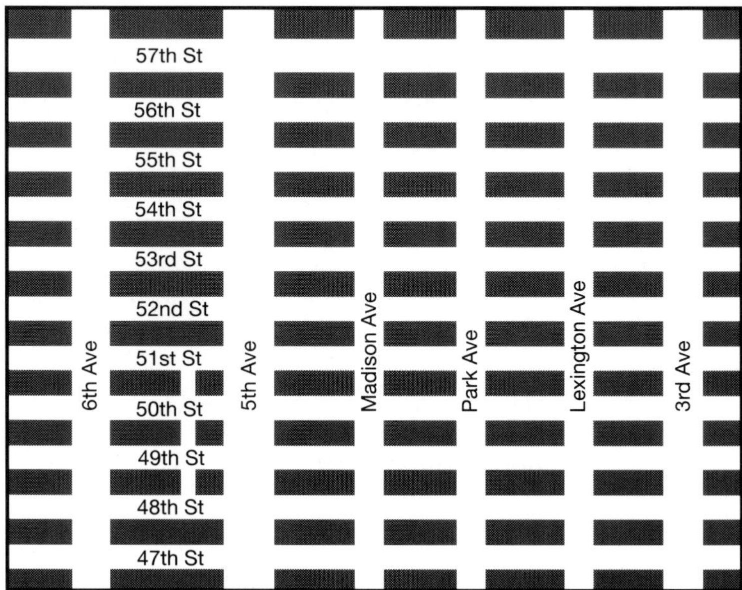


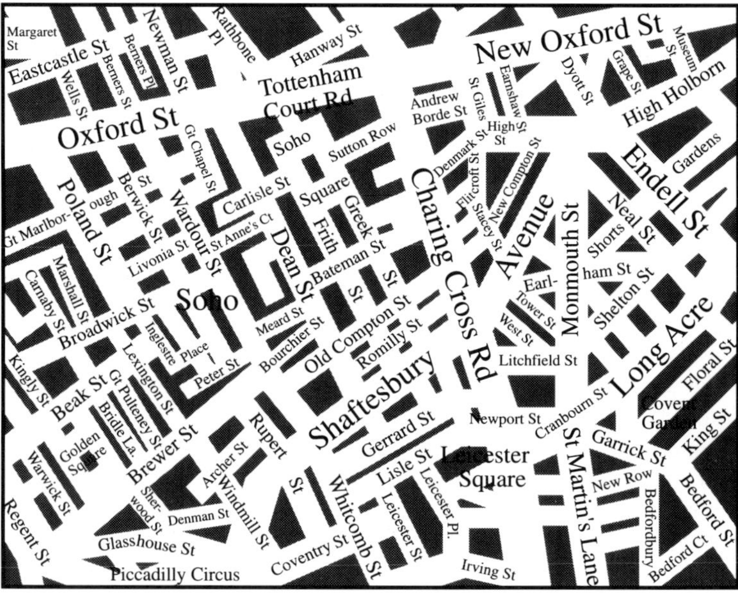


**Figure 4.1** *While Manhattan is an orderly city divided neatly into blocks of roughly equal size (top), London is a jumble of streets without any discernible regularity (bottom).*

In the structures of solid materials we can find an analogous situation. Some are like New York – divided into orderly "blocks" of atoms that recur in a regular manner. These solids are crystals, such as quartz, table salt or metals. Other solids have a Cockney character: their constituent atoms or molecules disport themselves at random, and no part of the material is ever exactly identical to any other. These are called amorphous solids, and include window glass and most plastics. It should come as no surprise that it is generally much more easy to describe the structure of crystals than of amorphous materials. Imagine a map-maker walking through New York. He finds that every time he has walked fifty yards or so down a street, he comes to a crossroads where an avenue intersects at a right-angle. Whether he takes one of these turnings or just continues along a straight course, he will find much the same thing repeating: a crossroads every fifty yards or so. Before long he decides it is safe to assume that this is what all of the city must look like. To complete his map, therefore, all he has to do is to fill in the outline of Manhattan island with a network of streets and avenues spaced fifty yards apart, a job that can be done over a Budweiser in a bar rather than by painstakingly plodding through the streets. While this will overlook anomalies such as Broadway, it will give a fair picture of the city's layout. But the poor cartographer of London is not so fortunate – she wanders on and on through the jumble of streets, scribbling as she goes, yet the map just goes on getting more convoluted, and she can't even hazard a guess at what the blank spaces will look like until she has visited them.

The distinction between a crystal and an amorphous solid is that between order and disorder. To be more exact, the distinction is between order and its absence over *large distances* (millions of atoms, say); a careful look at the structure of glasses and other disordered solids often reveals some degree of regularity amongst an atom's immediate neighbors. (We can identify something analogous over short distances in London: small islands of regularity sometimes crop up, and there are a few approximate rules apparent amidst the chaos, such as the fact that street corners are frequently not far removed from right-angles.) But the details of the structure of amorphous solids is a major topic in itself, and one that I shall not broach here.

A perfect crystal has an atomic-scale structure that repeats regularly and identically throughout (although few real crystals are so perfect). Scientists have learnt to exploit this fact in a technique that allows them to decipher the exact positions of the atoms within the crystal. The technique, which involves bouncing X-rays off crystals and measuring the patterns of "light" and "dark" in the reflected beam, is called X-ray diffraction, and represents the most powerful tool at the chemists' disposal for seeing the shape of a molecule directly. Because many biological molecules, including proteins, can be prepared in a crystalline form, X-ray diffraction provides a means of understanding the structure of these gigantic molecules, the complexity of which would pose tremendous difficulties for any other method of structure determination.

One of the principal subjects of this chapter is a novel class of solid which, when first discovered in 1984, seemed to turn the venerable field of crystallography – the study of crystal structures – on its head. These materials, called quasicrystals, have made necessary some careful rethinking of exactly what constitutes a crystal. Although arguments about the detailed atomic structure of quasicrystals still persist, the paradox initially posed by their existence is now largely resolved. In the course of discussing quasicrystals we will encounter some ideas about the concept of symmetry that have exercised artists, artisans and mathematicians alike.

## The benefits of constructive interference

### *The unit of order*

The regularity of a crystal's structure means, as our map-making New Yorker discovers, that one need deduce only what a part of it looks like to obtain a picture of the whole. The smallest region that needs to be mapped out in order for the entire solid structure to be known – the equivalent of New York's blocks – is called a unit cell. A perfect crystal consists of many billions of unit cells stacked together like boxes. The problem of working out how the many billions of atoms in a crystal are arranged thus reduces to that of figuring out the arrangement of a much smaller number of atoms – perhaps just half a dozen or so for a simple compound such as rock salt (sodium chloride) – in a single unit cell.

Rock salt provides an ideal example for illustrating the nature of crystal structures, not only because of the relatively simple way in which its constituent atoms (sodium and chlorine) are arranged in the unit cell, but because it takes nothing more than a stroll into the kitchen to reveal the consequences of the crystals' regular atomic structure. A close inspection of salt grains, assisted perhaps by a magnifying glass, will reveal many that are shaped roughly like a cube (Figure 4.2). This cubic symmetry can be traced all the way down to the shape of the unit cell, which itself has a cubic shape defined by the positions of sodium and chlorine atoms. Crystals of sodium chloride are constructed from stacks of the unit cell shown in Figure 4.3, each of which contains a total of four sodium and four chlorine atoms sitting at the corners, on the edges and faces, and on the inside of the cube. (There are in fact more than four of each kind of atom shown in Figure 4.3, because those on the edges, corners and faces overlap into adjacent unit cells. But if we add up all those segments that sit solely within the cube of the unit cell, the total number of each kind of atom equals four.) I should mention that in sodium chloride the atoms are in fact *ions*: that is, they bear electric charges, because the sodium atom gives up one electron to the chlorine, making the former a positively charged ion (called a cation) and the latter a negative ion (an anion, specifically a chloride ion in this case). The structure defined by the positions of the sodium ions is called a lattice – a regular, symmetrical array of identical points. The chloride ions occupy the corners of an identical, interlocking lattice.

While even simple ionic salts of this sort show a considerable variety of crystal structures, certain structures turn out to be common to several different compounds. The sodium chloride structure, for example, is shared by potassium chloride, copper oxide and magnesium sulfide; the metals occupy the positions of the sodium ions, and

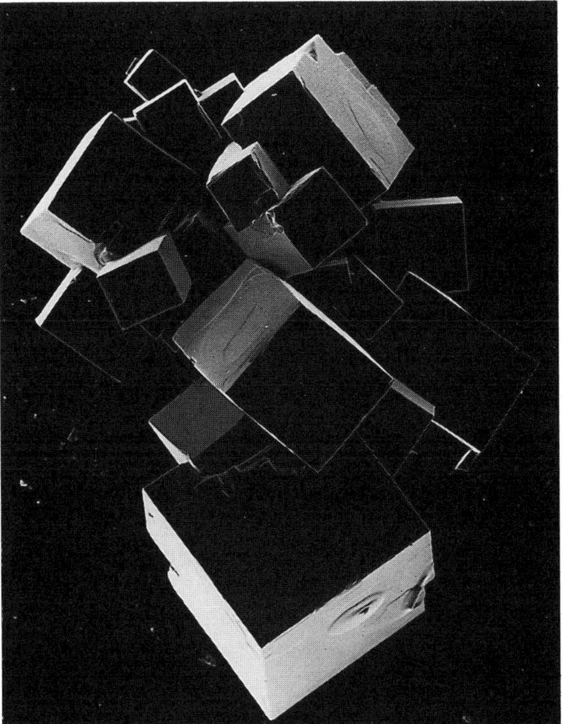

**Figure 4.2** *Table salt (sodium chloride) forms cubic crystals which reflect its regular packing of ions into cubic unit cells. (Photograph by Jeremy Burgess/Science Photo Library).*

the nonmetals that of the chlorides. This simplifies considerably the task of crystallographers because it means that compounds can be grouped together under generic structures. The unit cells of some other common crystal structures are shown in Figure 4.4.

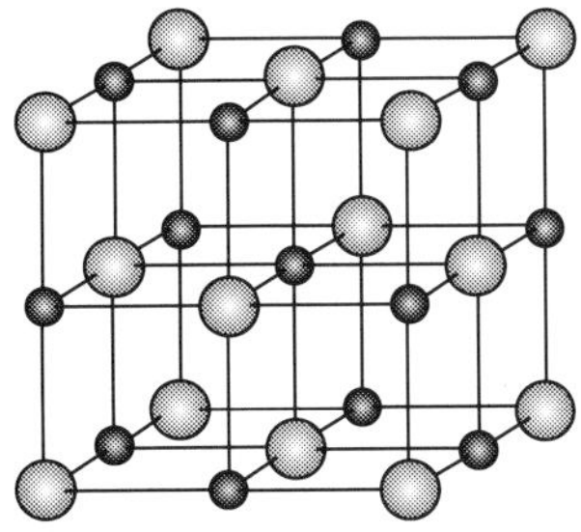

**Figure 4.3** *The cubic unit cell of crystalline sodium chloride. Large spheres represent chloride ions, and small spheres sodium ions.*

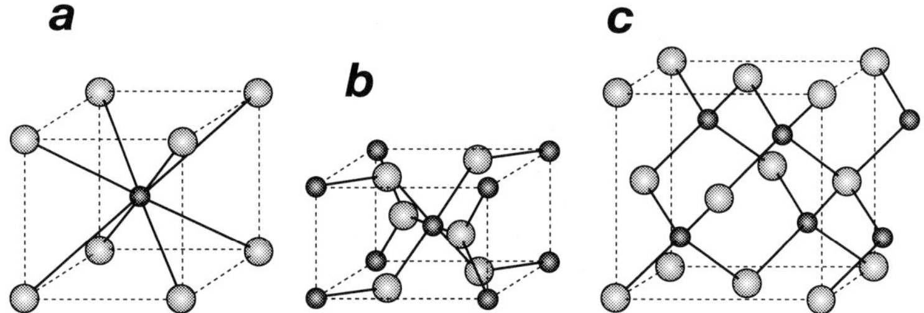


**Figure 4.4** *Certain packing arrangements of ions in crystals are shared by many compounds. The unit cells of some of the more common structures are shown here: the cesium chloride (a), titanium dioxide ("rutile") (b) and zinc sulfide (c) structures. In all cases, the small spheres represent the metal ions.*

## *Seeing with X-rays*

The structure of the sodium chloride crystal was one of the first to be deduced (in 1913) via the technique of X-ray diffraction. To investigate how electromagnetic radiation interacts with a regular array of "scatterers" such as atoms, in 1912 the German physicist Max von Laue fired X-rays at a crystal of copper sulfate and recorded the pattern of reflected beams on photographic plates. Von Laue found that in some directions there was a high intensity of reflected X-rays while in others there was virtually no sign of scattered radiation. The result was a symmetrical pattern of spots on the photographic plates (the pattern for zinc sulfide (Figure 4.4*c*) is shown in Figure 4.5). Only months later, W. Lawrence Bragg of Cambridge University showed that the pattern of spots constitutes a kind of coded image of the positions of atoms in the crystal. Bragg demonstrated how one could work backwards from the pattern to calculate the arrangement of the atoms and the distances between them. In collaboration with his father William Bragg, Bragg junior measured the patterns of X-rays reflected from several crystalline materials and translated these into pictures of their crystal structures.

Diffraction is the name given to the phenomenon in which bright spots are generated from a beam of X-rays bounced off a crystal. It originates in the wave-like nature of X-rays. This radiation, as pointed out in Chapter 3, is nothing more than a form of light with very short wavelengths (and consequently very high energy, since the energy in a quantum "packet," or photon, of electromagnetic radiation increases as its wavelength decreases). A beam of X-rays can be thought of as a bunch of undulating rays each with a series of peaks and troughs, corresponding to greater or smaller amplitudes of the electromagnetic field. When one wave encounters another, they disturb or "interfere with" each other. If two waves meet when they are both at a peak, they "add" together to form a single peak of twice the height. Conversely, if, at their intersection, one wave peaks while the other reaches a trough, the waves cancel each other out with the result that at the point of collision the intensity of the radiation is zero (Figure 4.6). The addition of two peaking waves is called constructive interference, and their cancellation corresponds to destructive interference. The effects of interference can be seen in water

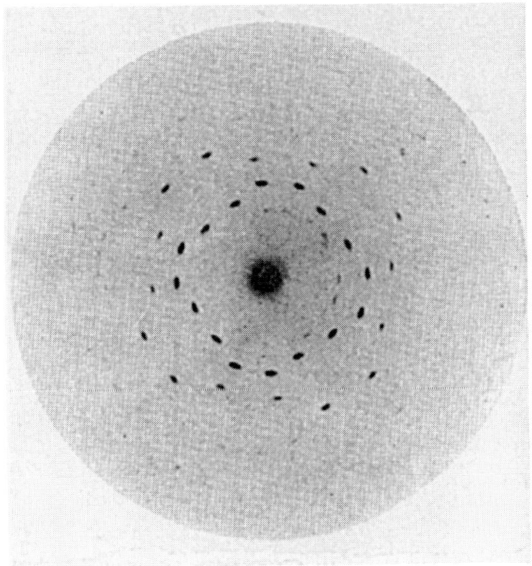

**Figure 4.5** *Max von Laue discovered that X-rays scattered off crystals form a regular pattern of spots, which can be recorded on photographic plates. The X-ray pattern shown here is that from a zinc sulfide crystal, and was one of the first determined by von Laue and colleagues in 1912. The fourfold symmetry evident in the pattern reflects a symmetry of the crystal structure, shown in Figure 4.4c. (From von Laue, 1961.)*

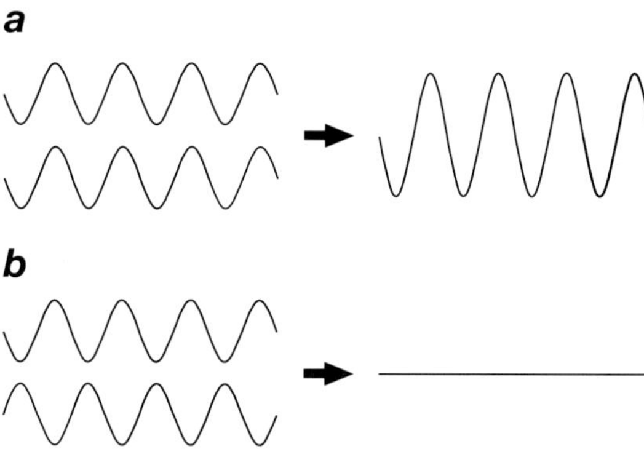


**Figure 4.6** *When two waves encounter each other, the oscillations can be enhanced or diminished by interference. In the one extreme where the two waves are perfectly in step ("in phase"), the peaks and troughs reinforce each other, and the result is a wave with twice the amplitude (a). This is constructive interference. In the other extreme, the waves are exactly out of step, and cancel each other out (b). This is destructive interference.*

waves: circular ripples radiating out from two points, like those produced when two pebbles are dropped into a pond, give rise to an interference pattern where they cross.

The image recorded on von Laue's photographic plates in 1912 was a picture of the diffraction pattern produced by the interference between X-rays scattered from the copper sulfate crystal. The X-rays bounce back off the atoms in all directions. Some bounce off the atoms in the first layer; others will miss these atoms but bounce off those in the layer below. Still others will be reflected by the third layer, the fourth, and so on. But an X-ray that bounces off the second layer has traveled a greater distance than one reflected from the first layer. This means that the peaks and troughs of the two rays can get out of step in the reflected beam, so that interference can occur. When the rays are half a wavelength out, interference will be destructive – the rays will cancel. If the mismatch is a whole wavelength, however, constructive interference is the result (Figure 4.7). If one of the rays has been reflected not from the second layer but from, say, the sixtieth, the difference can be many wavelengths. But whenever the difference in distance traveled by two interfering rays is a whole number of wavelengths, the peaks match up and interference will be constructive, while if the difference is a whole number plus another half of a wavelength, the rays cancel. The relationship between two reflected rays depends on the angle that they make with the atomic layers, and on the layer spacing. The reflected beams therefore produce a spatial pattern of high and low intensity, captured on the X-ray-sensitive photographic plates as bright spots. The regular, symmetric stacking of atoms in the crystal is echoed in the regular pattern of spots.

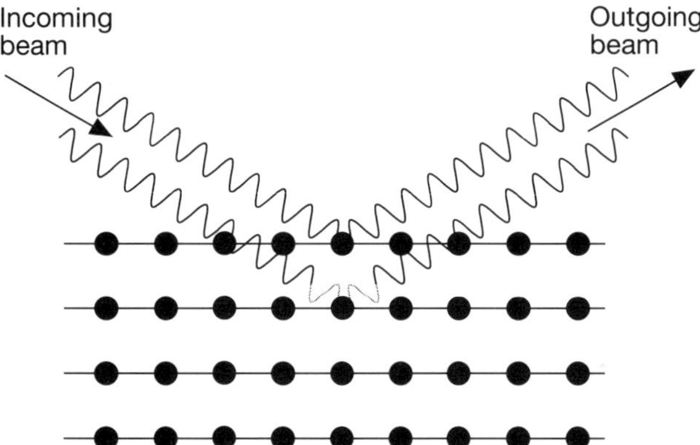


**Figure 4.7** *Parallel beams of X-rays reflected from different atomic layers in a crystal can get out of step and therefore interfere constructively or destructively. The nature of the interference is determined by the difference in path length of the two beams (here shown in gray), which itself depends on the distance between the layers and the angle of the incident beam. Thus the diffraction pattern formed from the interference of many beams from many different layers contains information about the layer spacing. Lawrence Bragg deduced a mathematical expression to extract this spacing from the positions of the bright spots due to constructive interference.*

Lawrence Bragg realized that, from the way that the scattered intensity of X-rays varies as the angle of the incident beam is changed, it should be possible to calculate the distance between atomic layers. He formulated an equation relating the layer spacing to the angles for which bright spots appear in the diffraction pattern. Bragg's equation helps to explain why we must use X-rays to obtain diffraction patterns from crystals, rather than more convenient and safe forms of radiation like visible light. The equation shows that for diffraction to be possible from a layered structure of any sort, the wavelength of the incident waves must be of the same kind of size as the distance between layers, which for crystallographic purposes means roughly the distance between neighboring atoms. In sodium chloride crystals, for example, the atoms of sodium are just under a third of a millionth of a millimeter from their neighboring chlorine atoms, which is comparable to the wavelength of 0.15 millionths of a millimeter for the X-rays typically used in diffraction studies.

Calculating the distance between layers of atoms in a crystal is not, however, the same as deducing exactly where each atom in the crystal is located. But a regular arrangement of atoms on a lattice amounts to much more than a single sequence of stacked layers: it is possible to identify in such an arrangement many different sets of layers in various orientations (Figure 4.8). X-rays incident on a given set of planes at a given angle will at the same time be striking another set at a different angle, and there will be a whole set of Bragg conditions for constructive interference from each stack of planes, determined by their interlayer spacing. As the angle of the incident beam is varied, we sweep through each of these diffraction conditions to obtain a complicated pattern of bright spots due to constructive interference from each set of planes (to do this, it is easier to keep the X-ray beam steady and merely rotate the crystal in the beam).

The full diffraction pattern contains information about the distances between atoms in the crystal in all three dimensions, not just the separation between a single set of planes. In principle, then, all the details of the crystal structure are coded within the pattern's spots (or "peaks"). But to decipher this code, the crystallographer has to be able to deduce which peaks come from which planes. This task, called indexing the peaks, demands some pretty hefty mathematics. There is one feature of the complex pattern that immediately tells us something about the crystal structure, however: the symmetry of the diffraction pattern echoes that of the crystal itself. We have seen that the unit

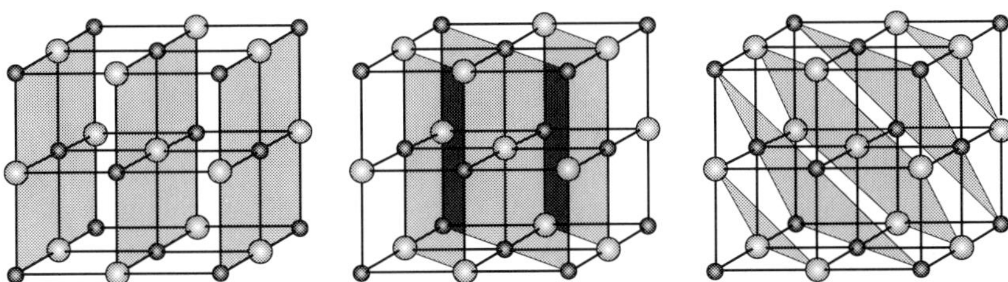

**Figure 4.8** *A crystal, such as sodium chloride shown here, contains a multitude of reflective atomic planes intersecting the lattice at different angles.*

cell of sodium chloride, for example, has the symmetry of a cube, which means in particular that it can be rotated by a quarter of a complete 360-degree rotation to end up in an identical position. Four such rotations bring the unit cell back to its starting position, so it is said to have a fourfold axis of symmetry. The diffraction pattern too has fourfold symmetry, analogous to that exhibited by the zinc sulfide pattern (Figure 4.5). Most simple crystals, such as pure metals, have fourfold or sixfold symmetries.

## *Thinking bigger*

If the unit cell contains a large number of atoms, the task of indexing the peaks is far from straightforward. Then, rather than trying to decode the diffraction pattern entirely from scratch, it becomes necessary to make an initial rough guess at the structure; the diffraction pattern that would be produced from this guessed structure can be calculated and compared with the one measured experimentally. If the guess was a good one, the patterns will be similar, and the trick is then to shuffle the atoms around until the match is close to perfect.

For the most complicated of unit cells, such as those encountered in crystals of biological molecules like proteins, even this guess-and-shuffle method may be to no avail. The chance of making a good initial guess in these cases is small, so that crystallographers may find themselves rearranging the atoms endlessly without any sign of targeting in on a calculated pattern that looks like the real one. Under these circumstances, it is best to forget about the positions of individual atoms altogether, and to depict the structure in a completely different way. When an X-ray is scattered from an atom, it is really the atom's electrons that are responsible for the scattering – the X-rays do not "see" the tiny atomic nucleus at all, but are deflected by the electron cloud that surrounds it. So the diffraction pattern is really a coded image of the distribution of electrons throughout the unit cell. As the electron clouds are generally centered around the positions of the atomic nuclei, the electron distribution gives a reasonable picture of the positions of the atoms themselves; but rather than dividing this distribution up into discrete atoms, it can be treated as a smooth, continuous map of smeared-out electrons, dense in some places and rarified in others.

The advantage of treating the structure in terms of an electron density map is that it allows one to bring to bear some mathematical tools that cannot be applied to a discrete "atomic" picture. Rather than pushing atoms about in order to match a calculated diffraction pattern to the measured one, a continuous map of electron density can be "molded" like clay to the right shape, using a mathematical procedure derived from the work of the nineteenth century French mathematician Joseph Fourier. Once the fitting procedure is completed to the crystallographer's satisfaction, the electron density map is inspected for regions where the electrons are dense, which generally indicates the presence of an atom (Figure 4.9).

The Fourier method of decoding diffraction patterns has proved invaluable for elucidating the structures of organic and biochemical molecules by X-ray analysis of their crystals. In the early days of crystallography it was believed that molecules of this

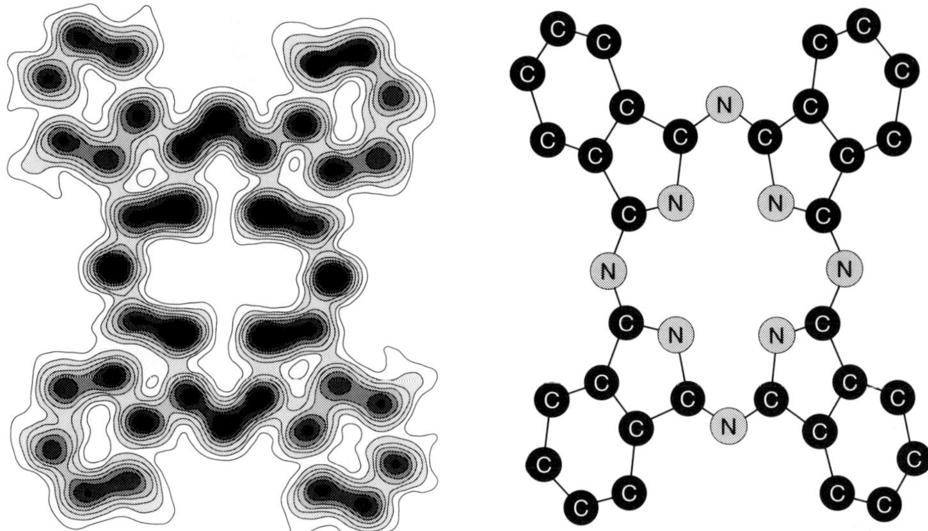


**Figure 4.9** *A map of electron density in crystals of phthalocyanine, obtained by Fourier analysis of the diffraction pattern, reveals the positions of the atoms in the molecule. The molecular framework is shown on the right.*

sort were too complicated for their diffraction patterns ever to prove intelligible. But the diligence of crystallographers showed otherwise. The structures of penicillin and vitamin $B_{12}$, molecules containing 41 and 177 atoms respectively, were deduced in the 1950s by Dorothy Hodgkin of the University of Oxford, for which achievements she was awarded a Nobel prize in 1964. Myoglobin, a relative of the hemoglobin that ferries oxygen through the blood, yielded the secrets of its structure to John Kendrew of Cambridge University in 1955, and the diffraction patterns of DNA measured by Rosalind Franklin of King's College in London provided the vital clues to the double-helical structure of DNA, deduced by Francis Crick and James Watson in 1953 (see Chapter 5).

The power of X-ray crystallography was greatly extended in 1953 with the discovery by Max Perutz that heavy metal atoms such as mercury or gold can be incorporated into a protein crystal without disturbing significantly the arrangement of the other atoms. Heavy atoms like this, with a very large electron density, show up loud and clear in the diffraction pattern and so provide a kind of peg around which the rest of the electron map can be constructed. The ability to study proteins opened the way to a new understanding of what molecules such as enzymes look like and how they function. Today huge and complex biological structures, such as that of the virus responsible for foot-and-mouth disease (Figure 4.10), can be solved more or less routinely. The foot-and-mouth virus is noteworthy in that it has fivefold (pentagonal) symmetry. In the crystal the viruses are stacked essentially as though they were spheres,

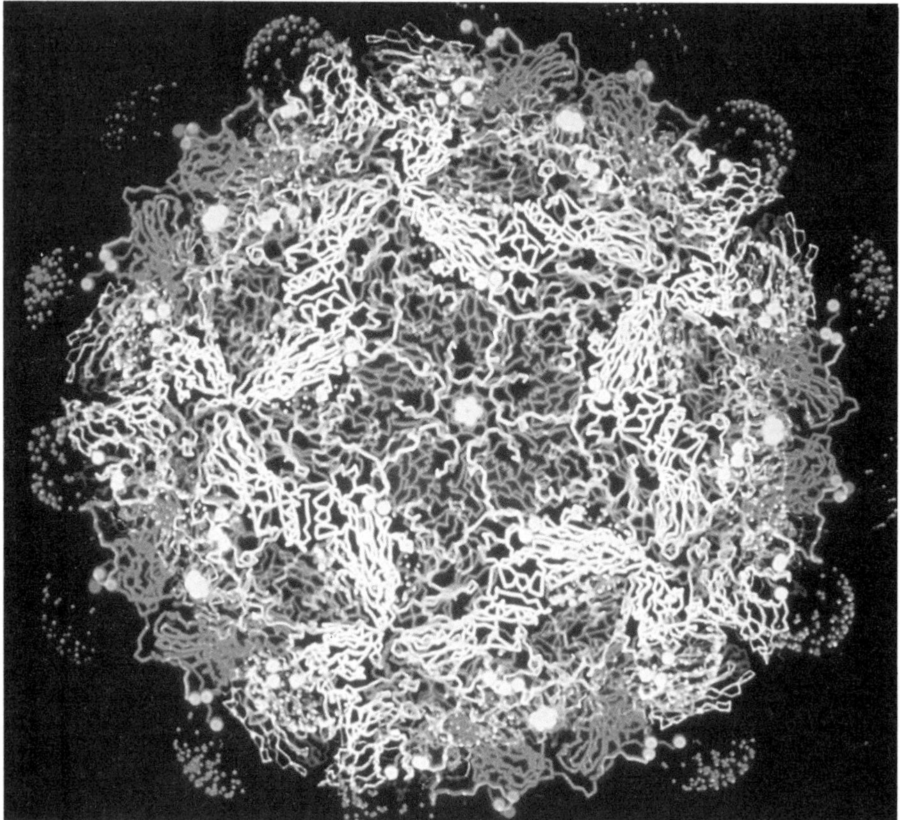

**Figure 4.10** *The structure of the foot-and-mouth virus, deduced in 1989, is immensely complicated. The virus contains more than a third of a million atoms, and while these cannot be distinguished individually, the general shape and structural features are clear. The virus has fivefold (pentagonal) symmetry. (Picture courtesy of David Stuart, University of Oxford.)*

so that the fivefold symmetry of the individual units is not reflected in the way that they are packed together. But as we shall see, fivefold symmetry is a very queer property as far as crystals are concerned.

## The paradox of quasicrystals

### *The forbidden symmetry*

In 1984 four researchers working at the National Bureau of Standards (NBS) in Gaithersburg, Maryland, discovered a material that seemed to break one of the most fundamental rules of crystal structure. Dan Schectman, Ilan Blech, Denis Gratias and John Cahn were studying alloys of aluminum and manganese formed by rapid cooling of a mixture of the two molten metals. By squirting a jet of the molten mixture onto a cold surface, the temperature of the liquid alloy could be lowered at a rate of about a million degrees Celsius per second. The researchers anticipated that this rapid cooling might freeze the mixture into structures substantially different from those obtained by slow cooling. Their "quenching" technique is an example of a non-equilibrium process of the type discussed in Chapter 9. But the structure that the NBS team observed when they measured the X-ray diffraction pattern of their quenched alloy was more than unusual – it was apparently impossible.

An example of the kind of pattern that they saw is shown in Figure 4.11. It consists of a symmetrical arrangement of sharp diffraction peaks, suggesting an orderly, crystalline structure (disordered, amorphous materials, in contrast, produce smeared-out scattering patterns, from which rather little can be deduced about the structure). There are ten bright spots in the ring around the bright central blob. I mentioned earlier that

**Figure 4.11** *The X-ray diffraction pattern from an aluminum/manganese alloy studied by Dan Schectman and coworkers in 1984 appears to be that of a crystalline material, since it contains sharp diffraction peaks. But it has a "forbidden" tenfold symmetry: note the ten bright spots in the innermost circle. This alloy is a quasicrystal. (Photograph courtesy of D. Gratias, Centre d'Etudes de Chimie Métallurgique, Vitry, France.)*

the symmetry properties of the diffraction pattern give an indication of that of the crystal. So in this case the implication was that the alloy was a crystal with tenfold (or possibly fivefold) symmetry. Rotating the crystal lattice by one fifth of a circle (72 degrees) would leave it unchanged.

The NBS researchers hardly needed their combined crystallographic experience to tell them at once that such a crystal is not possible. The diffraction pattern, looking deceptively like countless others produced from crystalline materials, was in fact a slap in the face for over a thousand years of geometry.

In this post-quantum-mechanical, post-relativistic world, scientists have learnt that there isn't much that we can take for granted. But one tenet that seems unassailable is that there are strict symmetry constraints on crystal lattices. While certain symmetries – threefold, fourfold, sixfold – are common, others – such as fivefold, eightfold, tenfold or twelvefold – are rigidly "forbidden". Any attempt to construct a lattice that repeats itself exactly throughout space while maintaining fivefold symmetry is doomed to fail. This is not due to a lack of ingenuity – one can show mathematically that it is simply not feasible. We can see this easily for the case of a two-dimensional (flat) lattice, where the task of devising a regularly repeating pattern with fivefold symmetry can be reduced to that of covering a flat plane with a regular pattern of fivefold-symmetric tiles, such as pentagons. Tiles with three-, four- and sixfold symmetries – equilateral triangles, squares and hexagons respectively – will all sit snugly side by side to fill the entire plane without gaps (Figure 4.12). But while we can certainly marry up the edges of adjacent pentagons, we quickly find that gaps between pentagons are unavoidable. Perhaps the gaps don't matter, though – perhaps they will turn out to repeat in a periodic manner? Not so. We can go on filling up most of the available space if we resign ourselves to a few gaps, but we can never produce an arrangement of pentagonal tiles and gaps that repeats itself regularly. In other words, unlike the three-, four- and sixfold tilings, the fivefold tiling has no *unit cell*. (To see how the same problem arises for two-dimensional lattices of atoms rather than arrangements of tiles, just imagine that each corner of a tile represents the position of an atom.)

For much of what follows, it will be more convenient to stay with these two-dimensional lattices rather than trying to visualize things in three dimensions. Everything said here for two dimensions applies equally to three, however. That is to say, one can stack atoms together in three-dimensional lattice arrays that have threefold, fourfold and sixfold symmetries, but fivefold-symmetric periodic stacking is not possible. A cubic lattice – a stacking of perfect cubes – exhibits not only the easily recognized fourfold symmetry but also threefold, when viewed along cube diagonals. But the two regular polyhedra with fivefold symmetry – the dodecahedron and icosahedron (Figure 4.13) – cannot be stacked together in a way that fills three-dimensional space without gaps, just as pentagonal tiles cannot be used to fill a plane.

The aluminum/manganese alloy of the NBS group, as well as the many other metal alloys that have since been found to produce diffraction patterns with forbidden five-, eight-, ten- and twelvefold symmetries, cannot be true crystals in the sense of possessing a genuine unit cell which repeats throughout the material. In other words, the positions

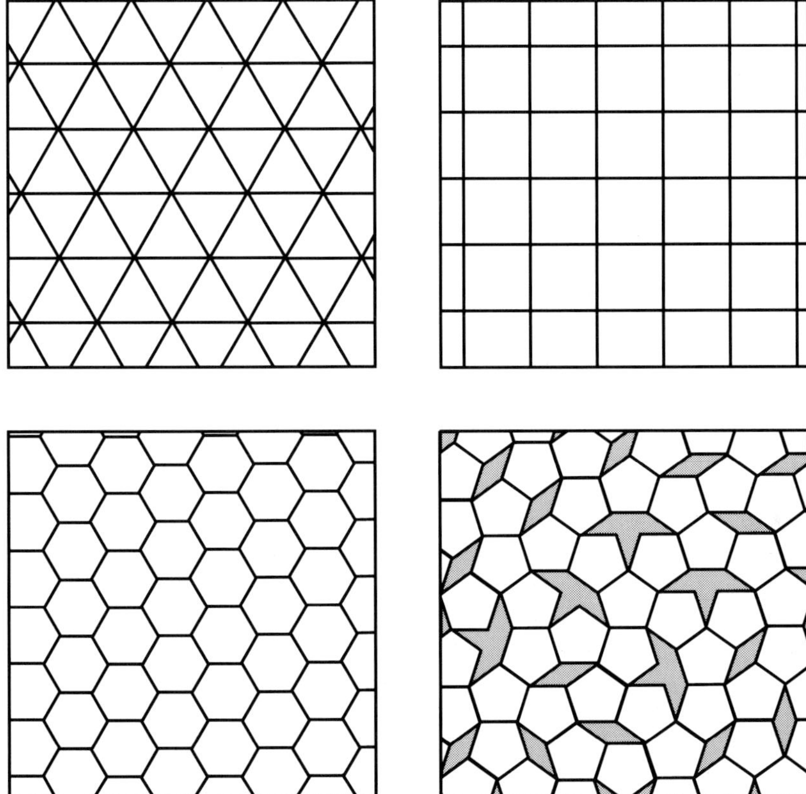

**Figure 4.12** *While it is easy to pack triangular, square and hexagonal tiles perfectly into periodic arrays, filling all of a two-dimensional plane, it is impossible to do the same for pentagonal tiles. The gap-littered pentagonal tiling can have no repeating unit cell.*

of their constituent atoms cannot be orderly over large distances. The geometric ban on crystals with these symmetries is absolute.

But the fact that these solids give diffraction peaks that can, in some cases, be every bit as sharp as those produced by perfect crystals, rather than the smeared-out diffraction patterns characteristic of amorphous solids, means that they must possess some kind of crystal-like characteristics which give rise to interference of reflected X-rays. These alloys, which appear to be neither crystalline nor wholly noncrystalline, have been given the name "quasicrystals".

The aluminum/manganese alloy made by Schectman and his colleagues exhibits icosahedral symmetry, which means that the structure possesses six axes about which there is fivefold rotational symmetry. There are, in other words, six axes around which rotation by a fifth of a full circle superimposes the structure on itself. Small groups of

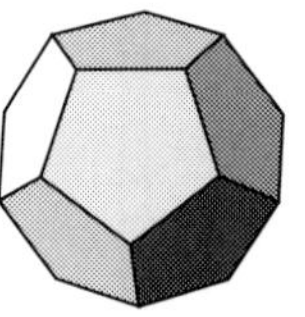


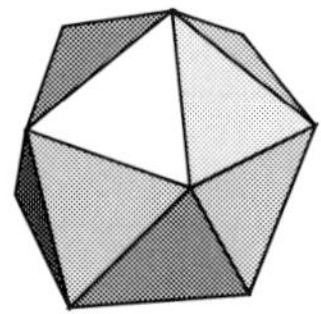


**Figure 4.13** *While cubes can be stacked together to fill three-dimensional space without gaps, giving periodic lattices with threefold and fourfold symmetries, the two regular polyhedra with fivefold symmetries – dodecahedra and icosahedra – will not fit together in a space-filling way, and there are no periodic lattices with fivefold symmetry.*

atoms packed with icosahedral symmetry had been seen long before these experiments. In 1952, Charles Frank of the University of Bristol proposed that clusters of atoms with icosahedral symmetry might be formed in liquids cooled below their freezing point (for small degrees of such "supercooling" it is possible to preserve the liquid state if the cooling is done carefully enough). In small clusters of atoms, icosahedral packing is actually quite favorable since it allows each atom to have a greater number of neighbors (on average), and thus more favorable atom-to-atom interactions, than does packing into clusters with fourfold or sixfold symmetries (Figure 4.14). The atomic-scale structure of these undercooled liquids can be preserved when they eventually freeze if the cooling is performed too rapidly to allow rearrangement of the atoms into crystalline structures. But experiments on rapidly quenched liquids in the 1970s showed that, while this icosahedral symmetry may indeed be maintained in the solid on a local scale, there is no regularity of structure over large distances. The liquid freezes into a disordered glass containing small icosahedral clusters oriented more or less at random. The diffraction patterns from these "icosahedral glasses" therefore exhibit no sharp peaks at all, let alone any sign of fivefold symmetry. It is no coincidence, however, that rapid cooling is also a feature of the NBS experiment, albeit at a rate that far exceeds those used to form icosahedral glasses.

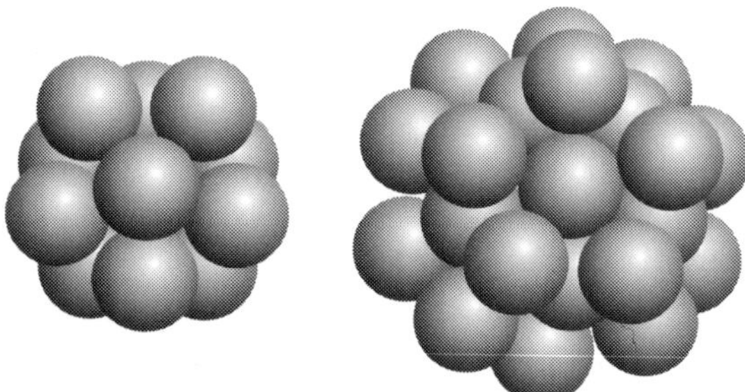

**Figure 4.14** *In supercooled liquids, atoms can come together in small clusters of icosahedral symmetry. When the liquid finally freezes, these clusters can be preserved in a glassy, essentially disordered state.*

## *The art of fivefold tiling*

Quasicrystals possess no unit cell but nevertheless manage to acquire some kind of long-range order based on fivefold symmetry. What kind of arrangement of atoms will generate these properties? The answer was not long in coming: for it turned out that, in one of those fortuitous convergences of theory and experiment, the tools needed to attack the problem had already been developed when the NBS team obtained their puzzling tenfold-symmetric diffraction pattern.

The symmetry properties of two- and three-dimensional lattices and tilings were investigated by Greek mathematicians over two thousand years ago, and by the latter half of the twentieth century it was assumed that they were relatively well understood. Some of the most interesting explorations in tiling patterns have come, however, not from mathematicians as such but from designers and artists. Two-dimensional symmetric patterns were close to the hearts of Moorish architects and designers throughout the ages, who had the somewhat Pythagorean notion that such patterns embodied a kind of divine perfection. The walls of their buildings, such as the Alhambra palace in Granada,

**Figure 4.15** *Fivefold and tenfold symmetries are much in evidence in the geometric patterns favored by Moorish artists.*

are decorated with complex geometric motifs (Figure 4.15). The Moorish designers were motivated not just by aesthetic ideals but by the more practical consideration that abstract designs were often the only option permitted to them: the depiction of living forms was generally frowned upon, if not actually forbidden, in Islamic tradition.

Nor was aestheticism the sole motivation of the artist whose works have introduced many nonscientists to the technicalities of two-dimensional tilings: the Dutchman Maurits C. Escher. Escher's illustrations tend to depict transformations of space and form: interlocking flocks of birds or a spiraling troop of lizards diminishing in size down a geometric progression (Figure 4.16). Escher's interest in tilings was stimulated by his trade as a ceramicist. He was himself influenced by Islamic tiling patterns, but he went on to discover of his own accord many of the mathematical rules that govern the filling of flat space with symmetrical shapes. He knew of the work carried out in this field by pure mathematicians such as George Polya and Heinrich Heesch, but preferred to find his own way around the topic. Thus he came to realize that the symmetry properties of repeating designs on a flat surface can be placed in one of just seventeen classes, a number that increases to forty-six if one distinguishes between two otherwise identical tile shapes by allowing them to have different colors. Having pursued these studies in the grim atmosphere of Nazi-occupied Holland during the Second World War, Escher found his work acknowledged in the 1960s by crystallographers. There can be no doubt that Escher would have delighted in the tiling schemes that, shortly after his death, were devised by the mathematical physicist Roger Penrose.

Penrose is amongst those scientists who have become captivated by Escher's work in the post-war years. But his own tilings represent an exploration of ways in which a plane can be filled completely by identical tiles *without* producing long-range order: that is, in an aperiodic manner. He found that one way of achieving this involved the use of just two kinds of rhombus-shaped tile. These can be thought of as squares that have been squashed so that the angles at each corner are no longer right-angles. Penrose used rhombuses with angles related to those between the edges of a pentagon, so that they can be assembled into objects with fivefold and tenfold symmetry (Figure 4.17). Unlike pentagons, however, Penrose's tiles can be used to fill up a plane without leaving gaps. Nevertheless, the pattern that results can never posses a repeating unit cell. To assemble the Penrose tiling, the two rhombuses are not arranged arbitrarily but according to a strict rule. If we imagine each tile to have single and double arrows along each edge (Figure 4.17), the rule is that equivalent arrows must always be matched up and must point in the same direction each time a new tile is added.

Even at a glance there seems to be something peculiar about the Penrose tiling. One can identify several types of fivefold- or tenfold-symmetric arrangements that recur frequently, but without any periodicity. Another curious feature comes to light on more careful inspection: in a very large tiling array, the ratio of fat to thin rhombuses is always more or less the same, being equal to about 1.62. In the limit of an infinitely large tiling pattern, this ratio will always be *precisely* the same; but its value cannot be written down exactly. The ratio is a number for which, like pi (the ratio of a circle's circumference to its diameter), the sequence of digits after the decimal point goes on

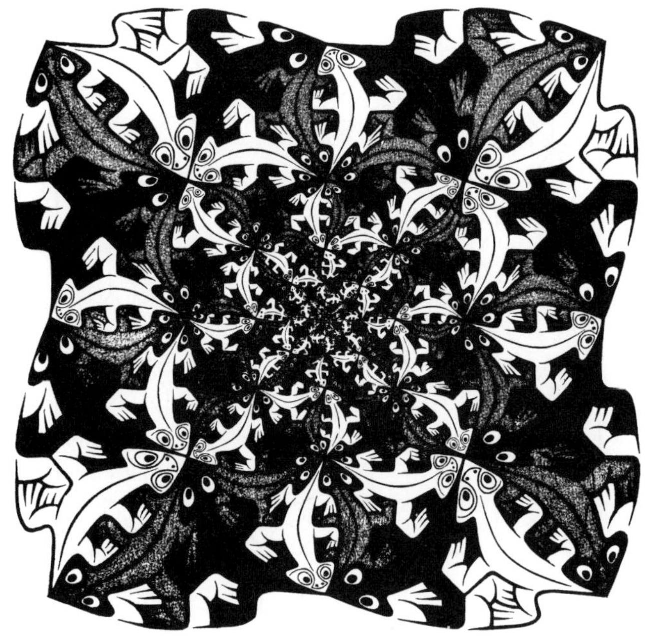

**Figure 4.16** *The designs of Maurits C. Escher display an astute awareness of the symmetry constraints on tiling patterns. (Pictures by permission of the M. C. Escher Foundation, Cordon Art BV, The Netherlands.)*

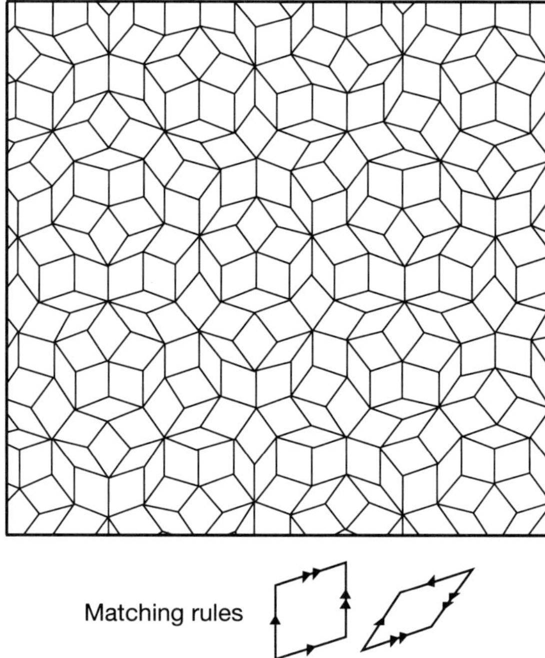


**Figure 4.17** *The tiling scheme developed by Roger Penrose uses two rhombuses to fill all of a plane without gaps, but also without a regularly repeating pattern – there is no unit cell. Nevertheless, certain shapes can be seen to recur that possess fivefold or tenfold symmetry. The tiling pattern is governed by strict "matching rules" between adjacent tiles: the arrows on adjoining edges must be matched up.*

forever without repeating or coming to an end. Numbers of this sort are called irrational. Simple fractions like 1/2 or 1/3, which can be written as the ratio of two integers, are called rational numbers: they can be defined completely either by a finite number of digits or a repeating sequence of digits after the decimal point ($1/2 = 0.5$, $1/3 = 0.3333\ldots$). But one can calculate pi to millions of decimal places (and indeed people have done this on computers) without ever finding that it becomes possible to predict what the ensuing digits will be. For the Penrose tiling, the ratio of fat to thin tiles is 1.618 to the first three decimal places; more precisely it is equal to half the square root of 5: $\sqrt{5/2}$. This number is called the golden mean; it crops up a little mysteriously in several contexts in mathematics, just as pi appears in situations that may seem to have nothing to do with the geometric properties of circles.

Penrose's explorations into fivefold tilings were conducted in the 1970s. In 1982, Alan Mackay of Birkbeck College in London calculated the diffraction pattern that would be produced by an array of atoms located at the vertices of a Penrose tiling. For this two-dimensional "quasilattice," the diffraction pattern is an array of spots with tenfold symmetry. Recall that this was still two years *before* such a pattern had been observed experimentally at NBS. But placing the atoms "by hand" on a Penrose lattice was one thing – it was far from clear to Mackay or anyone else that real atoms could be persuaded to adopt this kind of complicated arrangement, with its strict matching rule for assembling the tiles, in a three-dimensional solid.

In 1984 Peter Kramer and Reinhardt Neri at Tubingen and Don Levine and Paul Steinhardt at Pennsylvania independently succeeded in extending Penrose's two-dimensional tiling scheme to three dimensions. When, in the same year, Dan Schectman and his colleagues published the diffraction patterns of the aluminum/manganese alloy, Levine and Steinhardt made the connection immediately and proposed the three-dimensional generalization of Penrose tiling as a model for the alloy's structure. The building blocks of three-dimensional tilings are solid versions of the flat rhombuses, called rhombohedra, which are like cubes that have been sheared (Figure 4.18). Again, a thin and a fat rhombohedron are all that is needed to fill three-dimensional space without leaving holes, and matching rules are observed between adjacent edges. In the structure that results, one can identify objects with icosahedral symmetry; and once again, the ratio of thin to fat rhombohedra is related to the golden mean. If one places atoms at each corner of the rhombohedra, the diffraction pattern calculated for this arrangement of atoms agrees well with that obtained from quasicrystalline alloys.

## *Shadow crystals*

Thus both two- and three-dimensional Penrose tilings can be generated by packing the two types of tile or block "by hand" according to matching rules. These rules apply only on a local scale, however, so it is not clear how they can produce any kind of crystal-like long-range order, as seems to be required for sharp diffraction peaks. But there is another way of generating the tilings which provides a hint about their relationship to true crystals. To visualize this approach, it will be easiest if we first consider *one*-dimensional quasicrystals.

A one-dimensional solid can be thought of as a series of atoms arranged along a straight line. For a crystal, the arrangement is periodic: in the simplest case, the distance

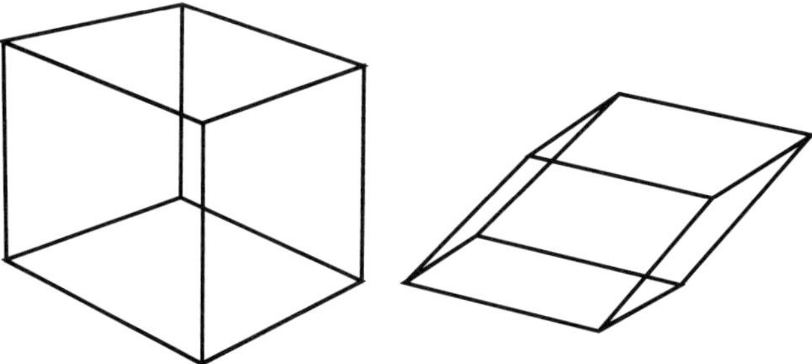

**Figure 4.18** *A three-dimensional version of the Penrose tiling can be devised that uses two types of rhombohedra as the structural elements.*

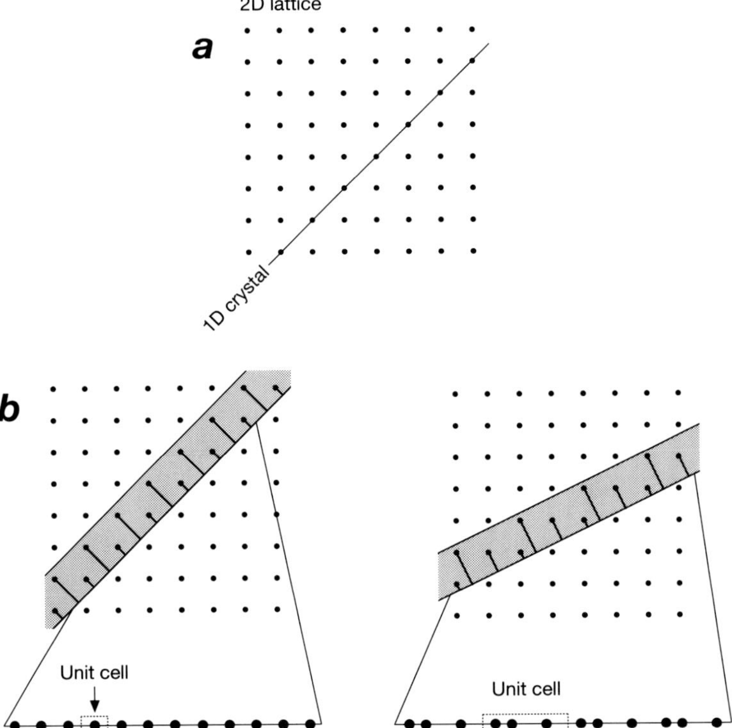


**Figure 4.19** *A one-dimensional crystal consists of a linear array of atoms which repeat in a regular pattern; in the simplest case, all of the atoms are equidistant. Such crystals can be obtained by taking a "cut" through a two-dimensional lattice (a). To avoid missing atoms entirely, we can consider instead taking strips through the lattice and projecting the points within the strip down onto a line (b). The kind of one-dimensional array that results depends on the angle of the strip.*

between each neighboring atom is equal. The unit cell for this structure contains just one atom. An amorphous one-dimensional solid would consist of an irregular array in which the distance between neighboring atoms is random.

A one-dimensional crystal can be generated by taking a "slice" through a periodic two-dimensional lattice (Figure 4.19*a*). A slice parallel to this one but slightly displaced might miss all the atoms, however; so more generally we can generate a one-dimensional crystal by considering every atom contained within a broad strip, and projecting the positions of the atoms down into one dimension – that is, onto one edge of the strip (Figure 4.19*b*). The nature of the one-dimensional crystal that we generate then depends only on the angle at which the strip passes through the two-dimensional lattice; strips at different angles produce crystals with different unit cells.

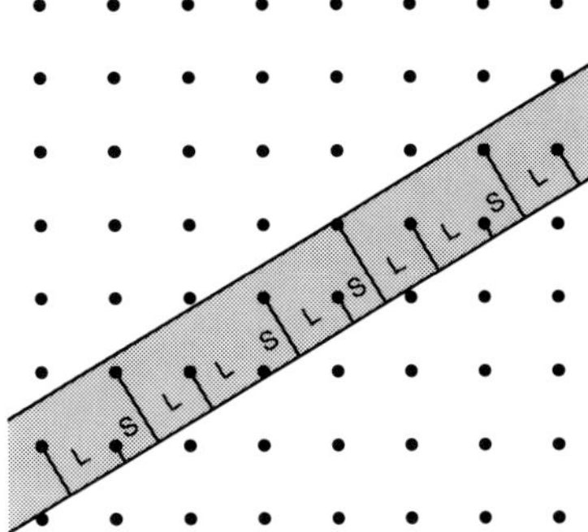


**Figure 4.20** *A one-dimensional quasicrystal can be generated from the projection onto a line of the points within a strip through a two-dimensional periodic lattice, when the strip lies at an angle that is an irrational number. In the quasicrystal shown here, the distance between atoms can either be long or short, but these two alternatives recur without any regular pattern.*

At first glance, it might appear that strips at any angle will generate a periodic one-dimensional arrangement of atoms. But it is not so. In fact, there are an infinite number of angles that will *not* generate one-dimensional crystals. It turns out that periodic one-dimensional arrays will be produced whenever the slope of the strip is equal to a rational number. But if the slope is equal to an irrational number, such as the golden mean, the pattern will be quasiperiodic; in other words, a one-dimensional quasicrystal.

What do one-dimensional quasicrystals look like? A strip passing through a square lattice at a slope equal to the golden mean will generate a linear array of atoms between which the distance can take one of two values: sometimes long, sometimes short (Figure 4.20). Clearly, this is very different from a one-dimensional array in which the atom-to-atom distances are simply random (that is, a one-dimensional glass). But on the other hand, there is no regularity in the way that the long and short steps alternate. The structure is not periodic; but it is nevertheless more orderly than that of a glass.

Quasicrystals of higher dimensionality can be generated in just the same way – that is, by taking a slice through a periodic lattice of still higher dimensionality. A two-dimensional quasiperiodic tiling can be generated by projecting down onto a plane the pattern formed by a cut through a three-dimensional cubic lattice at a slope equal to an irrational number (Figure 4.21). And three-dimensional Penrose quasicrystals, like those envisaged by Kramer and Neri and by Levine and Steinhardt, result from the projection into three dimensions of a slice through a periodic lattice of *six* dimensions. (If you experience some difficulty in picturing a six-dimensional crystal, rest assured that it is easy enough to generate one mathematically.) This relationship between quasicrystals and real crystals might make us feel less uneasy about finding sharp diffraction spots from a material that has no unit cell.

On the other hand, it might not. After all, while *we* may feel content with the notion that quasicrystals are a kind of "shadow" of perfectly crystalline objects of higher dimensionality, how on earth is a beam of incoming X-rays to be expected to recognize this recondite relationship to perfect crystallinity? The "projection" model does not

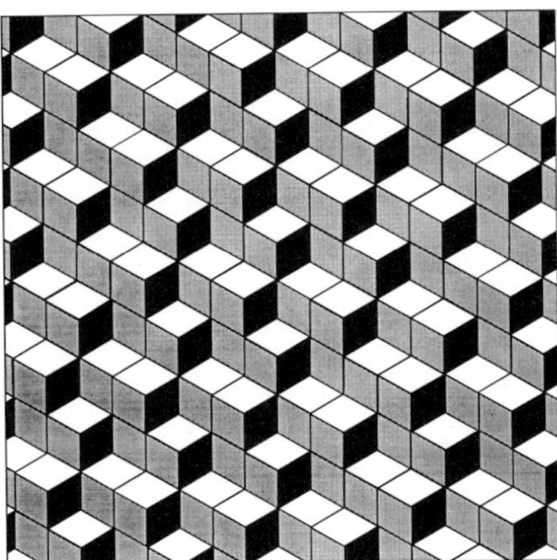

**Figure 4.21** *A two-dimensional quasiperiodic tiling can be obtained by projecting a slice through a three-dimensional periodic lattice onto a plane. Here I show a projection through a stack of cubes, using shading to bring out the relationship between the two-dimensional pattern of rhombuses and the underlying three-dimensional cubic lattice.*

seem to get around the fact that we need regularly spaced layers in order to produce constructive and destructive interference; yet in a quasicrystal there can be no such periodic layers, since this would imply the presence of long-range order. The resolution to this problem lies in the fact that a quasicrystal possesses what we might call periodic "quasiplanes."

This is best illustrated by returning to the two-dimensional Penrose tiling formed from fat and thin rhombuses. Imagine selecting a tile at random and then going through the array highlighting all those tiles that have a side parallel to one side of the chosen tile (in fact, these matching sides will come in pairs, since each rhombus has two pairs of parallel sides). What we find is not a haphazard scattering of highlighted tiles, but a pattern in which they form continuous rows which, while being somewhat kinked rather than flat, remain on average parallel and evenly spaced (Figure 4.22). There are five possible orientations for the rhombus that we choose to begin with, and so there are five sets of these parallel "quasi-rows." In a three-dimensional Penrose tiling, the analog of the quasi-rows are bumpy but roughly periodic quasiplanes.

The existence of quasi-rows does not imply long-range order, because displacing an entire row by the average distance between them will *not* superimpose the row perfectly on the next. But this displacement *will* superimpose on top of the quasi-row one that is roughly similar and which does not diverge off in some new direction for any significant distance. Quasiplanes in three-dimensional Penrose-type quasicrystals scatter X-rays to give a fivefold diffraction pattern. The kinkiness of the planes might lead us to wonder whether the diffraction pattern will get a little blurred by the disorder; but we should recall that even in a perfect crystal the atoms in the diffraction planes are being constantly displaced from their periodic arrangement by thermal vibrations.

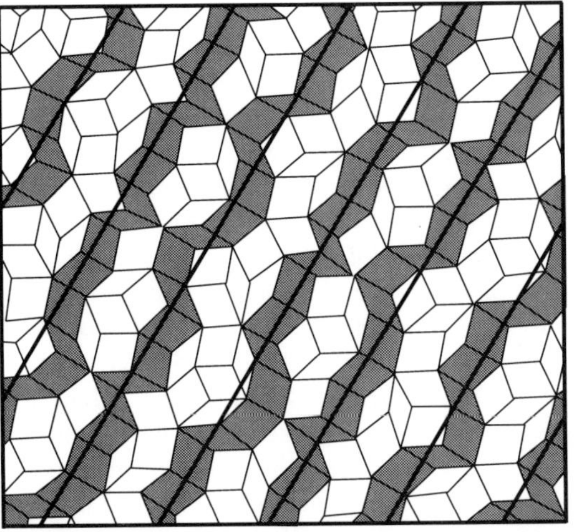

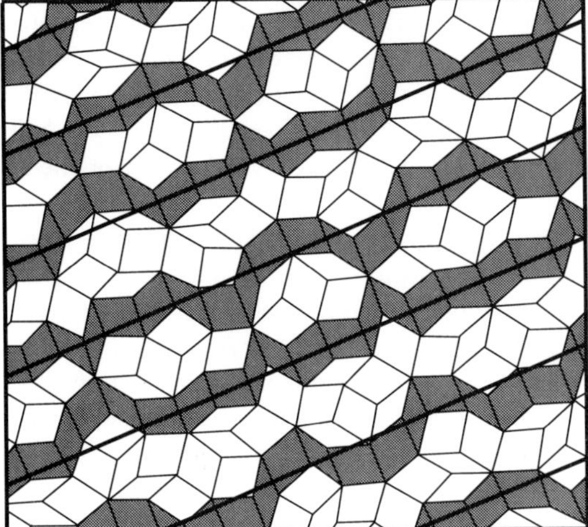

**Figure 4.22** *There are five "quasiplanes" in a Penrose tiling. Diffraction from these quasiplanes gives rise to a pattern with fivefold symmetry. Here I show two of the analogous quasi-rows in a two-dimensional Penrose tiling; the bold lines show the mean direction of the rows.*

So a quasicrystal, with its fivefold kinky quasiplanes, is not so different from a crystal after all.

## *Where are the atoms?*

Penrose tilings might seem to tie up the question of quasicrystal structure rather neatly. But there are some awkward questions that I have so far ignored, and which still continue to obstruct a full understanding of these unusual materials. We have seen that a quasicrystal's "lattice" can be represented by the corners of the Penrose tiles: place atoms at those corners and you have a quasicrystal. (I use quotes around the word "lattice" here because, strictly speaking, the term refers only to a periodic arrangement of points.) All of the quasicrystals discovered so far, however, are alloys of at least two (and often more) types of atom. The composition of these alloys can generally be specified very precisely: the best studied aluminum/manganese quasicrystals have the formulae $Al_4Mn$ and $Al_6Mn$, while other quasicrystals may exhibit more complicated but similarly well defined compositions, such as $Al_6Li_3Cu$ and $Al_{78}Cr_{17}Ru_5$. It is hard to see how these precise compositions can be maintained if the atoms are just distributed at random amongst the various "lattice" points; why, in that case, should the material care about preserving a specific and reproducible ratio of elements? And is it reasonable to ignore the fact that the different types of atom may have considerably different sizes?

For a crystal, the problem of figuring out how to arrange the atoms throughout the lattice so as to maintain the correct overall ratio is much simpler, because one need do this only for a single unit cell – as this repeats exactly throughout the material, so the ratio of atoms is guaranteed to be the same for a billion unit cells as for one. But for quasicrystals it is not sufficient to specify an arrangement of atoms with the correct ratio in just one small region of the structure, because this will not be repeated throughout the solid. And for the Penrose tiling models, there is no way to arrange the atoms on the tiles (that is, on the three-dimensional rhombohedra) to give exactly the right overall composition, because the ratio of the different types of atom in the bulk material is a rational number whereas the ratio of the two types of rhombohedra is irrational.

This problem can be circumvented if we allow the quasicrystal to have defects – places where the specified arrangement of atoms on each rhombohedron doesn't quite hold. There may be vacant sites, for instance, where one would normally expect an atom, or even regions where the rhombohedra are distorted. Even true crystals are invariably laced with defects in the periodic arrangement of atoms, so it is not difficult at all to imagine that a quasicrystal will contain them too. The defects provide some leeway which allows a compound with a rational ratio of atoms to form a quasiperiodic "lattice" that would, if perfect, require this ratio to be irrational. It is then possible to describe the structure in terms of rhombohedra "decorated" with atoms. Thus, for example, atoms arranged as shown in Figure 4.23 on the two icosahedral rhombohedra will give essentially the right atomic ratios for the quasicrystal $Mg_{32}(Al,Zn)_{49}$) when the rhombohedra are packed into a three-dimensional Penrose tiling with a small proportion of defects.

A more important and more difficult question than where the atoms are in a Penrose tiling model of a quasicrystal is how they got there. While we have been building our models by applying strict matching rules to the packing of decorated rhombohedra, or by envisaging a projection from six dimensions into three-dimensional space, the atoms in a rapidly solidifying molten alloy do not know these rules, nor do they have this kind of global overview of the structure; they know only what is happening in their own local environment. Imagine two tilers constructing Penrose-style floor tilings in opposite corners of a hall. Even though they might apply the matching rules assiduously

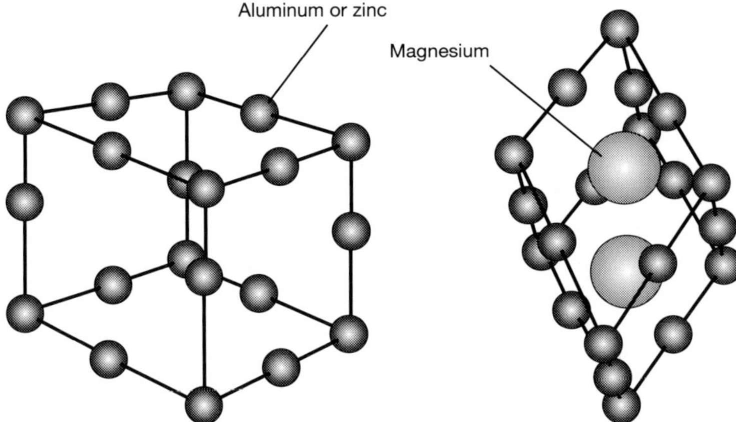


**Figure 4.23** *Decorating the rhombohedral structural units of a three-dimensional Penrose tiling with magnesium and aluminum or zinc atoms produces an atomic array that reproduces the diffraction pattern and the composition of the alloy* $Mg_{32}(Al,Zn)_{49}$, *provided that one allows for some defects in the structure.*

as they go, they are likely to find when they meet in the middle that their designs are not compatible: the edges of the tiles cannot be matched up in the correct way. Of course, we would then be left with two fairly big regions of the perfect tiling, and we might think that the flaws in the middle don't matter too much. But what if the hall contains fifty tilers all working independently on their own little patch? The problem is that the matching rules are local, in the sense that, each time a tile is to be added, one need consider only those in the immediate vicinity. But the only way to guarantee a perfect large-scale tiling is to ensure that there is a lot of foresight and communication between tilers working on widely separated regions. It is hard to see how, during the growth of a quasicrystal, this kind of long-range "communication" between atoms is possible. Indeed, the rapid growth conditions seem to make the problem even worse: it is as if the tilers have been set a very short time limit, with no latitude for taking apart and building again parts of the pattern that are found not to match.

In fact, real quasicrystals *do* show some signs of being significantly disordered, in the sense of departing from the kind of perfect quasiperiodic structure that the Penrose model predicts. For instance, whereas crystalline metals and ordered alloys should in general be excellent conductors of electricity, disorder in a material's structure can hinder the flow of current. In metals, defects are one of the principal causes of electrical resistance. Perfect quasicrystals are predicted to mimic crystals so well that they should be good conductors; but in practice, real quasicrystals turn out to conduct poorly.

Another signature of disorder in an otherwise crystalline solid is a smearing out of the diffraction peaks. I mentioned earlier that disordered materials such as glasses give diffraction patterns that are little more than a blur, with no sign of peaks at all. The diffraction peaks from quasicrystals are rendered somewhat fuzzy by the lack of perfect order (although curiously this fuzziness is *not* like that caused by defects and other sources of disorder in true crystals). The Penrose model, however, predicts peaks that are rather too sharp.

Thus there are some difficulties with the ideal Penrose tiling picture of quasicrystal structure. Not only does it predict a structure that is just a little *too* perfectly quasiperiodic, but it also fails to explain how a growing quasicrystal will be able to adhere to the strict matching rules, with their requirement of consistency between distant regions. These considerations have led some researchers to propose different models of quasicrystal structure. Shortly after they discovered the first quasicrystal, Dan Schectman and Ilan Blech suggested that the alloys were composed of icosahedral clusters of atoms (like those predicted for supercooled liquids by Charles Frank) connected together more or less at random, with scant regard for whether or not gaps were left and with no constraints imposed by matching rules. This idea has evolved into the so-called icosahedral glass model. A two-dimensional version of this model (Figure 4.24*b*) can be created using pentagonal tiles with gaps between them; the same system can be used to generate a Penrose tiling (Figure 4.24*a*). Clearly the model requires much less careful arranging than does the Penrose tiling; indeed, it is not obvious that it will contain any quasiperiodicity at all, but might be expected instead to combine short-range order of atoms in the individual clusters with a purely random structure over longer distances. But the surprising thing is that, with a little modification and fine tuning, the icosahedral glass model can be shown to generate a diffraction pattern with five- or tenfold symmetry. Whereas the Penrose tiling predicts peaks that are too sharp relative to those observed in practice, the icosahedral glass model predicts peaks that are too smeared-out, because of the rather high degree of disorder in the structure.

Since the Penrose tiling model predicts quasicrystals with apparently too much order while the icosahedral glass model contains too much disorder, the obvious compromise is to create a model that contains elements of both. Such a model, called the random tiling model, has been developed by researchers at Carnegie-Mellon University in Pittsburgh, amongst others. The matching rules of the Penrose tiling model ensure that there are no ill-fitting mismatches between tiles, whereas the icosahedral glass model pays no heed to their presence. In the random tiling approach, meanwhile, there are no matching rules governing the orientation of adjacent tiles but nevertheless mismatches and gaps are seen as a bad thing. These conditions are consistent with the physically reasonable picture of quasicrystal growth being governed only by rules that apply on a local scale: there is no call for the mysterious long-range communication that seems to be required for a perfect Penrose tiling, but all the same the material does its best to avoid too many defects. Each atom in the growing quasicrystal will therefore try to fit most comfortably with its neighbors but will not be concerned with what more distant atoms are doing. A surprising feature of this model is that, even though there are no matching rules, the predicted diffraction peaks may be as sharp as those of the Penrose model. However, it is fairly easy to "tune" the amount of disorder in the structure, and thereby the sharpness of the peaks, simply by varying the extent to which

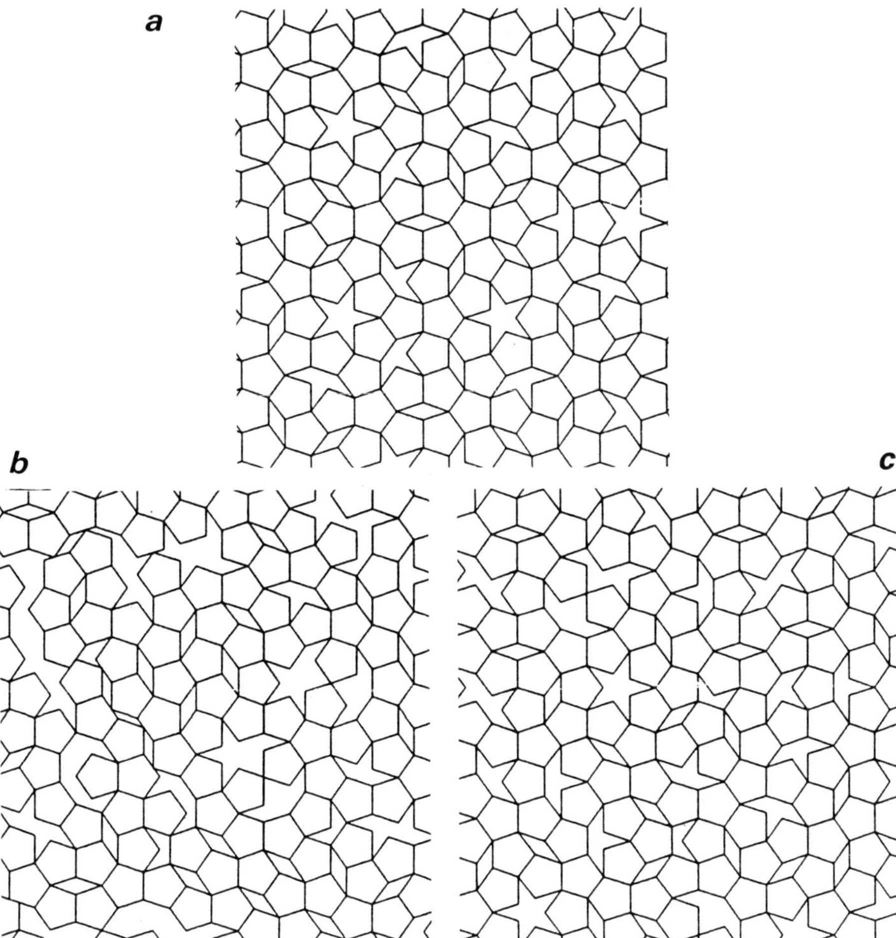


**Figure 4.24** *Three models of quasicrystal structure (shown here for two-dimensional quasicrystals, for clarity): (a) the Penrose tiling; (b) the icosahedral glass; (c) the random glass. The icosahedral glass model now looks unlikely to prove a viable model for quasicrystal structure, but the other two can, with some adjustments, account for the detailed features of quasicrystal diffraction patterns. (Pictures courtesy of Peter Stephens, State University of New York at Stony Brook.)*

gaps between tiles are excluded. It is therefore possible to generate diffraction patterns that correspond very closely to those measured.

The random tiling picture also turns out to have another surprising characteristic. Experimentally, most quasicrystals are formed by some variant of the rapid cooling technique used by the NBS team. This catches the alloy out, freezing it into a quasiperiodic structure before it has time to order fully into a perfect crystal. The quasicrystal is therefore not the alloy's preferred choice of structure; if the atoms were

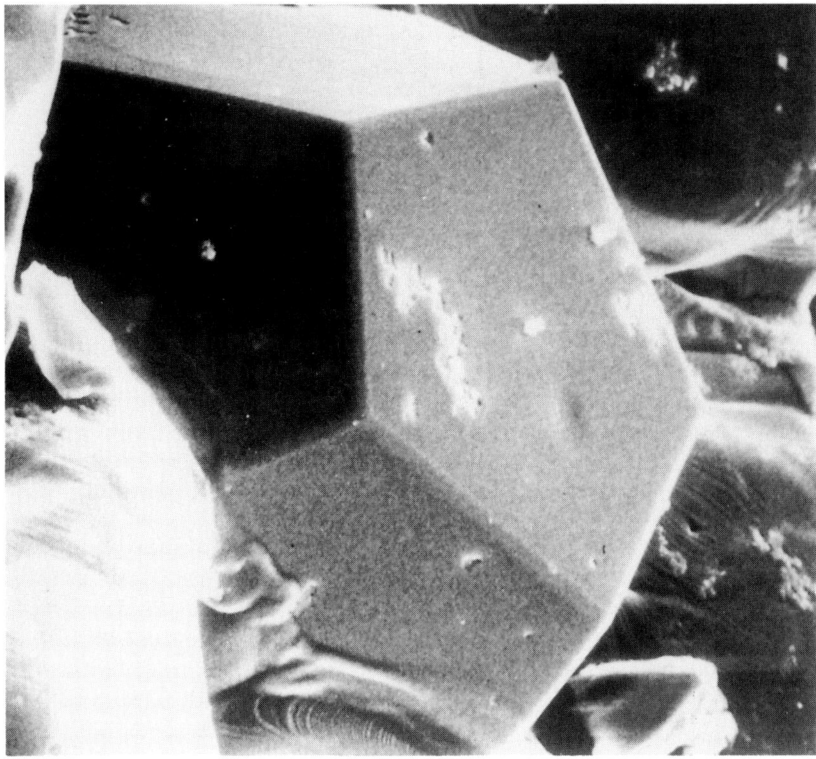

**Figure 4.25** *Thermodynamically stable quasicrystals can be grown more slowly than "metastable" ones, allowing large "pseudo-crystals" to be obtained. These can have shapes that reflect the forbidden symmetries of the atomic structure. In this picture, the pseudo-crystal has a dodecahedral shape. (Photograph courtesy of Kenji Hiraga, Tohoku University.)*

mobile, they would gradually rearrange themselves into a regular crystal. The quasicrystal is only *provisionally* stable (the technical term is "metastable"): melting it to get the atoms moving, followed by slow cooling, would produce a crystal. It turns out, however, that in some circumstances the disorder inherent in the random tiling model can help to make the quasicrystal *more* stable, especially at high temperatures. It is even possible for the quasicrystal to become more stable than a regular, crystalline structure.

More surprising still is the discovery that this stability can be seen in experiments too. Some of the new families of quasicrystalline alloys, such as aluminum/zinc/magnesium, retain their structure right up to the melting point rather than rearranging to crystals as the atoms become more mobile. This means that to grow such materials it is no longer necessary to cool the molten alloy rapidly; the cooling can be more gradual, which allows the growth of "ideal" quasicrystals with faceted, crystal-like shapes. The

**Figure 4.26** *Fivefold symmetries are not as uncommon in nature as might once have been supposed. Here they can be seen in the pattern formed by streamlines of fluid flow close to the point at which turbulence develops. (Picture courtesy of G. Zaslavsky, New York University.)*

fivefold symmetry of the atomic structure is then strikingly preserved in the bulk material (Figure 4.25).

Generally speaking, even these carefully grown materials show signs of the kind of disorder that is an inevitable consequence of the icosahedral glass model. But more recently, icosahedral quasicrystals have been grown that appear to be more or less free from this disorder. These new alloys therefore leave just the random and Penrose tiling schemes as viable models of quasicrystal structure.

After the initial excitement generated by the NBS discovery, work on quasicrystals has become pervaded with a feeling of deflation. There is no doubt that these materials represent a new and unexpected kind of solid, but we now have a reasonably good understanding of their structure and properties, and they don't, after all, force us to abandon cherished notions about symmetry. Much of the research now focuses on how the unusual structure affects properties such as electrical conductivity or magnetism, while a few enthusiasts still hope that quasicrystals will find some *practical* niche in the materials world. Perhaps one of the most stimulating aspects of the discovery of quasicrystals, however, is that it has led to a new appreciation of the importance of fivefold-symmetric objects in many spheres of science, from the shape of viruses and flowers to the patterns of fluid flow close to the onset of turbulence (Figure 4.26). Five is no longer a dirty number!

*Part II*

# New Products, New Functions

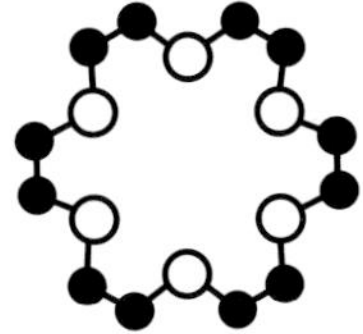

# 5

# Perfect Hosts and Welcome Guests

## Molecules that recognize each other and build themselves

*Sometimes it seems as if self-knowledge brought about the union, sometimes as if the chemical process were the efficient cause.*

Carl Gustav Jung

The human body is often compared to a well-oiled machine, controlled by billions of molecular components that intermesh with clockwork precision like so many cogs. There are certainly virtues in this image; but molecular biologists today tend to paint a less mechanical, more anthropomorphic picture. They describe the body as a veritable community in which individual molecules scurry about like ants, each performing its designated task so as to ensure the well-being of the society as a whole. Some gather food, for instance. Some build the structures in which they all dwell; others seek out and repel foreign invaders. Of course, each of these molecular workers is driven not by some autonomous consciousness but by the principles of chemistry.

This society generally assigns to each of its members a specific task. Because they exhibit little specificity, the molecules of the inorganic world are therefore of very little use to the community (except perhaps as raw materials). For example, biochemical catalysis – which is essential if the body's chemical reactions are to proceed efficiently at close to room temperature – does not employ inorganic catalysts such as transition-metal surfaces, since these tend to promote all manner of reactions. Instead, we saw in Chapter 2 that protein-based enzymes constitute the highly selective catalysts of biology.

In short, the body's molecular laborers must be very choosy about the individuals with whom they interact. They must be able to pick out their designated targets amidst a vast horde of other molecules, many of whom look almost identical. Chemists call this kind of phenomenon "molecular recognition" – the specific recognition of (and interaction with) one molecule by another. A capacity for exquisite molecular recognition, fine-tuned by millions of years of evolution, ensures that most biomolecules perform just a single chemical operation, which is generally no more than a small step in a complicated assembly-line process. In the biological world, there are very few jacks-of-all-trades; specialization is the name of the game.

This specificity dictates that the molecules be carefully "preprogrammed" for their task. In living organisms it is the DNA molecule, the carrier of the organism's genetic blueprint, that ultimately supplies this preprogramming. In other words, the information that tells a protein how to exhibit molecular recognition is encoded within DNA. How this coding works at the molecular level is one of the primary questions tackled by molecular geneticists.

Understanding the principles that lie behind molecular recognition should therefore help to unravel the complexities of much of the body's chemistry. But from a chemist's point of view, biomolecules are dauntingly huge (often containing several thousand or even several million atoms), so that the prospect of trying to deduce which parts of the molecule are responsible for its molecular-recognition capability is a rather frightening one. The chemist's approach is consequently to build simple model systems that mimic just one or a few of the characteristics of biomolecules. The insights gained in this way are of great value for the rational design of synthetic drugs.

The payoffs are by no means for pharmacology alone, however. The ability to perform highly selective chemical reactions has opened chemists' eyes to some of the breathtaking possibilities of their craft, allowing them to construct molecules of hitherto unimagined complexity in very elegant one- or two-step synthetic procedures. The consequences range from the creation of molecular-scale mechanical machinery and electronic devices to a greater understanding of what constitutes life itself.

## The chemistry of life

### *The gene library*

Molecular recognition plays a central role in the structure of the DNA molecule, in the translation of DNA's genetic code into protein molecules, and in the biochemical action of these proteins. It therefore lies at the heart of genetics and molecular biology. As this is a book about chemistry, I shall be able to offer no more than a brief glimpse at these biological systems, which I shall do simply to demonstrate the importance of molecular recognition in the natural world. For an introduction to genetics, you must look elsewhere.

Modern genetics has developed out of the study of heredity: that is, how characteristics of living organisms are passed on from one generation to the next. Many of the rules of inheritance were deduced in the mid-nineteenth century by an Austrian priest named Gregor Johann Mendel, who carried out extensive cross-breeding experiments on peas to understand how traits such as plant height, flower and seed color and shape were passed on between generations. Mendel gave the name "particulate factors" to the entities within the organisms that were responsible for passing on inherited traits, but could say nothing about what they consisted of. He deduced that offspring inherit two copies of these factors, one from each parent.

The importance of Mendel's work became apparent only after the turn of the century, when biologists acquired the ability to distinguish under the microscope the components of individual cells. They observed that all the cells of creatures less primitive than bacteria contained several X-shaped objects (typically between about eight and eighty, depending on the organism) (Figure 5.1). When a cell divides, these objects also split, half going into each of the two new cells. The X-shaped objects were termed chromosomes. It was also noted that the specialized cells involved in reproduction – ova and sperm – contain just half the usual complement of chromosomes. The union of an ovum and a sperm cell then restores the full chromosome count, half coming from each parent. Clearly this transmission of chromosomes from one generation to the next paralleled the transmission of Mendel's particulate factors. In 1903 Walter Sutton and Theodor Boveri proposed independently that the Mendelian factors (christened "genes" in 1909 by W. L. Johannsen) are molecular structures that reside on the chromosomes.

The chromosomes are gene "libraries" in which each gene is allocated a specific location. Some (but not all) of these genes determine the manifest characteristics of the organism, such as its gender. Defects in the molecular structure of a gene may in some

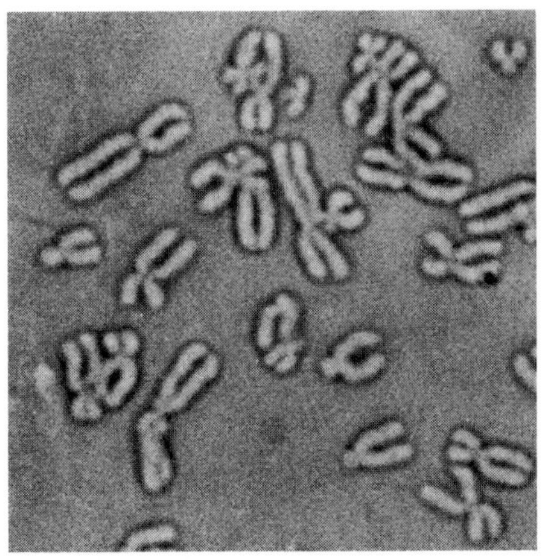

**Figure 5.1** *Human chromosomes. Each human cell contains forty-six of these X-shaped entities. (Photograph courtesy of William Earnshaw, Johns Hopkins University.)*

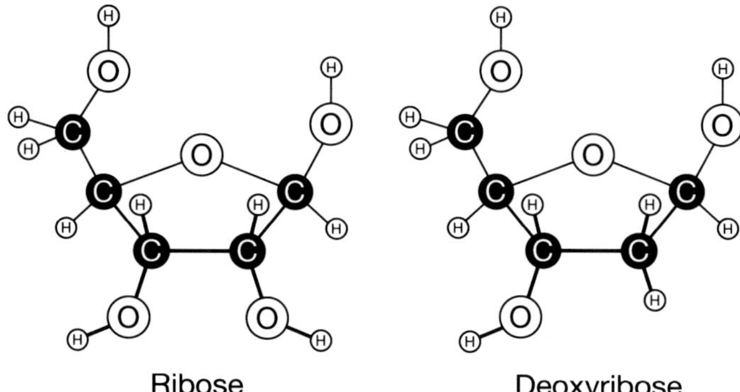


**Figure 5.2** *The sugars ribose and deoxyribose, which are components of the nucleic acids RNA and DNA, respectively.*

instances cause physiological malfunctions or render the organism susceptible to certain diseases. With the exception of primitive organisms such as bacteria, the chromosomes reside within a central compartment of each cell, called the cell nucleus. In the early part of this century it was widely believed that the genetic material in the chromosomes was comprised of protein. But analyses of the chromosomes' chemical composition revealed that, while indeed they contained protein, another component was also present which contained the sugar molecule deoxyribose (Figure 5.2). This component was named deoxyribonucleic acid, or DNA (the "nucleic" denoted its origin in the cell nucleus). A similar compound was also discovered in the nucleus, although not in association with the chromosomes; it contained the sugar ribose (Figure 5.2), and was named ribonucleic acid (RNA). (In the cells of those simple organisms that have no nuclei, DNA and RNA swim around unchecked amongst the other components of the cell.)

In 1944, O. T. Avery and colleagues from the Rockefeller Institute in New York provided compelling evidence that it was DNA, not proteins, that carried the genetic information. The molecular basis of this information storage became apparent in 1953, when Francis Crick and James Watson published in the journal *Nature* an inspired guess (guided by X-ray crystallography) at the molecular structure of DNA.

The compound is a chain-like polymer built up from smaller units called nucleotides, each of which contains three components: the deoxyribose sugar molecule, an ionic phosphate ($PO_4$) group, and a "base" (Figure 5.3). The nucleotide bases come in four varieties: in DNA they are adenine, guanine, cytosine and thymine (abbreviated to the symbols A, G, C and T), whereas in RNA the fourth base is uracil (U) rather than thymine (Figure 5.4). Adenine and guanine are members of a class of molecule called purines, while cytosine, thymine and uracil belong to the family of pyrimidine bases.

Crick and Watson's structure is based on the pairing of nucleotide bases via weak bonds called hydrogen bonds. This kind of bond is of great importance not just to the structure of DNA but to the interactions between biological molecules more

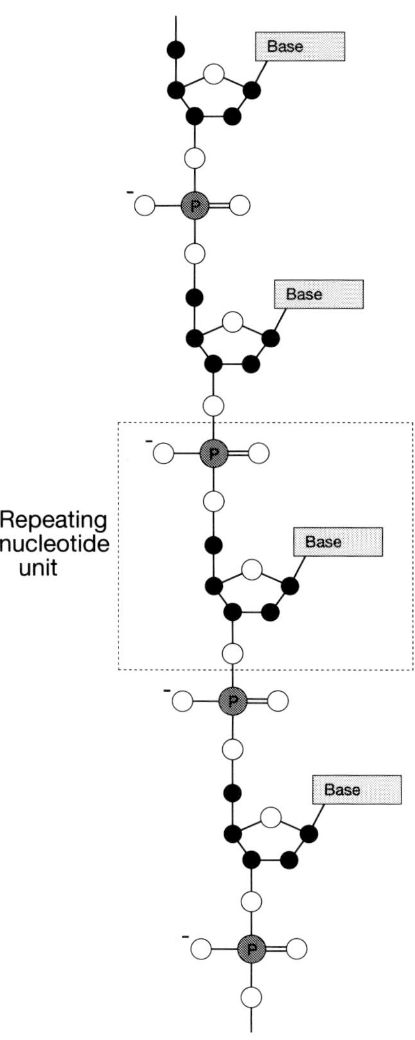


**Figure 5.3** *DNA and RNA are polymers built from units called nucleotides, each of which contains a cyclic sugar molecule – deoxyribose and ribose in DNA and RNA, respectively. These are linked via phosphate groups to those in neighboring nucleotides. To each sugar molecule in this sugar–phosphate backbone is attached one of four bases.*

generally. It is a consequence of the existence of lone pairs, described in Chapter 1. Nitrogen and oxygen atoms in molecules often possess one and two lone pairs respectively, which can bind weakly to hydrogen atoms attached to other molecules, or to other parts of the same molecule. When covalently bound to nitrogen or oxygen atoms, hydrogen atoms acquire a slight positive charge, to which the electrons of a lone pair can become attracted. It is hydrogen bonding that gives water many of the properties responsible for its unique role in the chemistry of living organisms. In liquid water hydrogen bonds are forever being formed and broken; in ice, they join water

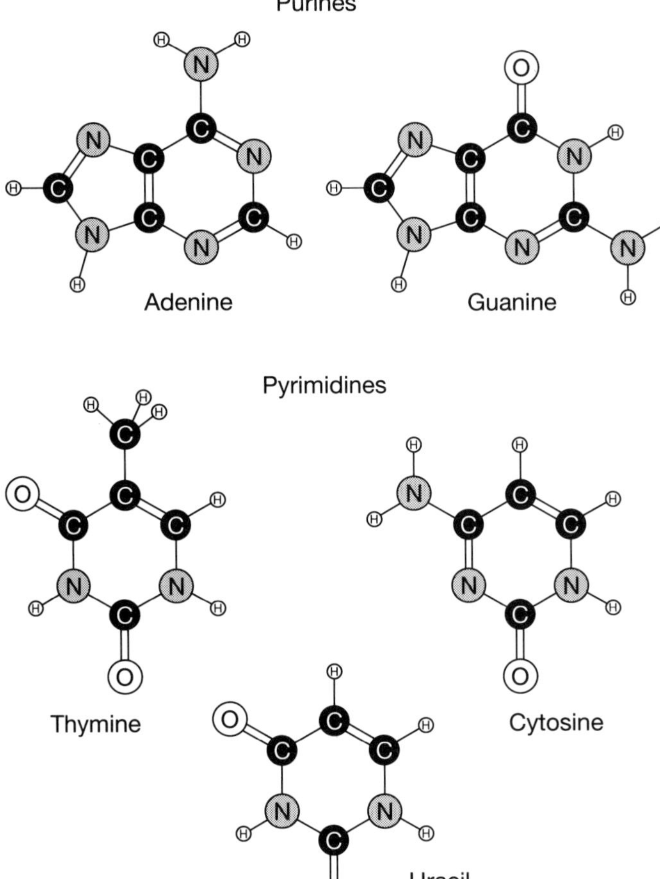


**Figure 5.4** *The bases in nucleic acids. Thymine occurs in DNA but not RNA, and uracil in RNA but not DNA.*

molecules into one huge crystalline network. Hydrogen bonds also play an important role in determining the shapes of proteins.

The DNA bases contain the ingredients for hydrogen-bond formation: they have both nitrogen and oxygen atoms with lone pairs, and hydrogen atoms attached to nitrogens. Crick and Watson proposed that each base has a complementary partner, with which it can lock together via hydrogen bonds like two pieces of a jigsaw. Adenine binds to thymine via two hydrogen bonds, and cytosine to guanine via three (Figure 5.5).

X-ray crystallography of DNA by teams at Cambridge and King's College, London, suggested that the nucleotide chains of DNA formed a double-stranded, helical (coiled)

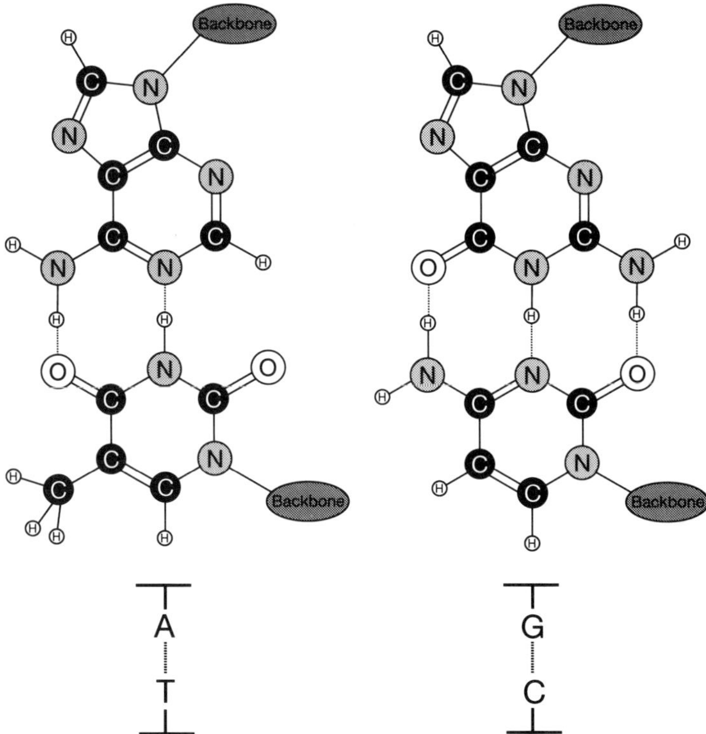


**Figure 5.5** *The DNA bases form complementary purine–pyrimidine pairs via hydrogen bonding. Adenine fits snugly with thymine (forming two hydrogen bonds, shown by dashed lines), and guanine with cytosine (forming three bonds). These complementary pairs are the same size.*

assembly. In the model of Crick and Watson, this is held together by complementary pairing of the bases on each strand, which are attached to the phosphate–sugar backbones of the two DNA strands. The resulting structure resembles a spiral staircase with A—T and C—G pairs for rungs (Figure 5.6). The two strands of the double helix are not identical; rather, one is like a "negative" for the other. Given just one strand, the structure of the other can be deduced (or built!) by attaching to each base the nucleotide containing its complementary base. Thus it was immediately apparent from this model how DNA might produce replicas of itself – this replication process must take place each time a cell divides. The two strands unwind, and each provides a template on which a new strand can be formed by assembling nucleotides according to the rules of complementary base pairing: wherever an A appears in the template, for example, a T is added to the new strand. In this way two identical DNA molecules will emerge from the unravelled original. Crick and Watson noticed straight away the possibilities in their model for a mechanism of DNA replication, and with devastating nonchalance they

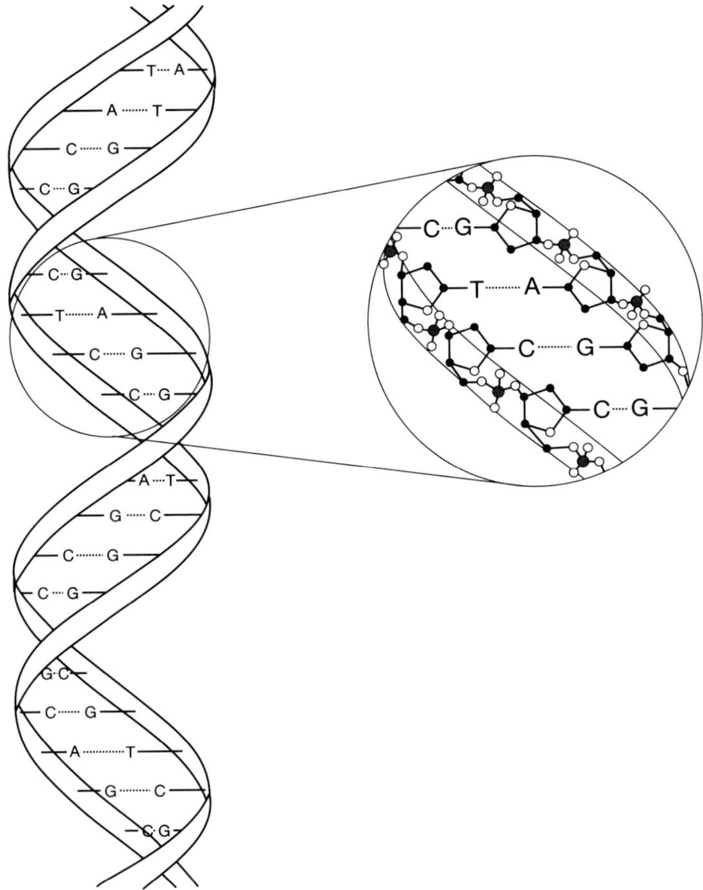


**Figure 5.6** *The DNA molecule is a double helix in which two polymeric chains of nucleotides twist around one another. The strands are held together by hydrogen bonding between complementary base pairs, which means that the sequence of bases on one strand must be matched by a complementary sequence on the other.*

pointed the way to the future of genetics: "It has not escaped our notice," they commented in the *Nature* paper, "that the specific pairing mechanism we have postulated immediately suggests a possible copying mechanism for the genetic material."

Complementary base pairing provides an excellent example of molecular recognition. During DNA replication, each base will pair specifically with its complementary partner in preference to other bases: it *recognizes* its complement. Although complex enzymatic machinery exists to ensure that the pairing rules are observed, ultimately the recognition is due to the geometry of the base pairs and the shape of the double helix: a wrong link-up, between two A's for example, will produce an uncomfortable bulge in the helix.

## *The code we live by*

A cell's full complement of DNA – that is, the organism's complete genetic blueprint, called its genome – does not reside all on one huge, double-helical molecule, but is divided up into several sections, each which is contained within a separate chromosome. In human cells there are forty-six of these discrete fragments. The chromosomes are a composite of DNA packed tightly together with protein "packing" molecules, called histones.

The function of the genes is to carry the information required to manufacture the enzyme proteins that orchestrate the body's chemistry. As the geneticist François Jacob puts it: "The gene gives orders. The protein executes them." In essence, each gene carries the plan for a protein. (This is not precisely true; some genes carry other kinds of information instead. Genes that represent "protein plans" are called structural genes.)

This information resides in the gene in a coded form. We can write down the chemical structure of a protein as a sequence of its constituent amino acids. DNA, meanwhile, can be represented as a sequence of base pairs linked by the sugar and phosphate components of nucleotides, which are simply part of the scaffolding. Thus we might imagine setting up a code in which specific base pairs on DNA correspond to specific amino acids on a protein. The sequence of base pairs within a gene on the DNA molecule can then represent a protein molecule in a coded form.

Yet there are twenty different amino acids that crop up in proteins, but only four DNA bases. So we can set up a complete DNA–protein code only by taking *groups* of bases to represent each amino acid; a code based on "one base to one amino acid" will not work. Morse code works on the same principle: it provides a code for all twenty-six letters of the alphabet based on just two symbols, by using groups of these symbols to represent each letter. A DNA code in which the characters are groups of just two bases will give us only $4 \times 4 = 16$ different elements – not enough to encode all of the amino acids. But with groups of three bases, there are $4 \times 4 \times 4 = 64$ possible characters, which is more than enough. The DNA–protein code must therefore use at least three bases to represent each amino acid.

This is indeed the system that DNA employs to record its protein plans. The amino acid sequence of the protein encoded in a gene can be deduced by reading along the base sequence of that gene in groups of three at a time. Experiments on bacteria have enabled this code to be cracked, allowing each natural amino acid to be assigned a corresponding triplet of bases. All living organisms employ this same genetic code. Because there is some redundancy in the code – there are sixty-four different base triplets available in the DNA cipher to represent just twenty different amino acids – some amino acids are encoded by more than one triplet. Some triplets, moreover, do not represent amino acids at all, but are instead "control codes" which signify the end of the protein-coding sequence in the gene. The full genetic code is listed in Figure 5.7.

Clearly, a DNA molecule, replete with meaningful information about the host organism's structure, cannot be created by the random assembly of nucleotides, any more than a random sequence of letters is likely to reproduce Shakespeare's works.

| First position | Second position: U | C | A | G | Third position |
|---|---|---|---|---|---|
| U | Phe | Ser | Tyr | Cys | U |
| | Phe | Ser | Tyr | Cys | C |
| | Leu | Ser | Stop | Stop | A |
| | Leu | Ser | Stop | Trp | G |
| C | Leu | Pro | His | Arg | U |
| | Leu | Pro | His | Arg | C |
| | Leu | Pro | Gln | Arg | A |
| | Leu | Pro | Gln | Arg | G |
| A | Ile | Thr | Asn | Ser | U |
| | Ile | Thr | Asn | Ser | C |
| | Ile | Thr | Lys | Arg | A |
| | Met | Thr | Lys | Arg | G |
| G | Val | Ala | Asp | Gly | U |
| | Val | Ala | Asp | Gly | C |
| | Val | Ala | Glu | Gly | A |
| | Val | Ala | Glu | Gly | G |

**Figure 5.7** *Sequences of DNA bases encode the information required for synthesis of proteins from amino acids. The blueprint for a single protein is encoded in a stretch of DNA called a gene. The code requires that each amino acid in a protein correspond to a group of three DNA base pairs. This genetic code holds for all organisms, and was cracked by studies of bacterial DNA. The shorthand notation for amino acids is as follows: Phe = phenylalanine; Ser = serine; Tyr = tyrosine; Cys = cysteine; Leu = leucine; Trp = tryptophan; Pro = proline; His = histidine; Arg = arginine; Gln = glutamine; Ile = isoleucine; Thr = threonine; Asn = asparagine; Lys = lysine; Met = methionine; Val = valine; Ala = alanine; Asp = aspartate; Gly = glycine; Glu = glutamate; and Stop is a codon that does not encode an amino acid at all but instead represents an instruction for protein synthesis to stop.*

Rather, the process of assembly must be directed with great control and precision. If an error is made in the DNA assembly process, it is akin to a spelling mistake: it might remove any meaning from the sentence, or it might even alter the meaning in an insidiously plausible way. Mistakes in DNA assembly do indeed occur during replication, despite the existence of an astonishingly efficient array of molecular machinery for checking and "editing" the molecule as it is formed. It is these mistakes that allow a species to mutate and evolve by Darwinian selection; that is, to develop by chance new traits that might turn out to be either beneficial or detrimental to its chances of survival.

When the base sequence of a gene gets converted to the amino acid sequence of a protein, the information in the gene can be considered to have been translated into a different code on the protein. The amino acid sequence on most proteins is sufficient in itself to determine the shape of the protein – that is, the way in which the chain folds up to give the compact, enzymatically active form (see Chapter 2). The rules that govern protein folding are not yet completely understood, however – this is one of the major unsolved problems facing biochemists today.

## *Translating the message*

Translation of the genetic code from base sequences to amino acid sequences does not occur in a single step, but requires an intermediary. This is the role of the other nucleic acid in the cell, RNA. The RNA base uracil is the same size as thymine and has the same capacity to bind specifically to adenine. RNA therefore has the potential to encode information in just the same way as DNA. But the RNA molecules in a cell are shorter than DNA, and can be found in places where DNA is absent. It is the task of RNA to create a copy of the information in individual genes and then to act as the scaffolding on which this information is converted into protein molecules.

The progression from gene to protein involves a molecular assembly line of remarkable efficiency. The first step is to produce an RNA version of the gene encoded on the DNA molecule, where the sequence of A, T, C and G nucleotides in the gene is reproduced as a sequence of complementary U, A, G and C nucleotides in the RNA. The relevant gene-bearing portion of the DNA double helix is unravelled and one of the single strands acts as a template for the construction of the RNA molecule (Figure 5.8). This process is called transcription. The RNA molecule so formed is known as messenger RNA (mRNA), in distinction from the other types of RNA that become involved in protein building at later stages. The mRNA is then detached from the DNA strand, bearing the protein plan in the form of the sequence of bases (still with each group of three bases representing a single amino acid in the protein). Another type of RNA, called transfer RNA (tRNA) then sets about the task of foraging throughout the cell for the amino acids required to build the protein and ferrying them to the mRNA. The operation of using the coded information on mRNA to put together the protein is called translation.

Proteins are assembled on the mRNA template one amino acid at a time. Each three-base group on mRNA corresponding to an amino acid is called a codon. The tRNA molecules have one end that becomes anchored to a specific codon; this anchoring region on tRNA consists of a triplet of the base pairs that are complementary to those

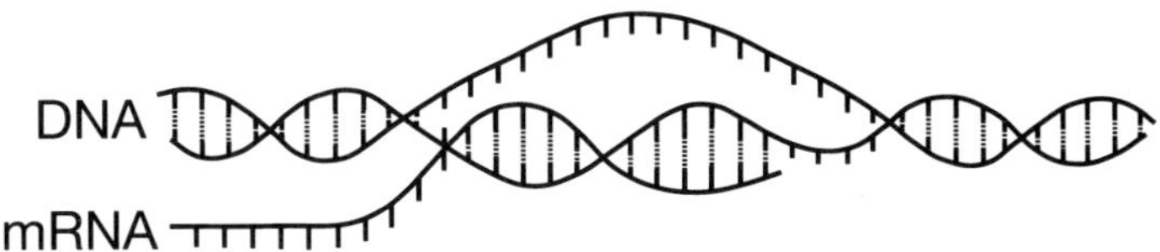


**Figure 5.8** *The genetic code is translated from DNA sequences to amino acid sequences via the intermediation of RNA. RNA molecules containing the information in a single gene, called messenger RNA or mRNA, are constructed by unwinding sections of the DNA double helix and using one of the unwound stretches as a template. The sequence of nucleotides in the RNA strand is determined by complementary base pairing with the DNA bases. This process of transcription can be considered to be driven by biochemical molecular recognition.*

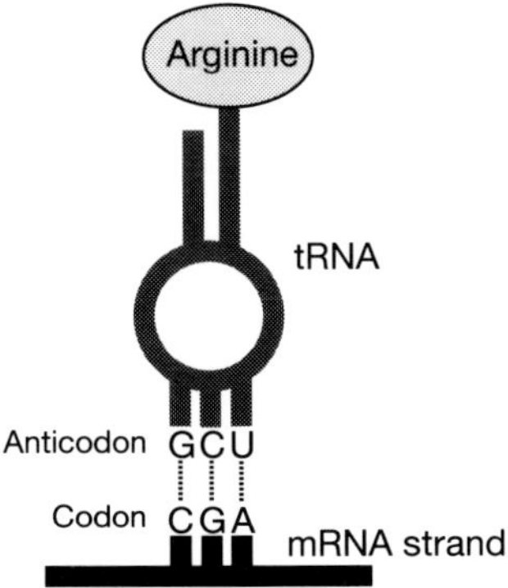


**Figure 5.9** *Amino acids are captured and brought to the mRNA "protein template" by transfer RNA or tRNA. One end of the tRNA molecule binds the appropriate amino acid through a complex molecular recognition process; the other end contains a sequence of three bases called an anticodon, which docks into the complementary codon sequence on the mRNA molecule. Here the codon is that for arginine (CGA; see Figure 5.7).*

on the mRNA codon, and it is therefore called an anticodon. For example, to the mRNA codon with the base sequence CGA will become anchored the tRNA anticodon GCU, via the complementary pairs C---G, G---C and A---U (Figure 5.9). The other end of the tRNA molecule binds specifically to the amino acid corresponding, in the genetic code, to the mRNA's codon. In the example above, the sequence CGA on mRNA, produced from the sequence GCT in a gene on DNA, corresponds to the amino acid arginine (see Figure 5.7). Thus the tRNA molecule responsible for putting arginine into its place in the protein chain has the anticodon GCU at one end, and binds arginine at the other.

The codon–anticodon recognition process works according to the principle of complementary base pairing. But the other of these two recognition processes – the picking up of the amino acid – is less well understood. It requires the help of an enzyme, called aminoacyl-tRNA synthetase. There is a different enzyme of this sort for every amino acid, and its task is to help the amino acid to form a bond between its acid part and the five-membered ribose ring of an adenine nucleotide on the end of the tRNA. Regardless of the nature of the tRNA's anticodon, the amino-acid-binding end of the molecule always ends with the base sequence CCA, so that the amino acid–tRNA bond is always to the A nucleotide. Given that this end group is always the same, the enzyme must somehow be able to "feel" the rest of the tRNA to make sure that it is attaching the right amino acid. Molecular biology abounds with complex recognition processes of this sort that are yet to be understood.

The linking together of the tRNA-borne amino acids at the mRNA template requires the assistance of a third kind of RNA, called ribosomal RNA or rRNA, on conjunction with other enzymes and proteins. Several rRNA molecules are bound together with many more proteins into a very complex entity called the ribosome, whose job it is to oversee this linking process. The ribosome binds to the mRNA and first facilitates the docking of the tRNA anticodon onto the mRNA codon. As successive tRNAs bring their respectively amino acids to the mRNA for stitching into the protein molecule, the ribosome shunts along the mRNA chain one codon at a time, so that it is always ready in the right place to receive the next tRNA in the sequence. The ribosome holds in

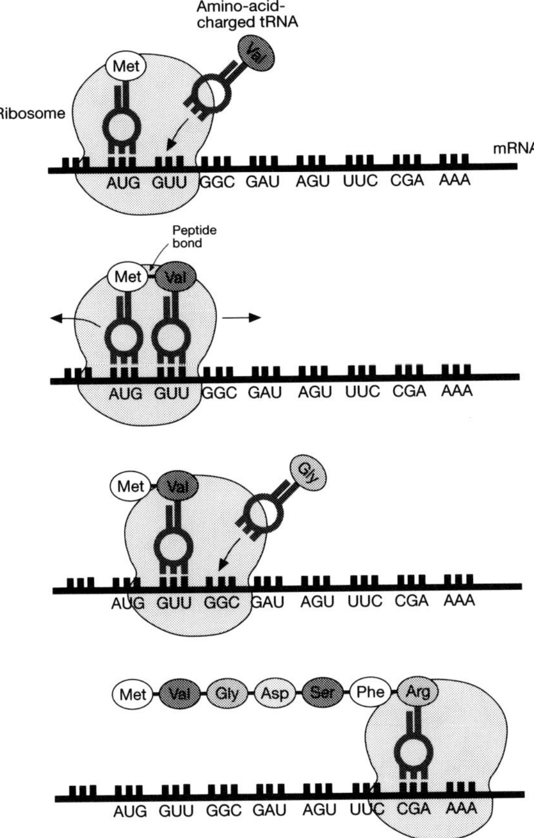


**Figure 5.10** *Protein synthesis on the mRNA template is a carefully orchestrated process. A molecular unit called the ribosome, which incorporates ribosomal RNA and several proteins, facilitates the docking of amino acid-charged tRNA onto the mRNA codon. The amino acid on the tRNA is then linked to the growing protein chain via a peptide bond, and the ribosome clicks one codon further along the mRNA chain. The tRNA is then discharged and destroyed.*

place two successive tRNAs at a time. To one of these will be attached the growing protein chain, while the other carries the next amino acid to be inserted, via formation of a peptide bond, into the chain (Figure 5.10). The ribosome holds together the end of the polypeptide chain and the next amino acid in just the right position for a peptide bond to form. The formation of this bond transfers the chain to the new tRNA, whereupon the old one is released by the ribosome, which then shunts along to the next codon and is ready to receive the next tRNA. There are base sequences at either end of the mRNA that do not correspond to codons, but instead act as signals to tell the ribosome where to begin and where to end protein synthesis. Once the protein chain is completed, it is detached from the ribosome/mRNA complex, and the mRNA, having done its task, is then mercilessly broken down by enzymes. So easy are these apparently complicated tasks of translation and transcription for the body's molecular machinery that it can afford to throw away the carefully constructed messenger after having used it just once!

## Learning to recognize

### *Lessons from life*

The vast majority of biochemical reactions involve a high degree of molecular recognition. An understanding of how these recognition processes work should not only provide insights into the chemistry of life, but may also guide the design of artificial enzymes which can carry out novel tasks or replace defective natural counterparts. Protein-based enzymes are, as we have seen, generally very large and complicated molecules, but once the essential aspects of their function are understood it may be possible to design smaller, simpler molecules that retain the most essential features and so can perform the same task. The pharmaceutical industry therefore stands to gain much from studies of molecular recognition in simple, model systems. Moreover, Chapter 2 showed how enzymes are becoming used increasingly as catalysts for industrial processes, in place of traditional, less selective catalysts. Here too, artificial molecules that mimic natural enzymes could prove to be of considerable value, especially as it can be a slow and costly task to isolate and purify large quantities of natural enzymes for industrial use.

But artificial enzymes and new drugs provide by no means the only motivation for investigations into molecular recognition in synthetic chemical systems. It has become clear in recent years that the organic chemist's conventional approach to putting together large and complex molecules is not necessarily the most efficient. Organic syntheses have tended to follow long, arduous routes involving many small steps. The long-winded nature of these strategies is the consequence of a lack of specificity in the reactions, which often results in a need to add "protecting groups" to parts of the molecule under construction so that a reagent targets the correct region without disrupting the rest of the molecule. How much better it would be to design reactants that interact with each other in a highly organized and specific way so as to form the product in just one or two steps. Such reactions would be driven down the right pathway by molecular recognition between the reactants, which locks them together in such a way as to ensure a single outcome. Chemists have found that by exploiting molecular recognition in this way, complicated molecules can be left to assemble themselves simply by throwing the reactants together in a rather cavalier fashion. "Self-assembly" has become one of the catchphrases of the discipline.

Organic chemists are not alone in benefiting from this new approach to chemical synthesis. As we shall see, molecular recognition is now providing the means to assemble molecules into remarkable superstructures, just as if one were dealing with the components of a microscopic construction kit. In a sense, this is engineering on the molecular scale. Molecular self-assembly processes provide perhaps the most promising avenue for research in the very young scientific enterprise known as nanotechnology – technology on the scale of nanometers (a millionth of a millimeter, or roughly the size of a $C_{60}$ molecule). The potential of molecular-scale engineering for electronics and materials science is vast and, as yet, scarcely tapped.

## *Hints of recognition: molecular hoopla*

While it can be argued that chemists have been learning lessons from the natural world for as long as the science has existed, molecular recognition as a field of study began in earnest only in the 1960s. At that time the American chemist Charles Pedersen was engaged in work for the petrochemical industry which involved studying the way that metal ions affect the chemical properties of rubber products. Pedersen came across a class of molecules that were able to attach themselves to specific metal ions while ignoring others. This kind of selectivity is rare in the world of inorganic chemistry – in general, metals tend to exhibit generic, rather than individualistic, modes of chemical reactivity. Even organic-based molecules that can bind to metal ions generally do so with little selectivity. An enzyme, in contrast, will remain aloof to any chemical species other than its designated target. If this were not the case, biochemistry simply would not function.

Pedersen's molecules, however, could be designed to attach themselves to just one type of metal ion and no other. They might, for example, bind to potassium ions but not to sodium ions (even though these have the same electrical charge and similar chemical properties) or to silver ions (which are of similar size to potassium). This finding was particularly appealing for two reasons. First, the fact that the molecules showed a binding selectivity for metals from the first and second groups of the Periodic Table (the so-called alkali and alkaline earth metals), rather than the transition metals, was surprising, since the latter usually show a greater sensitivity to the shape of their environment because they have preferences for particular geometric arrangements of their bonds. The ions of alkali and alkaline earth metals are little more than charged spheres, whereas the transition metals bristle with empty or partly filled electron orbitals sticking out in specific directions. The second interesting point was that the alkali and alkaline earth metals, particularly sodium, potassium, calcium and magnesium, are involved in many physiological processes, such as the responses of nerve cells. So it was possible to imagine that molecules with the ability to recognize individual members of this group of metals might have useful pharmaceutical applications.

What were the molecules that Pedersen had discovered? They were members of a class called cyclic ethers, which contain rings of carbon and oxygen atoms. In the ether group, an oxygen atom forms a bridge between two carbon atoms ($—CH_2—O—CH_2—$). The oxygen atom possesses two lone pairs, just as it does in water, so the molecule can form hydrogen bonds, or dative bonds to metal ions. For this reason, many metal ions are soluble in ethers. Cyclic ethers contain one or more ether groups joined in a ring. Pedersen's molecules typically had rings of between nine

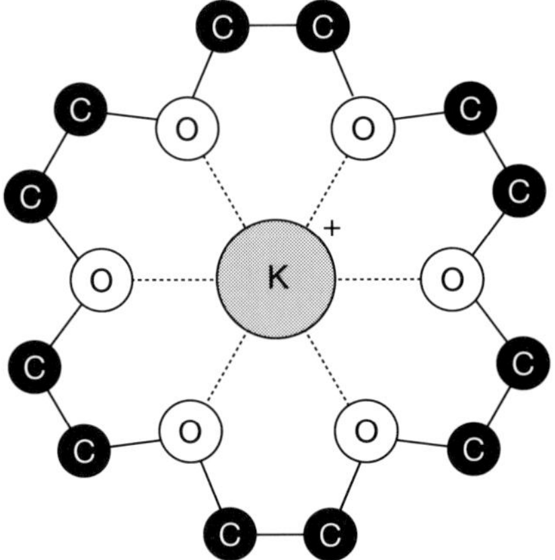


**Figure 5.11** *Crown ethers are cyclic molecules containing ether groups, the oxygen atoms of which can bind metal ions (here potassium ions) via their lone pairs. The strength of the binding depends sensitively on the size of the ring. Here and in all the following figures, hydrogen atoms attached to carbons are omitted.*

and eighteen atoms, in which an oxygen atom alternates with two $CH_2$ groups. The links between atoms in the ring are all bent in these molecules: the rings are puckered, resembling a crown. Pedersen therefore called these molecules crown ethers (Figure 5.11).

By uniting crown ethers with metal ions, Pedersen was playing a kind of molecular hoopla. The crown ether encircles the ion, each of the oxygen atoms using its lone pairs to bind the metal so that it ends up trapped securely in the center of the ring. The selectivity for different ions depends very sensitively on the ring's size. If the metal is a little too big, the hole in the ring will not be able to accommodate it; if it is too small, the oxygen atoms will be too far away to bind the ion effectively. Even the small difference between the radii of a potassium and a silver ion can cause pronounced differences in a crown ether's capacity for binding them.

### *Becoming more cagey: the cryptands*

It was the physiological significance of these metal ions that drew Jean-Marie Lehn's attention towards crown ethers. Lehn, a French chemist working at the Université Louis Pasteur in Strasbourg, was studying the role of sodium and potassium ions in nerve signaling. The electrical signals that nerve cells send to the brain originate from the way in which these ions are distributed on either side of the cell membrane. One way in which ions can be transported from one side of a membrane to the other is by becoming bound to molecules called ionophores, which ferry them through the membrane. Ionophores are usually ring-shaped molecules that exhibit a high degree of selectivity for different ions. The ion cannot pass through the membrane by itself because it is not soluble in the fatty compounds of which the membrane wall is composed. The ionophore gives the metal ions a fat-soluble coat, then releases them again on the far side of the membrane.

Many natural ionophores use lone pairs on oxygen or nitrogen atoms to bind metal

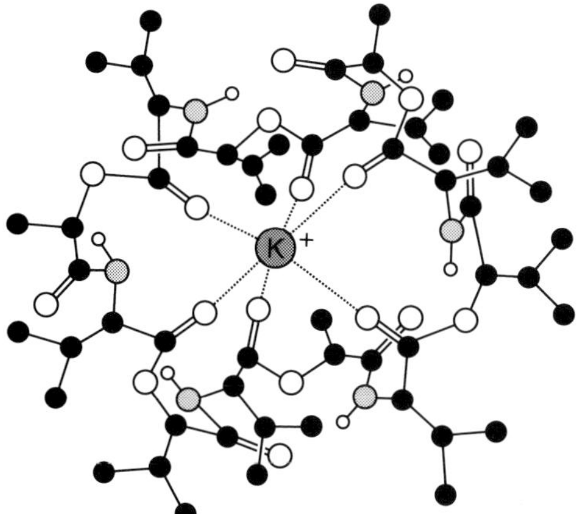


**Figure 5.12** *Compounds that can bind metal ions and ferry them through cell membranes are called ionophores. Natural ionophores are often macrocyclic (large-ring) molecules containing oxygen or nitrogen atoms that bind metals much as do crown ethers. The ionophore valinomycin, shown here, forms particularly stable complexes with potassium ions. It can act as an antiobiotic, by virtue of its ability to promote changes in the chemical content of cells and the functioning of their membranes.*

ions (Figure 5.12). Pedersen's crown ethers were therefore candidates for simple models of these compounds which might permit studies of the way that ion transport occurs. It was conceivable that they might even provide the basis for new drugs that mimicked the functions of real ionophores: certain of the latter can be used as antibiotics. Lehn realized that stronger and more selective binding of the metals might be achieved by making crown ethers with more than one ring, thereby imposing more severe space constraints on the cavity into which the ions fit. These molecules would be rather like molecular baskets.

The simplest molecule of this kind can be created by adding a bridge across the ring, producing a "bicyclic" species with two rings that are partly fused along one edge (Figure 5.13). The bridging link is made by incorporating into the ring two nitrogen atoms in place of oxygens. While the oxygen atoms form just two bonds, the nitrogens can form three, and so they can hold together the junction of three chains. Nitrogen-linked molecules of this sort are called azacrown ethers or simply azacrowns. To the bicyclic azacrowns (or "diazacrowns") Lehn gave the evocative name of "cryptands," from the Greek *krypt*, meaning hidden. The cryptands did indeed show the anticipated ability to hide metal ions within their central cavity. By varying the length of one or more of the ether chains, Lehn could make cryptands that were highly selective in their binding affinity towards different alkali metals, achieving a degree of discrimination comparable to that of natural compounds.

Yet there was no need to stop at bicyclics. In 1976, Lehn's group produced a tricyclic azacrown or triazacrown (Figure 5.14). The cavity of this molecule is truly cage-like, and it extended still further the possibilities for carrying out selective ion-binding. The potassium ion and the ammonium ion ($NH_4^+$) are of somewhat similar size, and a bicyclic cryptand will ensnare either with more or less equal avidity; but a tricyclic cryptand shows a marked preference for the ammonium ion. This is because in the latter case, selectivity is determined not only by size but also by geometric factors – by shape. Whereas potassium ions are just charged balls, the ammonium ion has a well-defined

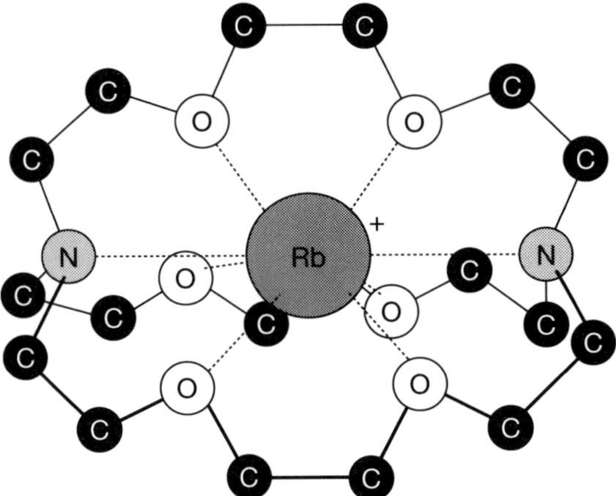


**Figure 5.13** *Bicyclic crown ethers contain two partially fused ether rings. Nitrogen atoms permit the junction of three ether chains in these molecules, and also supply lone pairs for binding metal ions (here rubidium). Because of their ability to hide ions in their cage-like cavities, Jean-Marie Lehn has named such molecules "cryptands."*

shape in which the four hydrogens sit at the corners of a tetrahedron. The tricyclic cryptand has just the right arrangement of nitrogen and oxygen atoms to form directional bonds to $NH_4^+$ (Figure 5.14).

In this latter example, molecular recognition is more than a case of simply wedging a peg in a hole of the right size: the hole has to be the right shape too. It is this sensitivity to molecular shape that characterizes most biological recognition processes. The molecule that is doing the recognition, which chemists generally call the host, has

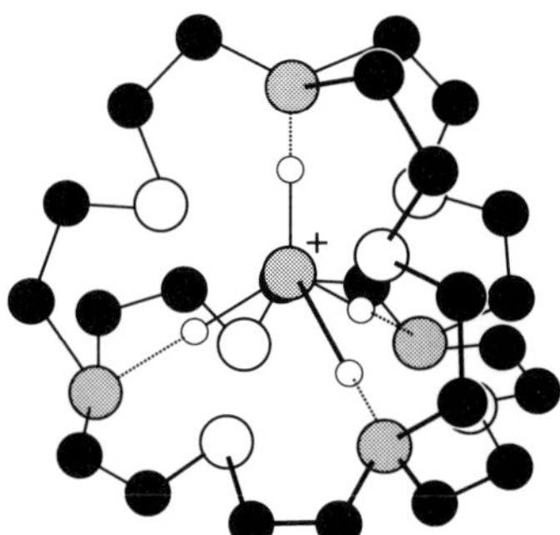


**Figure 5.14** *The cavities of tricyclic azacrown ethers have roughly spherical shapes. That shown here is particularly well suited to forming a complex with ammonium ions ($NH_4^+$).*

a shape that is complementary to that of the molecule that it captures (called the guest), just as the cavity of a lock has a shape complementary to its key. The "lock-and-key" concept in molecular interactions was first invoked by the German biochemist Emil Fischer in 1894. The assemblies of molecules created by this fitting together of complementary shapes are examples of so-called supermolecules, the formation and behavior of which is the topic of supramolecular chemistry. Supermolecules are distinguished from mere large molecules by the fact that their components are not linked together by strong covalent bonds but instead by weaker interactions such as hydrogen bonding, which can in principle be prised apart to regenerate the constituent molecules. What one encounters in these assemblies is precisely the same kind of phenomenon that gives rise to complementary base pairing in DNA, or to the binding of biomolecules to enzymes. In biology, the host molecule is known more commonly as the receptor, and the guest is called the substrate. This terminology has become popular in supramolecular chemistry too, and to some extent I shall be using "host/guest" and "receptor/substrate" as synonymous terms.

## *All shapes and sizes*

Combinations of size and shape selectivity can be used to design receptor molecules that will recognize substrates far more complex than metal ions. Lehn and his colleagues have used the principles of crown ether chemistry to make molecular "pencil cases" into which fit long, linear substrates of various lengths. The pencil-case molecules have azacrowns at either end, joined together by two linear "spacer" groups. Each end of this receptor has the capability to bind an ammonium group, just as cryptands can. The cavity is therefore the perfect receptor for a linear diamine molecule, a hydrocarbon chain with amine ($NH_2$) groups at each end. In acidic solutions, the amine groups pick up a hydrogen ion in just the same way that ammonia does, forming positively charged $—NH_3^+$ groups. When the length of the "diamino" ion matches that of the spacers holding the azacrowns together, it will slot into place in the receptor (Figure 5.15). Molecules of this sort, which have several binding sites arranged in such a way that only substrates with a certain geometry can become attached to them, come considerably closer to mimicking biological recognition.

An intriguing variant of the molecular pencil case has been devised by George Gokel and colleagues at the University of Miami. Gokel's receptor has an azacrown at either end for binding amino groups, but these two ends start out as separate entities. To one azacrown are attached two hydrocarbon chains capped with thymine bases; to the other are attached similar chains but with adenine bases at the ends. When these two components are mixed together, the adenine groups link up with the thymines by forming hydrogen bonds, creating precisely the same base pair as is found in DNA (Figure 5.16). The pencil case thereby *assembles itself* spontaneously from its components. In this way, the receptor is itself the product of molecular-recognition processes.

Gokel has also devised a variant of the crown ethers in which the substrate is bound with increased firmness by a flexible arm which swings around and clips it in place.

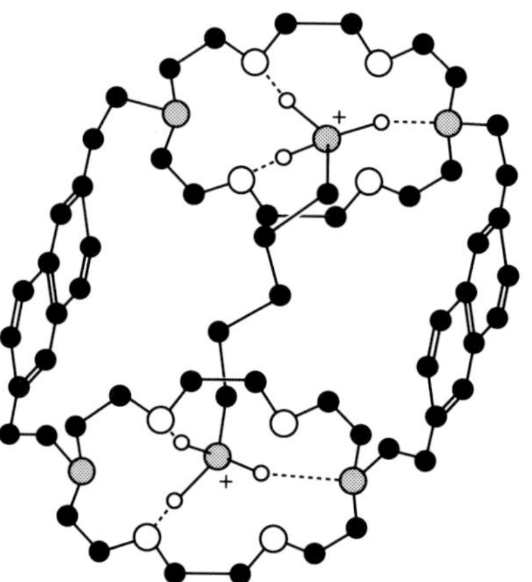


**Figure 5.15** *Two azacrown rings joined by hydrocarbon spacer groups form a slot-like cavity of variable length, into which long, thin "pencil-like" guest molecules can be bound.*

These molecules are azacrowns that have a linear hydrocarbon chain attached to the nitrogen atom, at the end of which is an additional ion-binding group (that is, one with a lone pair, usually an ether group). Because of their lasso-like shape, Gokel calls these molecules lariat crown ethers (Figure 5.17). The free arm on the nitrogen atom is relatively flexible, so that when a metal ion becomes bound within the ring, it can twist into a position in which the ether group at the end lies above the ion, binding it more

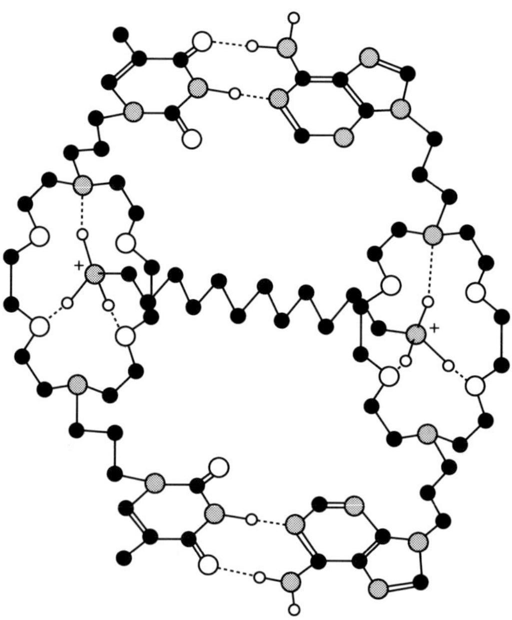


**Figure 5.16** *A self-assembling pencil case. The two halves link together spontaneously in solution via hydrogen bonds between the complementary nucleotide bases adenine and thymine.*

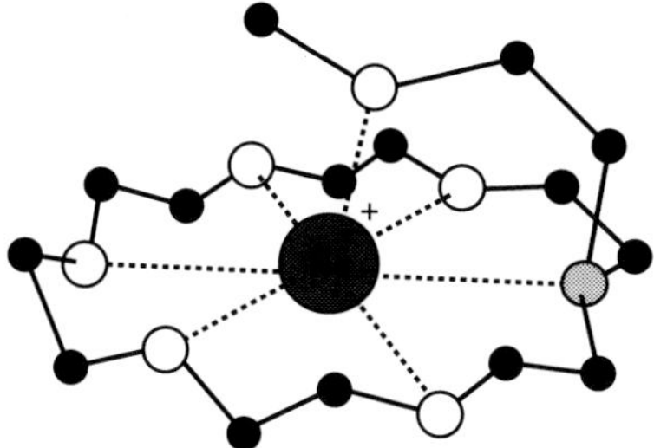

**Figure 5.17** *In lariat crown ethers, a flexible arm provides greater control of the affinity for binding guests, being able to swing round to clip the guest molecule in place or open up to release it.*

securely still. Their movable arm gives the lariat crown ethers a combination of flexibility and the ability to "cage" their substrates, so that while binding is effective and selective, the substrate need not become entrapped so tightly that it cannot be easily released again. You might notice that the two halves of Gokel's self-assembling pencil case are themselves nothing more than two-armed lariat crown ethers with nucleotide bases at their ends.

A group at Texas Tech University led by Richard Bartsch have designed a pair of molecular tongs (Figure 5.18) that can trap a variety of metal ions between its jaws. The molecule bites on lithium ions with relish, but larger sodium ions are too much of a mouthful to be captured effectively, making the tongs efficient at selectively picking lithium out of solution. Seiji Shinkai and colleagues from the University of Kyushu in Japan have found an ingenious way to obtain more control over the opening and closing of the jaws: they have made a pair of molecular tongs that snaps shut in response to light.

In Shinkai's tongs, the connection between the two crown ether rings is made with a group based on the azobenzene molecule. Azobenzene consists of two benzene rings joined together via two nitrogen atoms linked by a double bond. Because the nitrogens each have a lone pair, the molecule is crooked rather than linear. The benzene groups can sit either on opposite sides or on the same side of the double bond; the molecule therefore has two isomers. The configuration with opposite benzenes is called the *trans* isomer, and that with benzenes on the same side is the *cis* isomer. Because the double

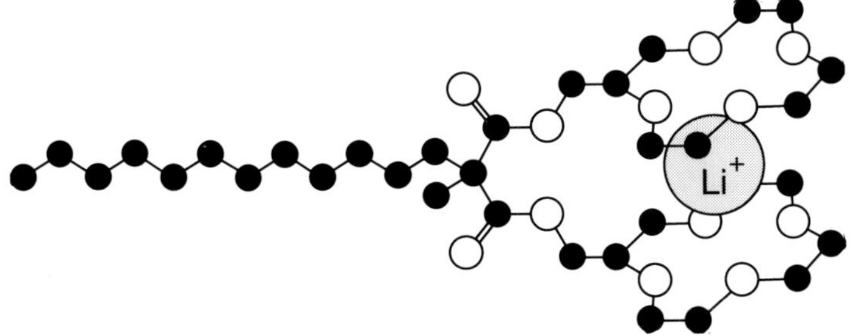


**Figure 5.18** *The jaws of these molecular tongs will close on lithium ions.*

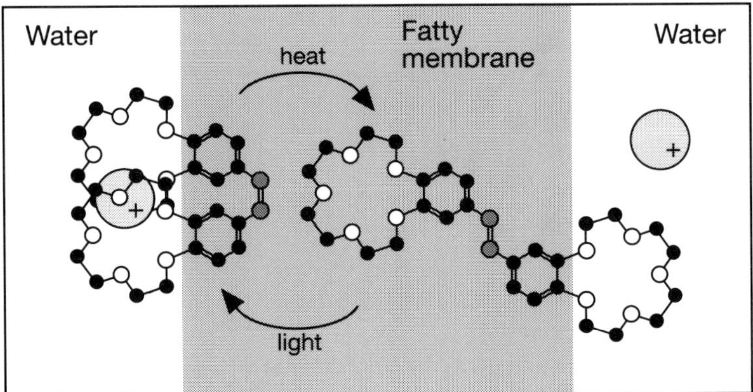


**Figure 5.19** *In the molecular tongs devised by Seiji Shinkai, the jaws can be closed by irradiation with ultraviolet light, which induces photochemical isomerization of the double bond between the nitrogen atoms. Heat will reverse the process, opening the jaws. In other related systems, irradiation with visible light or changes in acidity can induce opening. The consequent controlled trapping and release of metal ions can be used to bring about transport of metal ions across artificial membranes: the ions are trapped by the* cis *isomer (left), ferried through the membrane, and released by conversion to the* trans *isomer (right). Host molecules of the kind shown here are sometimes known, for reasons that should be apparent, as "butterfly" molecules.*

bond is rigid, the molecule cannot easily interconvert between the two isomers. But ultraviolet light can induce a photochemical rearrangement from the *trans* form (which is preferred normally because it keeps the bulky benzene groups furthest apart) to the *cis* form. When heated, the molecule returns to the *trans* arrangement.

Shinkai's group attached crown ethers to both benzene rings of an azobenzene molecule. In the *trans* form these rings are far apart, but in the *cis* isomer they lie adjacent to each other and so can trap a substrate between them. Shinkai has used the molecule as a set of molecular tweezers to transport potassium ions across a "liquid membrane" – a fatty organic liquid separating two aqueous solutions, which serves as a crude model of a cell membrane. The azobenzene remains in the liquid membrane all along. In response to irradiation with ultraviolet light, the *trans* isomer switches to the *cis* form, which enables it to snap up a metal ion at the interface between one of the aqueous solutions and the membrane. It carries this ion through the membrane to the other side, disgorging it when the system is warmed to regenerate the *trans* form. Membrane transport controlled by light might provide a means of storing solar energy, since it can pile up charge on one side of a membrane which may later be discharged as an electrical current.

The simpler cryptands also show an ionophore-like ability to effect cross-membrane transport of ions. The aim of most membrane transport processes, both in biology and in these artificial systems, is to build up an excess of ions on one side, thereby setting up a difference in electrical potential (that is, a voltage) across the membrane. But the system's natural preference is to have equal concentrations on both sides (this corresponds to the situation at thermodynamic equilibrium; see Chapter 2). So a source of energy will be needed to drive the process up the "thermodynamic slope." Shinkai provided this energy in the form of light and heat; in cells, it is commonly supplied by metabolic processes involving the energy-storing "battery" molecules of biochemistry, adenine triphosphate (ATP).

There is an entirely different way in which ions can cross cell membranes: through tunnels that have been created specifically for this purpose. These so-called ion channels in cell walls are rather like tubes with interior surfaces that interact favorably with the ions (rather than repelling them, as do the fatty compounds that constitute the membrane itself). Ion channels contain gates that can be opened and closed at appropriate times to direct the ion traffic. Channels obviate the need for carrier molecules to ferry the ions through the cell wall, but selectivity in the transport process is maintained by recognition processes between the ion and the channel – the gates, for example, might admit only one specific ionic species. Lehn imagines that it should be possible to build synthetic ion channels from crown ethers stacked like hoops, and other groups have explored this concept using "boland" molecules, in which several crown ether rings are

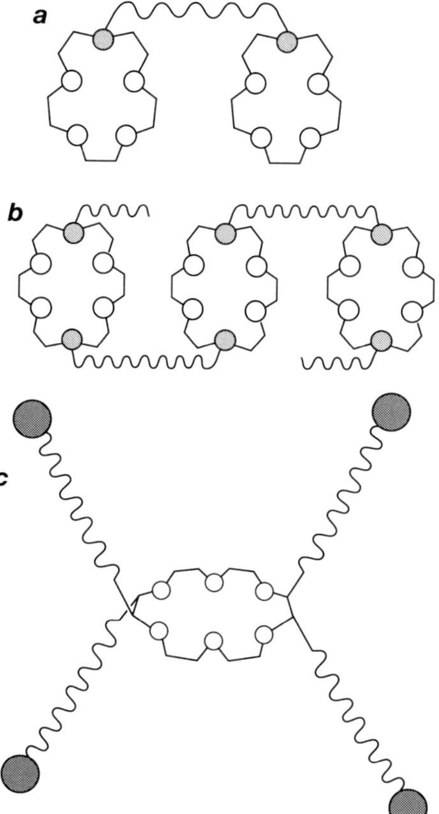


**Figure 5.20** *Crown ethers stacked one atop the other might constitute artificial "ion channels" which allow selective passage of metal ions through membranes. Linked crown ether rings called "bolands" will fold up to create such channels (a). A molecule possessing several linked rings can fold up to create a tunnel of "hoops," between which metal ions might jump (b). Jean-Marie Lehn has taken a somewhat different approach to the construction of artificial channels by attaching long tails to a central crown-ether ring (c). He calls these "bouquet molecules." The tails end in water-soluble groups (gray circles), which make the molecule compatible with those that constitute an artificial cell-like membrane. Note that these tails have a rather complex structure which I have not attempted to show in full here. Bouquet molecules will allow alkali metal ions to pass through such membranes.*

linked together by hydrocarbon chains. When incorporated into artificial cell-like membranes, these molecules can indeed effect the passage of ions from one side to the other (Figure 5.20). Lehn also envisages schemes in which gates can be introduced into these channels, activated by acidity, light or by electrochemical means. At present, these ideas have not led to anything remotely resembling an artificial nerve cell; but it is fair to say that many of the basic chemical principles for effecting selective ion transport across membranes are now well understood.

## *The inside story*

Pedersen's crown ethers and Lehn's cryptands function by incarcerating their target ions within a kind of molecular cage. But this cage is formed only as the guest is captured; when the host molecules are "empty" they are rather floppy, without a well-defined cavity at all. A crown ether, for instance, does not drift around by itself in the form of a rigid ring, but is more akin to a rubber band until it captures a metal. Natural enzymes, on the other hand, are folded up into highly organized, preformed shapes that present a ready-made keyhole into which the substrate will fit. It is likely that a recognition process will be relatively inefficient if it involves a good deal of rearrangement of the receptor's shape. So there is considerable interest in designing artificial receptors that have their cavities already formed into the shape appropriate for the intended substrate.

Donald Cram of the University of California at Los Angeles has been building molecules that are stiffer and have more clearly defined shapes than the crown ethers and their relatives. To the first generation of these molecules Cram gave the name

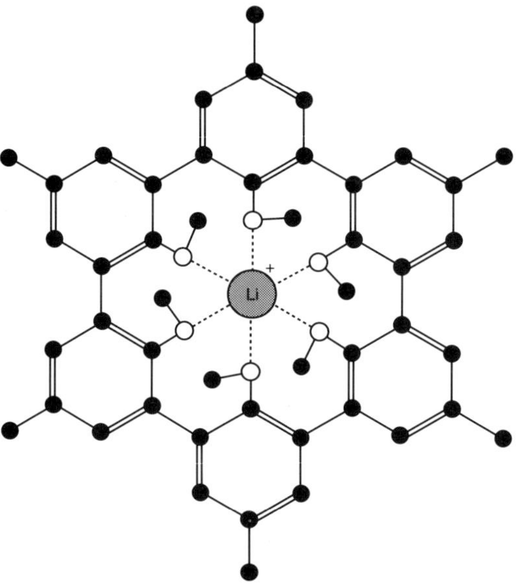


**Figure 5.21** *Donald Cram's collar-shaped spherands provide ready-made, rigid cavities for metal complexation.*

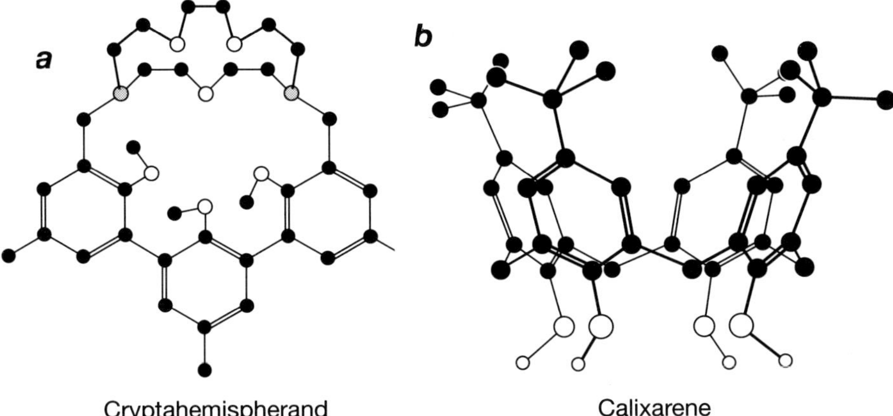


**Figure 5.22** *A combination of the component parts of cryptands and spherands generates a cryptahemispherand (a). The bowl-shaped calixarenes (b) represent another kind of rigid-cavity host.*

spherands; one such is shown in Figure 5.21. The six benzene rings form a more or less rigid framework with a cavity in at its center lined with oxygen atoms for binding metal ions. The cavity in Figure 5.21 is quite small – it will accommodate a lithium ion but not sodium or potassium. Because the molecule does not have to do any gymnastics in order to bind the lithium ion, it turns out to be much more efficient at doing so than is a crown ether.

Better still than these "hard" rings are bowl-shaped molecules, such as cryptaspherands and calixarenes (Figure 5.22). The benzene rings in these molecules hold their cavities rigid. Calixarenes in particular have attracted interest because the rim of their bowl-like cavities can be lined with chemical groups that determine which guests it will accept. For example, placing negatively charged groups such as sulfonate ($—SO_3^-$) onto the benzene rings along the rim will make the bowl repel other negative ions while appearing highly inviting to positive ones. Seiji Shinkai has used this idea to create a molecular capsule from two calixarenes. To the rim of one he attached positively charged groups, and to the other, negatively charged units. The two bowls lock together spontaneously, perhaps trapping small molecules inside (Figure 5.23). The capsule can be cracked open by making the solution more acidic, whereupon hydrogen ions attach themselves to the negative groups and neutralize their charge. Perhaps one day we may see these microcapsules used to deliver drug molecules into the body.

Donald Cram's molecules have been getting increasingly adept at confinement. One class of these compounds is called the carcerands, from the Latin word for prison. These are something of a cross between the bowl-like calixarenes and the spherands: the benzene groups are linked by ether chains into a net that almost closes in on itself to form a sphere (Figure 5.24). Small holes are left for molecules to get in, but once there they are effectively trapped.

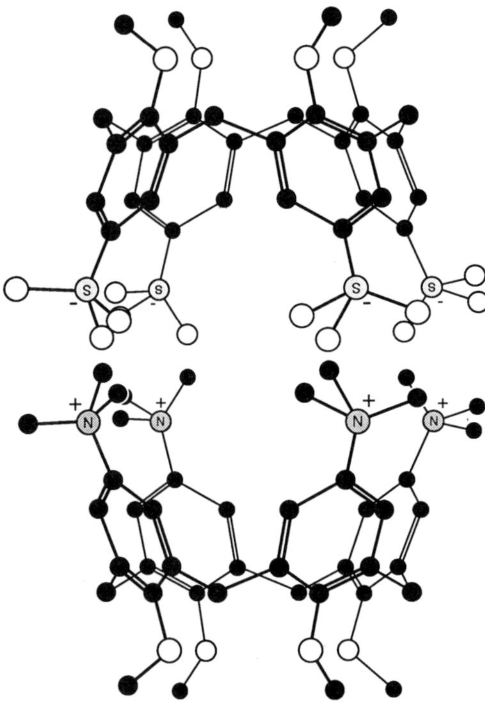


**Figure 5.23** *Calixarenes with rims lined with oppositely charged ionic groups will self-assemble into capsules, inside which guest molecules may be trapped.*

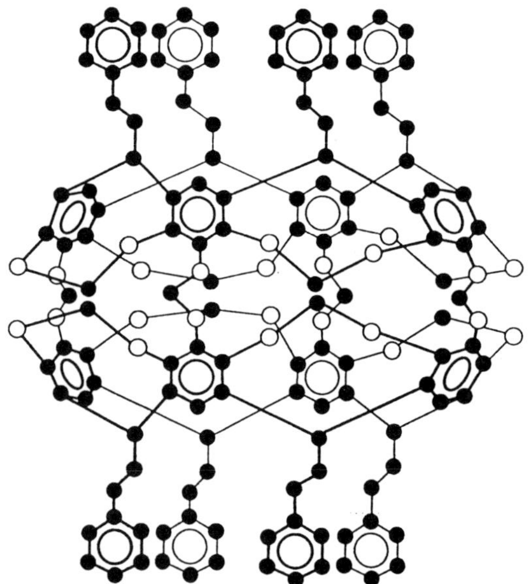

**Figure 5.24** *Some of the most uncompromising of rigid-cavity molecules are Donald Cram's carcerands, which are hollow, spherical molecules containing narrow openings to admit small guests. Note that here, for simplicity, I denote benzene rings using the common "ring" notation rather than showing the double bonds of the Kekulé structure.*

The spherical cage-like structures of these molecules have some interesting analogs. One is the football-shaped buckminsterfullerene molecule described in Chapter 1, which also contains benzene-like rings in its shell. The buckminsterfullerene shell is, however, closed completely, so if one wants to incarcerate molecules inside, one has either to ensure that they are around when the molecule is being formed or to cut open a hole in the framework. We saw in Chapter 1 that encapsulation of metals has already been achieved by the former route. Chemists are now working on ways to open up holes in the shell by using reagents that attack the carbon rings. It is widely hoped that buckminsterfullerene will provide a convenient short cut to carcerand-type molecules.

In Chapter 2 we encountered another analog of the carcerands – the porous crystals called zeolites. These materials contain molecular-scale cavities within an aluminosilicate network, accessible via narrow channels. The well-defined sizes and shapes of the cavities give the zeolites the ability to act as selective catalysts, a little like inorganic analogs of enzymes. Mindful of these properties, Cram has plans for his own molecular prisons that go beyond simple recognition and binding: he sees them as little flasks in which chemical reactions can be carried out. Some reactions, particularly those of a biochemical nature, can be hard to control when all the reactants are floating around in solution, particularly if the reaction products or the molecules formed at intermediate stages are highly reactive. The carcerands might supply the kind of isolated environment needed to perform delicate reactions of this kind.

In a demonstration that this is not mere fantasy, Cram has used one of his cage molecules as the beaker in which he has synthesized the very reactive compound cyclobutadiene. This latter molecule consists of a ring of four carbon atoms, each with a hydrogen attached, linked by alternating single and double bonds. Because the square shape induces rather severe bending of the bonds between carbons, cyclobutadiene is highly strained and breaks apart quite easily to produce either two acetylene molecules or, following the linking up of two fragments, an eight-membered carbon ring. To provide a protective environment in which cyclobutadiene can be created and maintained without breaking up, Cram and his colleagues have used a so-called hemicarcerand: a carcerand with a small gap in its shell, through which small guest molecules can pass when the system is heated (Figure 5.25). Cram's group admitted into the hemicarcerand's interior the starting material for the synthesis of cyclobutadiene, a molecule called alpha-pyrone. This can be converted into cyclobutadiene and carbon dioxide by exposing it to a flash of light. The strained product remains stable at room temperature provided that it stays inside the hemicarcerand. But opening the door to admit other small molecules, such as oxygen, allows the cyclobutadiene to react with the new guest while still inside. So unusual is the environment inside the hemicarcerand that Cram feels it deserves to be designated a new state of matter.

### *Threading the needle*

The Scottish chemist Fraser Stoddart has been putting supermolecules through ever more remarkable contortions. Stoddart envisions a time when it will be possible to

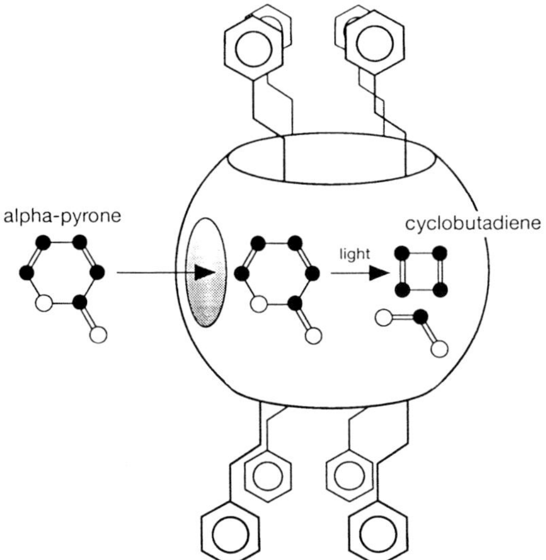


**Figure 5.25** *Carcerands can act as protective vessels in which delicate chemistry can be performed. Here an alpha-pyrone molecule is admitted into the interior of a hemicarcerand through the hole in its shell and is then converted photochemically to the highly reactive molecule cyclobutadiene. While it remains isolated from molecules outside, the product is stable.*

assemble molecules (or rather, to design them so that they assemble themselves) into just about any structure one might wish for, just as if they were no more than the struts, brackets and blocks of a toy construction kit. He refers to this as "molecular Meccano"; other researchers, meanwhile, are to be found fantasizing about molecular "tinkertoy" sets based on stiff, rod-like molecules.

Much of Stoddart's work has been concerned with molecules that self-assemble in such a way that one is threaded through a hole or ring in another. The principle behind these creations is similar to that of Lehn's pencil-case molecules except that the substrates are chosen to be too big for the slot, so that they lie at an angle to the ring with each end poking out. One such assembly can be created from a crown-ether-type molecule in which two ether chains are linked by benzene groups. This molecule (which has a complicated name that I shall represent as HY to signify its derivation from the compound hydroquinol) can act as a receptor for the molecular ion shown in Figure 5.26. The latter contains two benzene-like rings incorporating positively charged nitrogen atoms; I shall denote this species as $PQ^{2+}$, as it is related to the paraquat molecule. A supermolecule similar to this one can be constructed by "inverting" the roles of each molecule – that is, by joining two $PQ^{2+}$ ions into a ring, and making the substrate a linear version of the HY cyclic ether (Figure 5.26).

These assemblies can be prevented from coming unthreaded by capping the ends of the thread with big molecular groups – this is rather like tying knots in the ends. For instance, Stoddart's group threaded a linear HY molecule through a $PQ^{2+}$ ring and then added to each end of the thread a silyl group, a silicon atom with hydrocarbon groups attached (Figure 5.27). The ring was thereby trapped like a bead on the thread. These assemblies are called rotaxanes.

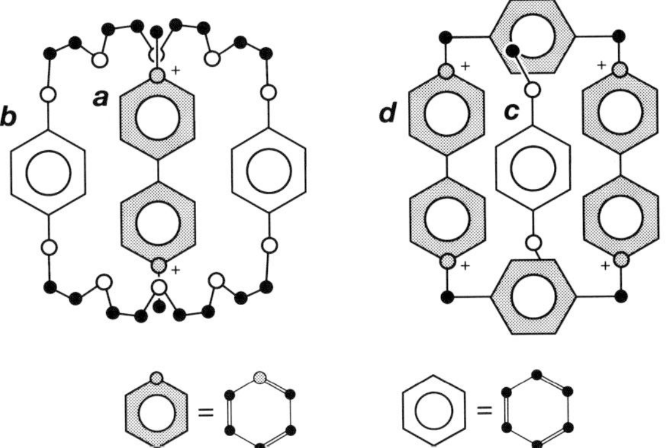


**Figure 5.26** *The charged paraquat-based molecule* a *will thread spontaneously through the hydroquinol-based crown-ether ring* b *in solution to form a host–guest complex (left). Similarly, a ring* d *based on molecule* a *will act as a host for the molecule* c, *which is based on* b *(right).*

The term "rotaxane" was first coined in 1980 by Gottfried Schill of the University of Freiburg, who constructed a rather different type of rotaxane from hydrocarbon rings (Figure 5.28*a*). A very similar molecule had been constructed 13 years earlier by Ian Harrison and Shuyen Harrison of Syntex Research in Palo Alto, California (Figure 5.28*b*), but they had preferred the name "hooplane." Schill's choice was derived from the resemblance of the assembly to a wheel (*rota* in Latin) on an axel. The crucial difference between these earlier rotaxanes and those made by Stoddart and his colleagues is that the former were the product of very delicate methods of traditional chemical synthesis, whereas by building into his bead and thread the capacity to recognize each other Stoddart was able to let the supermolecules assemble themselves.

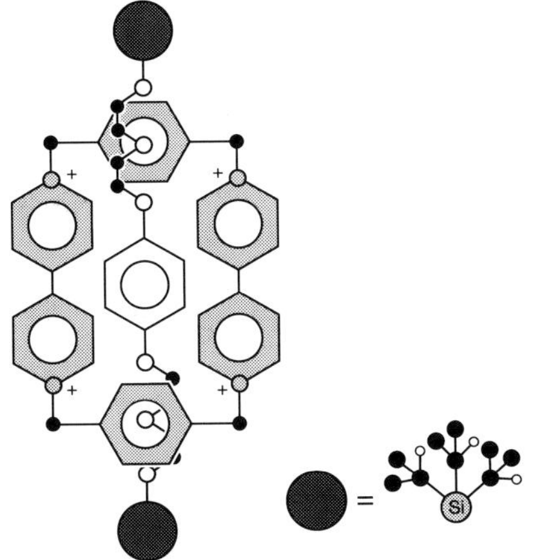


**Figure 5.27** *Threaded molecules can be prevented from slipping out of the ring by capping them with bulky groups. The result is a rotaxane.*

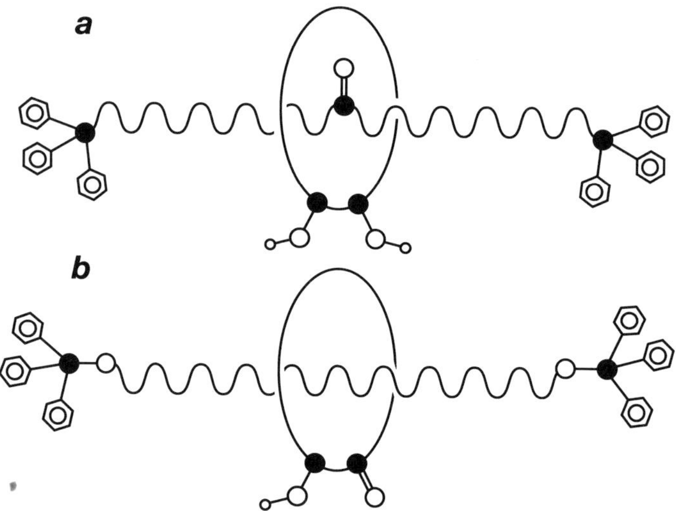


**Figure 5.28** *Gottfried Schill's rotaxane, incorporating a hydrocarbon-based ring and thread and formed via delicate chemical synthesis in 1980 (a), and the related hooplane constructed 13 years earlier (b). Fraser Stoddart's more recent rotaxanes have the crucial difference that they assemble themselves spontaneously, with no need for complex synthetic pathways.*

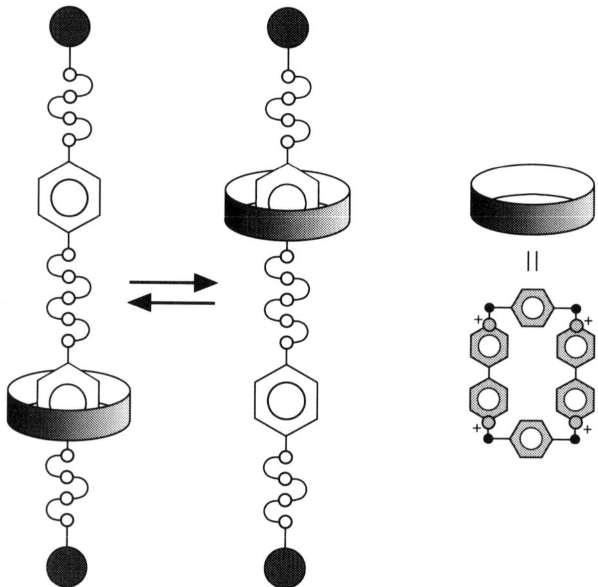

**Figure 5.29** *A rotaxane with two "sticking points" for the threaded bead becomes a molecular shuttle.*

The most exciting prospects for these curious supermolecules involve designs in which the threaded beads can move. For example, Stoddart has synthesized an HY—$PQ^{2+}$ rotaxane in which the HY thread contains *two* potential docking positions for the $PQ^{2+}$ bead (Figure 5.29). It is then possible for the bead to jump back and forth between these positions, like a shuttle. Stoddart's molecular shuttle has sparked off speculations about using these supermolecules for information storage: an array of them, each of which could be switched back and forth between two states, could be used to store data in binary form, just as computer memories do. Of course, it would be necessary to develop some way of controlling the switching process so that information could be fed in, and also some way of reading the information back out again; how this might be done is not obvious. These difficulties have not, however, dampened the enthusiasm with which speculations even more extravagant are made. How about a molecular abacus for nanotechnological calculations?

## *Molecular trains*

Given his success in threading both a $PQ^{2+}$ ion through an HY ring and vice versa, Stoddart reasoned that it should be possible to build an assembly that combines the two; in other words, an assembly in which two closed rings are threaded like the links of a chain. Stoddart's team were able to create this remarkable supermolecule by threading through an HY ring a double-$PQ^{2+}$ thread, and then adding a benzene ring across the two ends of the threaded molecule to close the loop (Figure 5.30). Although it might appear fiddly, this procedure is actually very straightforward because of the way that chemical interactions guide the thread into a position in which the two loose ends lie next to each other, ready to be capped. Clearly, if the host and guest are suitably designed, these topological feats of synthesis are no problem at all – molecular recognition does it all for you.

Linked molecular assemblies of this sort are called catenanes. Perhaps surprisingly, they have a history a little longer than the rotaxanes, the first catenane having been constructed in 1960 by Edel Wasserman of AT&T Bell Laboratories in New Jersey. Like the early rotaxanes, Wasserman's catenane was based on simple hydrocarbon rings

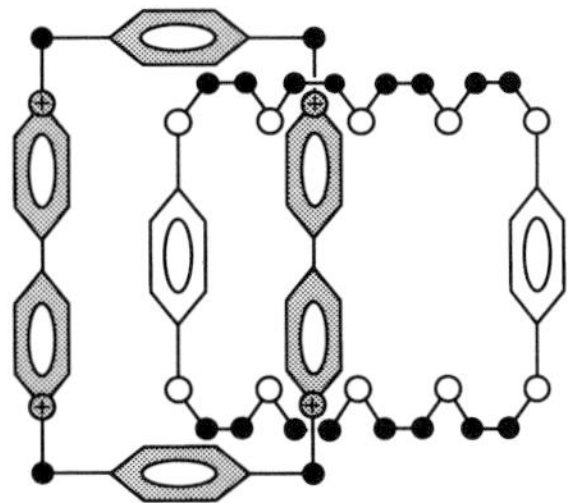

**Figure 5.30** *By sealing together the ends of a rotaxane, Fraser Stoddart has assembled a two-link molecular chain – a [2]catenane. The most favorable arrangement of the benzene-like rings is that in which they sit face-to-face, more or less perpendicular to the plane of the molecular rings.*

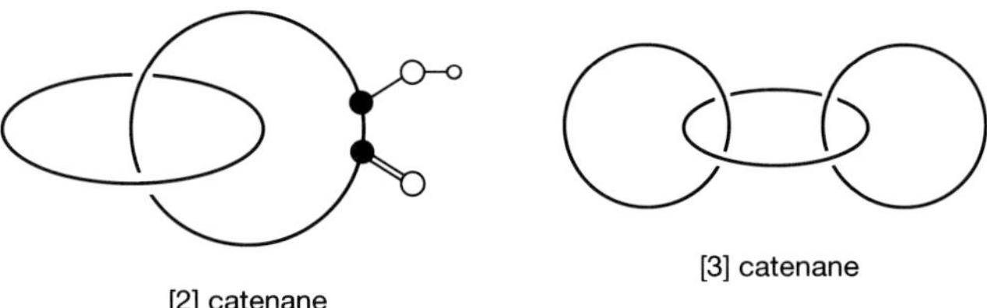


**Figure 5.31** *The earliest catenanes were, like the rotaxanes, hydrocarbon-based molecules, the synthesis of which involved no host–guest complexation. The [2]catenane shown here was made in 1960, and the first [3]catenane was synthesized in 1977.*

(Figure 5.31). A chain of two links is called a [2]catenane; Gottfried Schill went a step further in 1977 by making a hydrocarbon-based [3]catenane (Figure 5.31).

Jean-Pierre Sauvage and colleagues at Strasbourg have combined catenanes with the binding of metal ions to form metallo-catenanes (or "catenates") in which a metal ion is trapped between the two rings (Figure 5.32). The [2]catenate, created in 1983, can interchange copper, lithium and silver ions in the metal-binding pocket.

While at the University of Sheffield, Stoddart took up the challenge of making a [3]catenane from his own brand of supermolecules, using a longer double-$PQ^{2+}$ thread which could penetrate two HY rings. The $PQ^{2+}$ units in the thread were separated by two benzene rings, rather than just one, with the result that, having threaded through one ring, the molecule was left with one $PQ^{2+}$ group dangling too far away to interact with the same ring, so that it picked up another instead (Figure 5.33). Very recently, Stoddart has extended his self-assembling systems to the synthesis of a [5] catenane. If you look carefully at the symbol for the Olympic Games, you will appreciate why he has suggested that this supermolecule be christened "olympane".

If one loop of these catenanes is much longer than the other(s), the assemblies become molecular necklaces on which the bead(s) can slide around. Stoddart has built an assembly

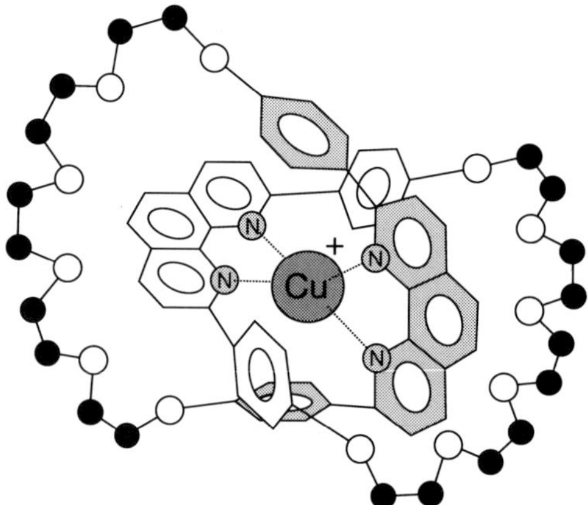


**Figure 5.32** *Catenanes based on crown ethers can bind metal ions in the cavity between the two rings. These complexes are called catenates.*

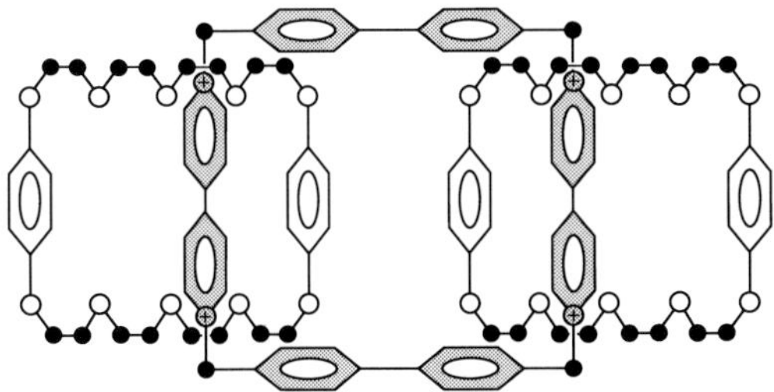

**Figure 5.33** *A [3]catenane can be formed by using a thread long enough to pass through two rings. The ends of the thread (here shown in gray) are then joined together.*

containing one bead and two docking points, like a looped version of the molecular shuttle. The bead then becomes a molecular "train" on a circular track. It runs freely around the track at room temperature, but when the system is cooled to minus 80 degrees Celsius the interactions between the train and the "stations" are strong enough to overcome the thermally induced cycling motion, bringing the train to rest. Stoddart has also built a longer track with four stations, and can watch either one train circle it (stopping at minus 60 degrees Celsius; Figure 5.34) or two (both of which stop at minus 40 degrees Celsius). In the latter case, there would seem to be the possibility of crashes between the two trains; but it appears that the rail network is a very

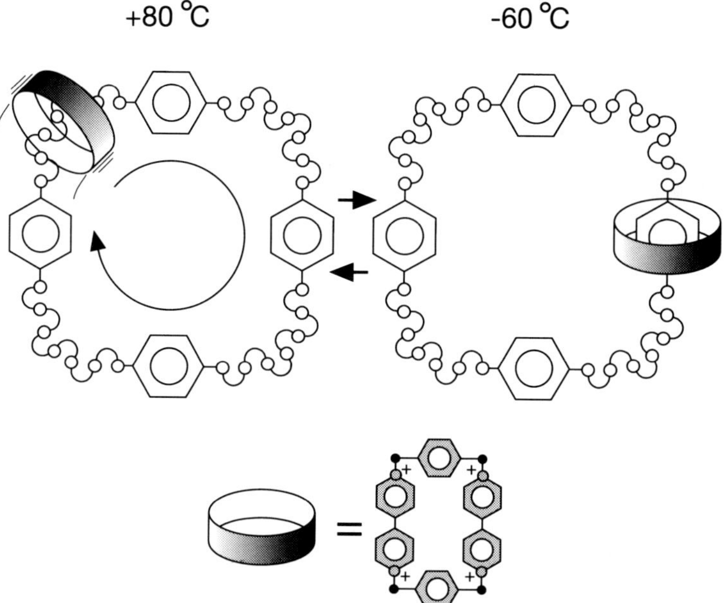


**Figure 5.34** *A [2] catenane in which one ring is much larger than the other provides a molecular analog of a train on a circular track. The train circulates around the large ring at room temperature. When the molecular assembly is cooled, the train stops at one of the HY stations.*

well-organized one, in which the trains always make sure that they keep one station between them. Stoddart hopes that he will be able to find a way to insert into the track sets of signals to provide control over the trains' motions. These extraordinary molecular assemblies show that, while nature remains in many ways a vastly more capable chemist than we are, she does not necessarily display a more vivid imagination!

## From recognition to replication

### *Carbon copies*

It may not now appear altogether unreasonable to suggest that cells of living organisms are nothing more than large and complex supramolecular assemblies. In Chapter 7 we will explore a little further the sense in which this is true. But what is it, precisely, that distinguishes the living from the merely complex? Most scientists agree that there are three fundamental properties that characterize living systems: metabolism, replication and regeneration. (Some scientists would include further criteria too: one might regard these three as necessary but perhaps not sufficient conditions.) Metabolism is the process of harvesting from the surrounding environment the substances needed for sustenance, energy and growth; for us, these substances come primarily in the form of carbohydrates. Replication is an obvious requirement – organisms must be able to reproduce and propagate. Multicelled organisms start their lives as a single cell, which must produce many replicas that "differentiate" at some stage to form the various bodily organs. And finally, regeneration – it is hard to see how any organism would get very far in the world if it were unable to repair itself when the need arose.

We've seen how the discovery by Watson and Crick of the structure of DNA provided the vital clue to understanding how an organism makes copies of its genetic information: each of the two complementary strands of the double helix can act as a template on which a new strand is constructed. This replication of genetic material must be carried out every time a cell divides, since each of the two new cells must have complete copies of the genome. The simplicity of the template concept should not, however, obscure the awesome magnitude of the task of replicating a strand of DNA; the process has to be very carefully orchestrated by enzymes called DNA synthetases. Nevertheless, this procedure is self-contained in the sense that the enzymes are themselves produced in the first place from the information encoded on DNA. In other words, the DNA molecule carries all the information needed for its own replication. This is a characteristic that any truly living system must possess. (We might note in passing, however, that it is by no means the case that all of the biomolecules within our bodies are nucleic acids or genetically encoded proteins.)

The "complementary template" idea is so simple in principle that chemists have been moved to wonder whether it might not be possible to apply the same idea to much less complicated chemical systems in order to develop models for studying replication. Julius Rebek, a chemist working at the Massachusetts Institute of Technology, created a considerable stir in 1989 when he described the synthesis of molecules that can generate copies of themselves from their component parts. Although at best these molecules satisfy just one of the requirements for a truly living system (and while it is not clear that they will be adaptable enough to acquire the others), Rebek felt entitled to claim that his molecules could be regarded as showing "a primitive sign of life."

An appealing aspect of Rebek's work is that, while his molecules share some properties in common with both nucleic acids and proteins, their replication operates according to novel kinds of molecular interaction rather than mimicking the complementary base pairing of nucleic acids. One could view this as an indication that perhaps DNA is not the *sine qua non* of life, so that one might conceive of organisms that "live" according to completely different molecular principles. The evolutionary biologist Richard Dawkins has suggested that Rebek's replicating molecules "raise the possibility of other worlds having a parallel evolution [to Earth's] but with a fundamentally different chemical basis."

Although the copying act of DNA has long been considered a remarkable feat, Rebek believes that replication in itself is not such a big deal. All it takes, he says, is a molecule with complementary parts that are prevented from binding to each other, say by the geometry of the molecule. To illustrate the general principle, consider a molecule that contains two complementary groups X and Y; these groups will lock together if allowed to do so. For example, X could be the adenine base found in nucleotides, and Y the thymine base to which it binds in the DNA double helix. If the part of the molecule linking X and Y is rigid, the X and Y groups on a single molecule may be unable to link up.

But if the X and Y groups are oriented in a suitable manner, it may be possible for a second, identical molecule to lock comfortably onto the first in an upside-down orientation, with its X group bound to the Y of the first and vice versa. Of course, this kind of pairing is not replication; if, however, we present the initial molecule not with an identical partner but with just the *unassembled components* of such a partner, these should lock into place so that the first molecule provides a template that holds the components together and makes it easier for them to join up into an identical copy. Suppose our XY molecules can be assembled by linking one half, containing the X group

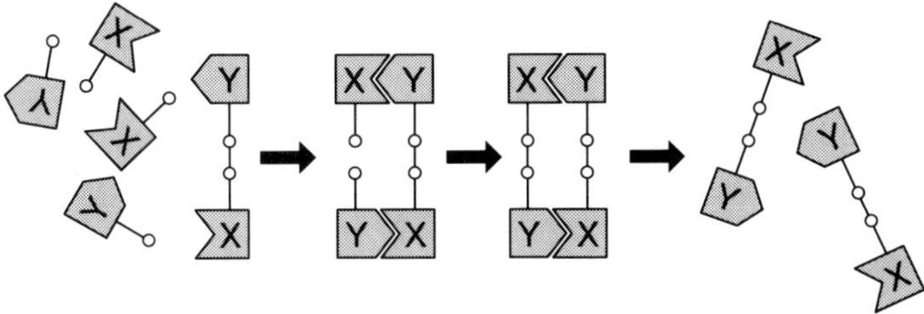

**Figure 5.35** *Molecular "copying" or self-replication is possible for molecules with complementary ends that are prevented from linking up (by the molecule's rigidity, for example). Here the X and Y groups are complementary. The XY molecule can act as a template on which X- and Y-containing fragments are assembled into a replica.*

and part of the connecting chain, with the other, containing the Y group and the rest of the chain (Figure 5.35). Then if we add these separate components to a solution of complete XY molecules, each component will bind to its complementary end on the XYs, bringing the two parts together so that the connecting chain can join up easily. The initial molecule is then assisting in its replication.

Of course, the constituent halves may be able to assemble themselves into complete XY molecules of their own accord, without help from the preformed XYs. But with an assembled molecule already present to act as a template, the assembly of the second molecule should be easier and therefore faster. The XY molecule can be said to catalyze its own synthesis. Such behavior represents an approximation to replication, since the molecule enhances the efficiency with which copies of itself are created.

To put these ideas into practice, Rebek synthesized the J-shaped molecule B in Figure 5.36, which has a highly reactive pentafluorophenyl ester group at the end of the long arm (shown schematically as a gray hexagon) and an imide group at the short end. To the latter he attached molecule A, which is similar to the adenosine nucleotide in DNA. The adenine base on this molecule forms hydrogen bonds with the imide, just as it does to thymine in DNA. The result is a U-shaped molecule, which Rebek denotes a "molecular cleft."

The two loose ends of the U lie side by side. The reactive ester group seizes the chance to react with the amine group on the other arm to form a peptide bond. But this linkage is severely kinked, generating a strained molecule like a young sapling bent over on itself. To relieve this strain the relatively weak hydrogen bonds between the imide and adenine break. The molecule swings open like a jacknife. It is this opened-up molecule ($R_1$ in Figure 5.36) that will supply the template for replication.

No doubt this sounds like a very complicated sequence of steps, but you needn't be too concerned about following them in detail. The critical consideration is the character of the molecule that we end up with. By forcing apart the bonds between complementary groups attached to the same molecule, Rebek succeeded in creating a molecule with complementary ends – a version of the XY molecule discussed above. If the opened-up molecular cleft is then mixed with its own unassembled building blocks, it will assemble a partner for itself, as shown in the final two steps of Figure 5.36.

The result of the template assembly process is two identical molecules clinging to each other by hydrogen bonds. They have a tendency to stay together, but in a dilute solution the molecules will eventually separate from each other, and both can then act as templates for the assembly of others. Adding just a small amount of the completed molecule to a solution of the uncombined components will therefore trigger replication until all of the components have been assembled.

Rebek suspects that it should be possible to design replicators from any components that exhibit specific mutual interactions, which is to say, components that recognize and bind to each other. In support of this hypothesis, he has created a second "species" of replicator which operates according to similar principles (Figure 5.37). In an exciting and particularly suggestive set of experiments, Rebek and colleagues investigated whether the two kind of replicator could be crossbred to form hybrids. This seemed

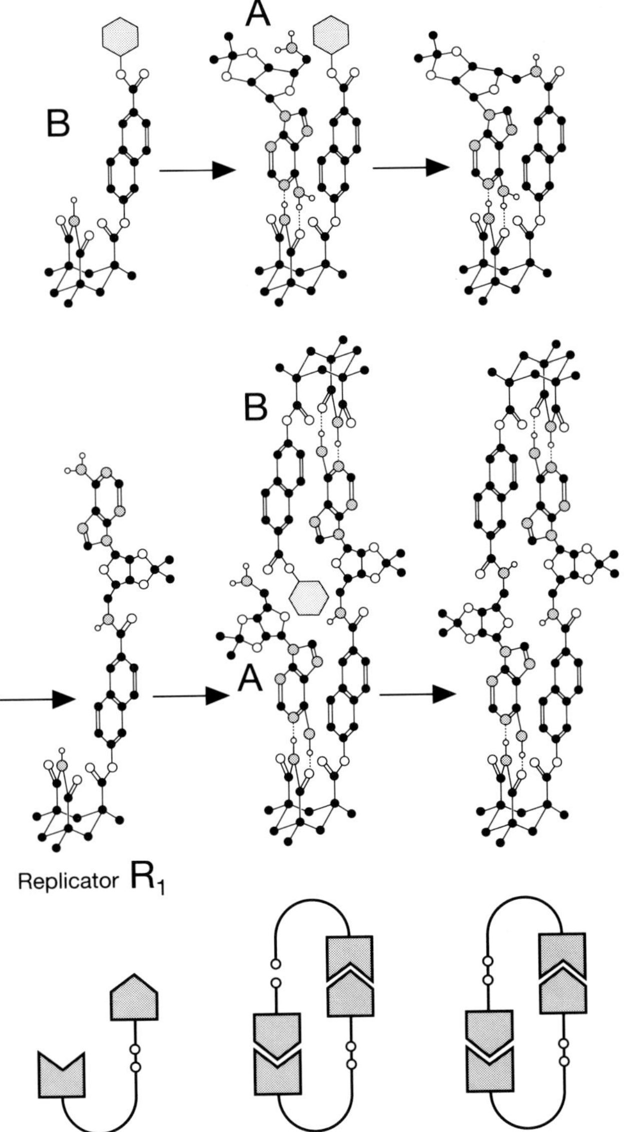


**Figure 5.36** *Julius Rebek's replicating molecule is based on two components A and B, which can bind together via hydrogen bonds. Once paired up, the two components form a covalent bond between their free ends, expelling the reactive pentafluorophenyl group (gray hexagon) in the process. The new bond is severely bent, and to relieve the strain, the composite molecule springs open, breaking the hydrogen bonds, to form the replicator* $R_1$*. This has complementary ends, each of which can bind a new fragment – the A end binds a B molecule, and vice versa. In this way, a new replicator is assembled on the template of the original one. The final three steps are shown in schematic form at the bottom.*

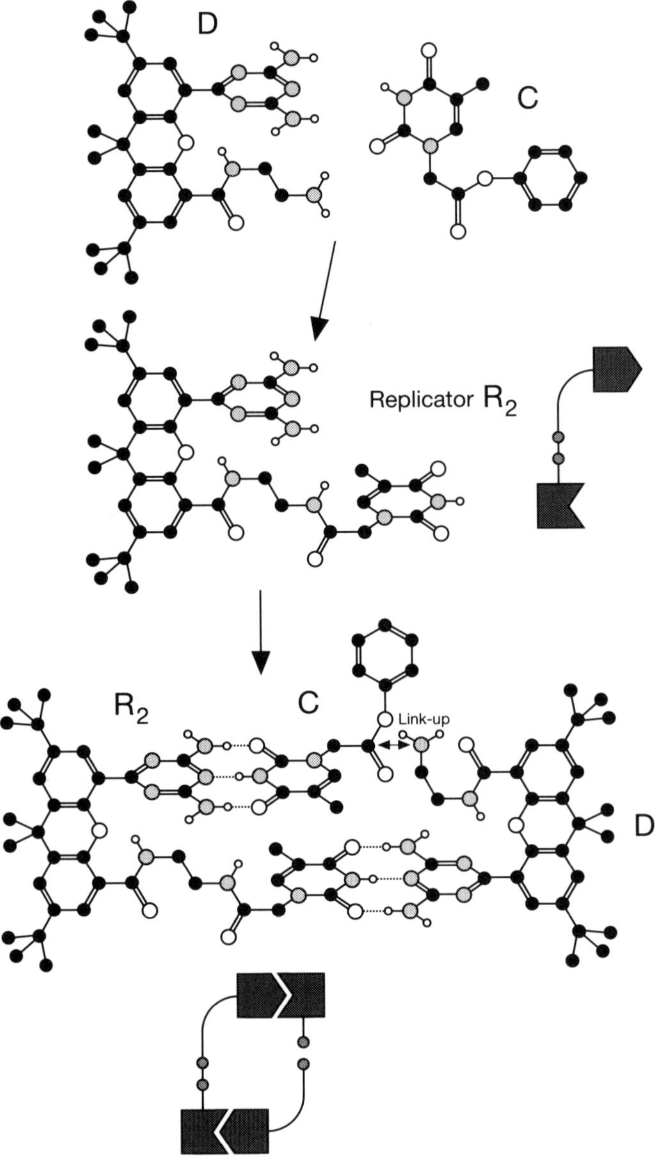


**Figure 5.37** *Rebek's second species of replicator employs different chemical interactions but the same principle of complementarity.*

possible in principle because the system of hydrogen bonds that holds together the complementary groups in the first kind of replicator is somewhat similar to that found in the second, so that the components of the two species should be able to interact with each other.

What Rebek did was to add the component parts of the first and second replicators to a solution containing both kinds of preassembled molecule. There are four possible products of such an experiment: both replicators will presumably assemble some copies of themselves, but hybrids might also be formed that comprise half of one and half of the other, of which there are two combinations (Figure 5.38). The hybrids may then act as

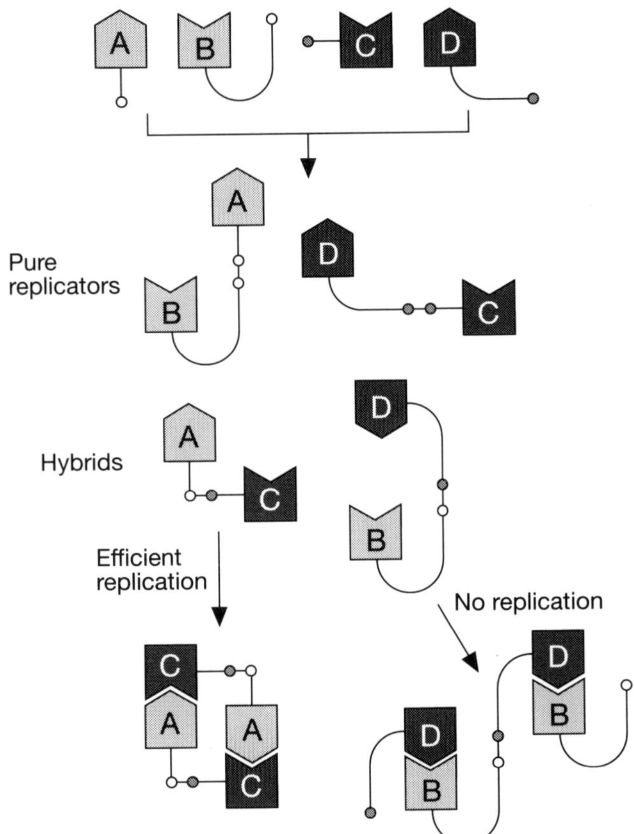


**Figure 5.38** *Hybrids of the two replicators $R_1$ and $R_2$ can be created by mixing their four components. One of the hybrids (A–C) is a better replicator than either of the "pure" species; the other is sterile, because its "C" shape prevents it from acting as a template. Instead, it twists into an "S" shape and binds the two complementary fragments to form a stable three-part complex.*

templates for their own replication. In effect, the test-tube becomes a battleground in which the four varieties compete with each other for the components that they need to reproduce.

Rebek did indeed find the two hybrids amongst the products of the molecular crossbreeding experiment, but they exhibited remarkably different properties. One hybrid was infertile, like a mule: it could not replicate. This is because it has an S-like shape, rather than the J shape of the "pure" replicators, preventing it from acting as a template for further assembly. The other hybrid, meanwhile, was a better replicator than either of the two pure varieties. If a mixture of the two pure replicators were to be fed continually with their component parts, this hybrid would soon take over in the battle for supremacy, appropriating all the parts in order to make copies of itself.

Rebek has been able to pursue further this idea of molecular "evolution" by making artificial replicators that can be mutated. He prepared versions of the first replicator $R_1$ to which were attached bulky substituents. Despite being lumbered with these impediments, the molecules were still able to catalyze their own replication as before. By shining ultraviolet light on the molecules, however, the bulky groups could be snipped off, mutating the replicators. (Mutations in DNA can similarly be induced by ultraviolet light.) The mutants act as templates for assembly of both the unmutated and the mutated varieties, and a competition ensues between the two replicators for the component parts that they need in order to replicate. Whichever of the two is the better replicator will eventually win out. In this instance, Rebek found that the UV-mutated replicators were the "fittest" for survival.

Molecular replication and competition is now being exploited in the creation of new variants on biological molecules which can function as drugs. Researchers are able to generate billions of mutant varieties of natural proteins or of molecules based on the nucleic acids DNA and RNA and to set these species competing against one another to fulfill the task that is required of the drug, thus fine-tuning the molecules' efficiency via chemical evolution. Such strategies are far less controlled and planned than Rebek's: they involve the use of the techniques of biotechnology (which employ enzymes rather than the simple principle of complementary templating) to create copies of the biomolecules. Subtle mutations are introduced at random, so that it is impossible to keep track of all of the mutant species produced. Those mutants that function best as drugs – for example, by binding efficiently a certain substrate – are separated and preserved; the rest of the "junk" mutants are thrown away. The capabilities of these proto-drugs are then further enhanced using them as starting materials for more mutants and repeating the cycle. A new breed of drugs based on DNA and RNA, rather than on proteins, is evolving in this way, since techniques to replicate and mutate the nucleic acids are well established in biotechnological research.

It may seem odd to be using the terminology of Darwinian evolution to describe chemical reactions. But it is highly likely that something just like natural selection did indeed take place amongst the primitive replicating molecules that most scientists believe must have presaged the appearance of life on Earth. It is hard to imagine how a process with anything like the complexity of DNA replication could have arisen simply from

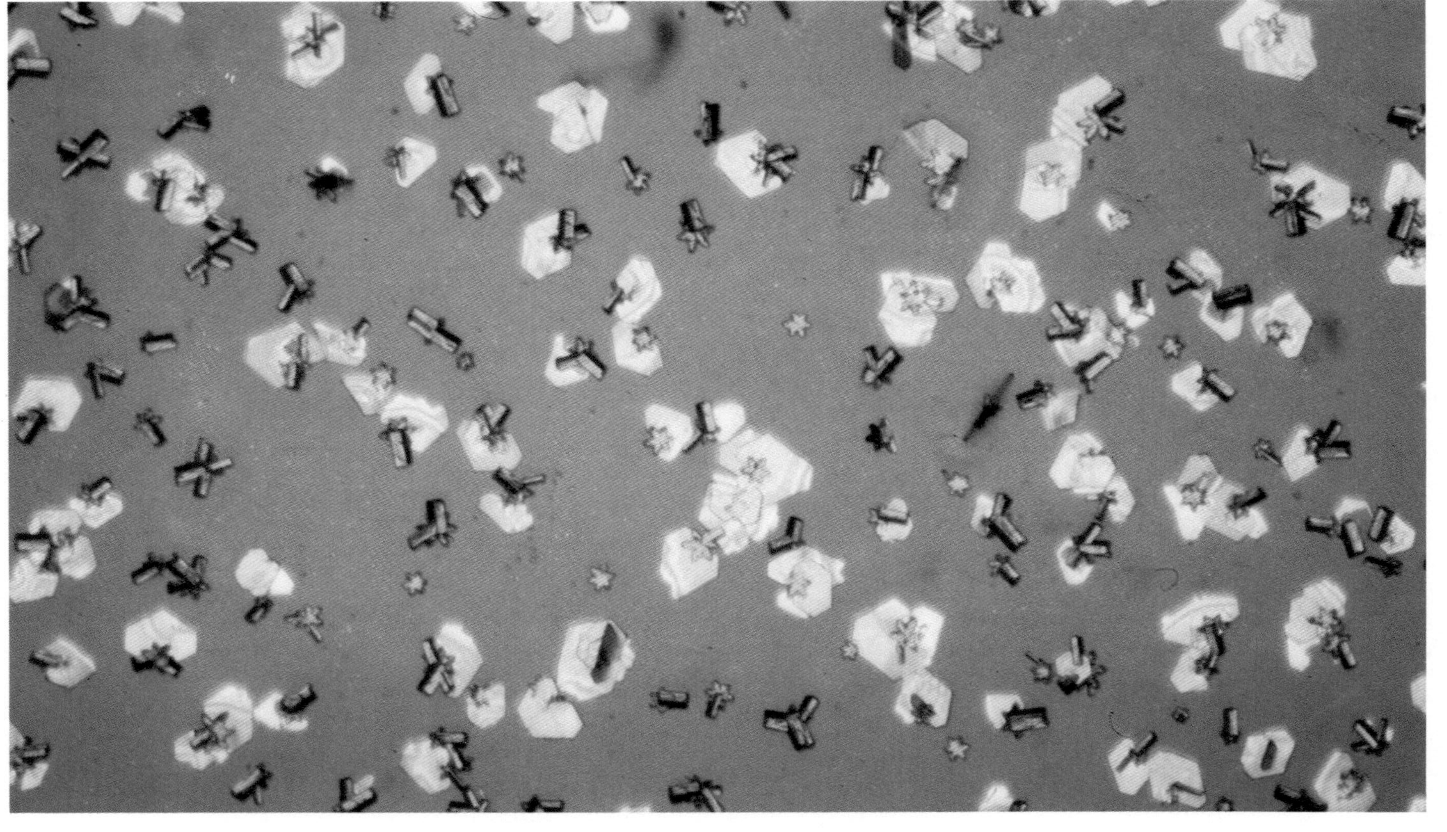

**Plate 1** *Fullerene crystals isolated and purified by Donald Huffman, Wolfgang Krätschmer and their students in 1990. They consist mainly of $C_{60}$, with about 10 per cent of $C_{70}$. (Picture courtesy of Donald Huffman, University of Arizona.)*

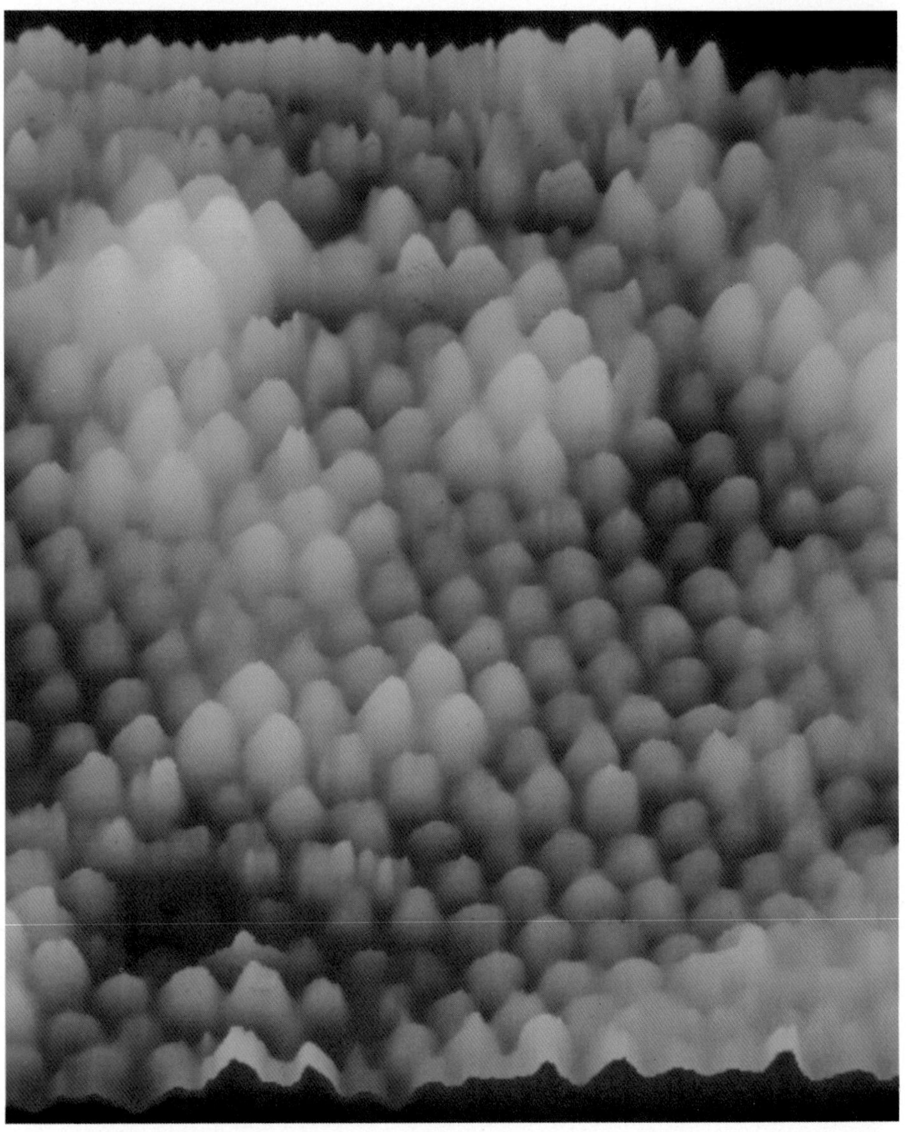

**Plate 2** *A layer of $C_{60}$ molecules on the surface of gold. The picture is obtained using a scanning tunneling microscope, which can reveal much finer detail than a light microscope. The spherical molecules appear as bright peaks; their hexagonal and pentagonal rings cannot be resolved here as the $C_{60}$ molecules are thought to be spinning rapidly. (Picture courtesy of Don Bethune, IBM Almaden Research Center, California.)*

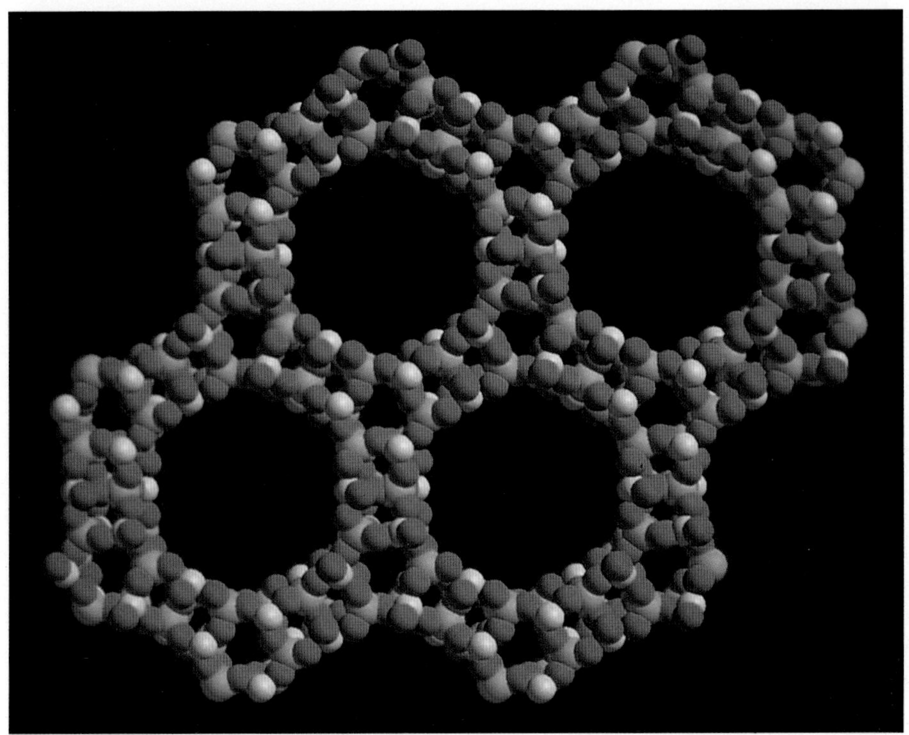

**Plate 3** *The aluminophosphate VPI-5, synthesized in 1988, has particularly large pores, the openings to which consist of rings of eighteen atoms. Here we are looking down onto the openings of the pores. Red atoms are oxygen, purple are aluminum and blue are phosphorus. (Picture courtesy of Mark Davis, California Institute of Technology.)*

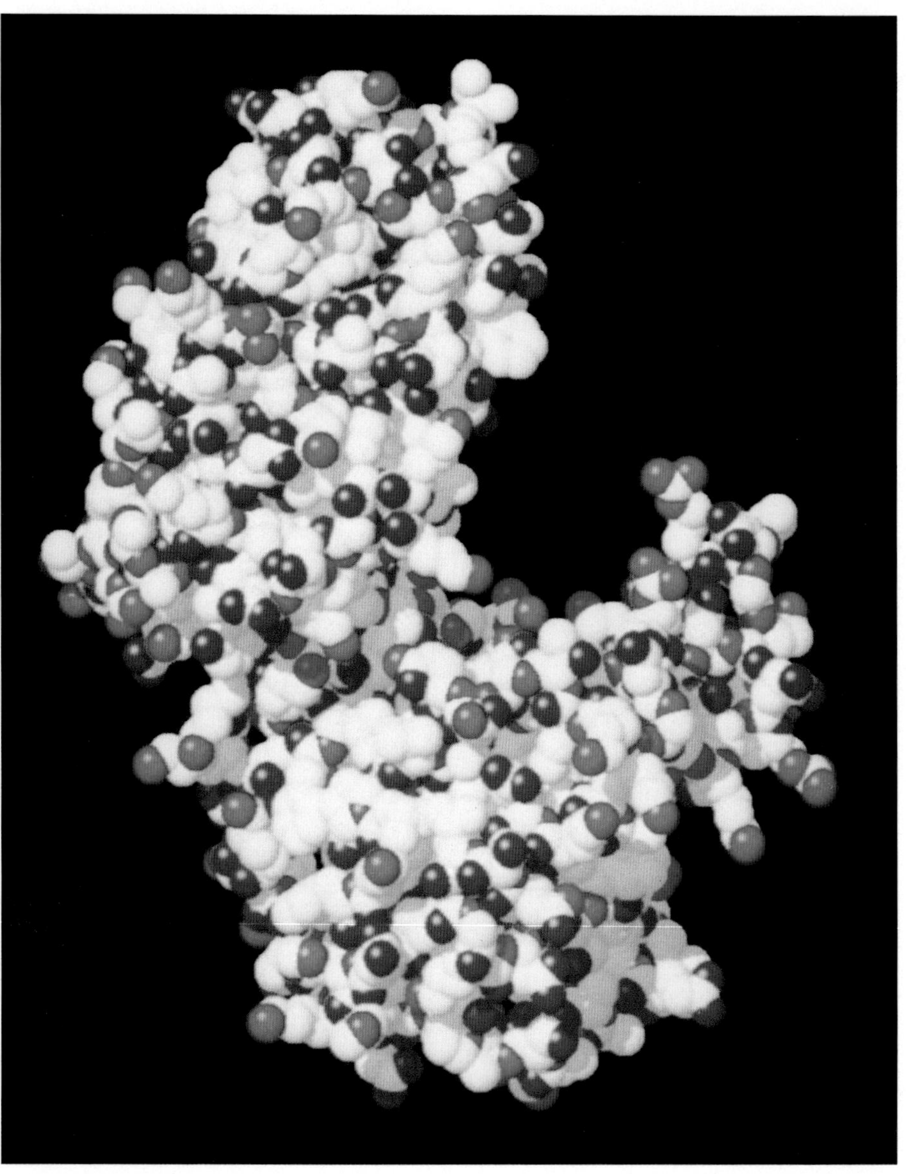

**Plate 4** *The enzyme phosphoglycerate kinase, which mediates a key reaction in the breakdown of glucose to provide metabolic energy. The reactant molecules are bound within the cleft, and the two lobes close around them to catalyze the reaction. Carbon atoms are shown in white, nitrogen in blue and oxygen in red. (Picture courtesy of David Goodsell, University of California at Los Angeles.)*

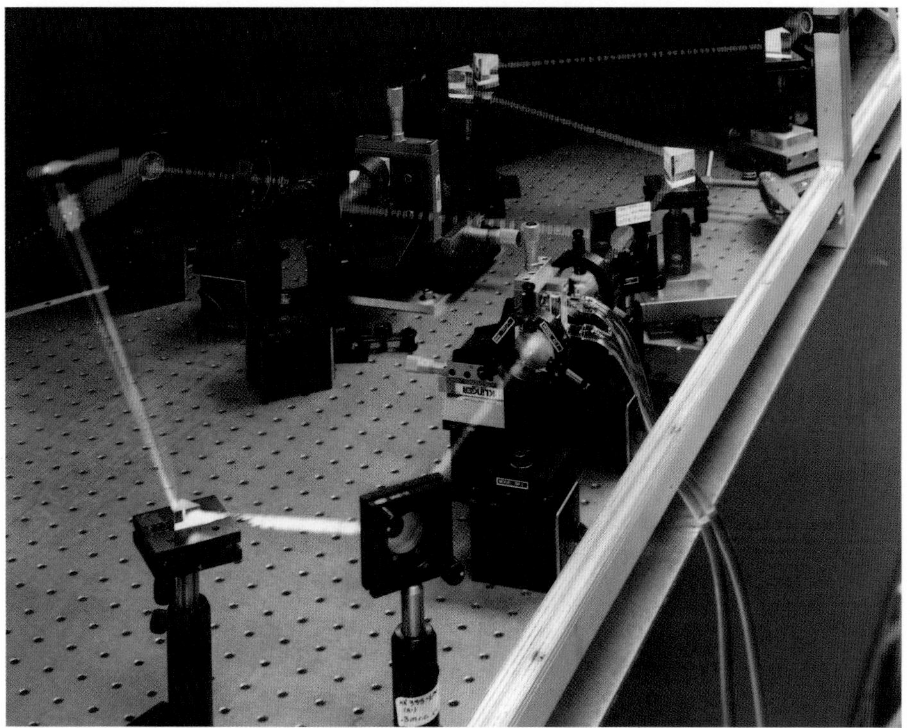

**Plate 5** *Ahmed Zewail's laser spectroscopy laboratory at the California Institute of Technology is a kaleidoscope of color. Some of these laser beams come in pulses so short that several million billion are emitted each second. (Photograph courtesy of Ahmed Zewail, California Institute of Technology.)*

**Plate 6** $K_{1.75}Pt(CN)_4$ *crystals grown from solution by an electrochemical process developed at the Xerox Corporation in Webster, New York.*

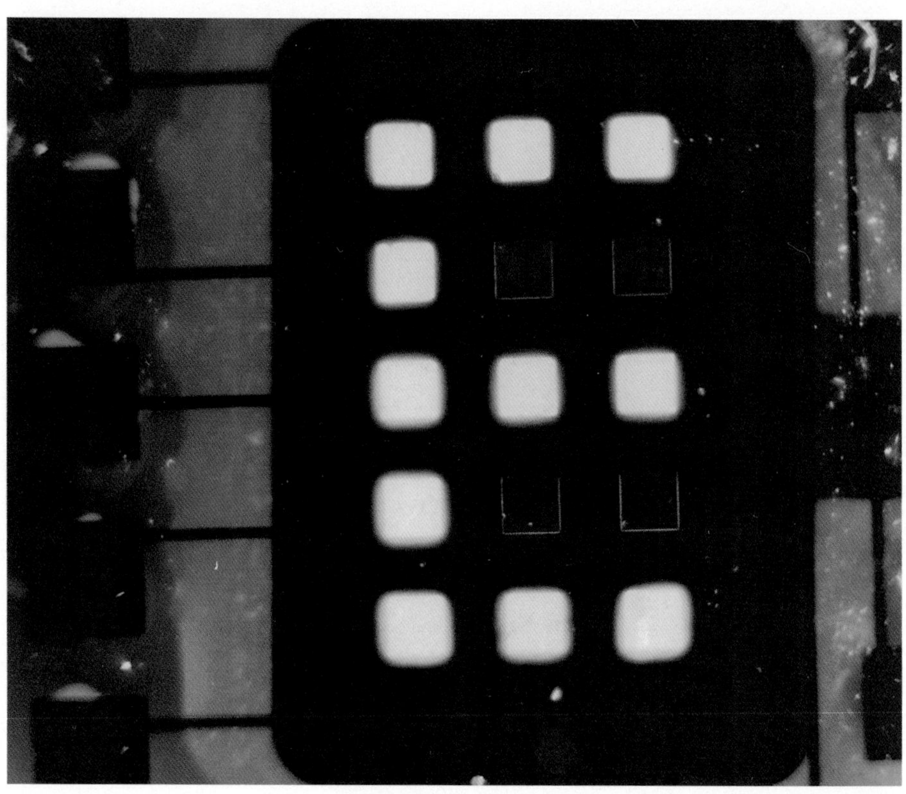

**Plate 7** *Plastic light-emitting diodes have been made from the polymer poly(paraphenylene vinylene) (PPV), which luminesces when electrons and holes are injected into it. The first polymer LEDs shone with yellow light, but tinkering with the chemical makeup of the polymer can change the color of the emitted light; red, orange, green and blue polymer LEDs have now also been made. (Photograph by the Photographic Department of the Chemistry Department, University of Cambridge.)*

**Plate 8** *Polyacetylene changes color with temperature. The lower, red part of the film shown here is kept cool by immersion in dry ice. The upper part of the film has turned blue on being warmed by a heating element. (Photograph courtesy of Richard Kaner, University of California at Los Angeles.)*

**Plate 9** *The structure of potassium-doped* $C_{60}$ *(*$K_3C_{60}$*), which is superconducting at 18 K. The* $C_{60}$ *balls are shown in blue, and the potassium ions in red and pink. (Picture courtesy of Richard Kaner, University of California at Los Angeles.)*

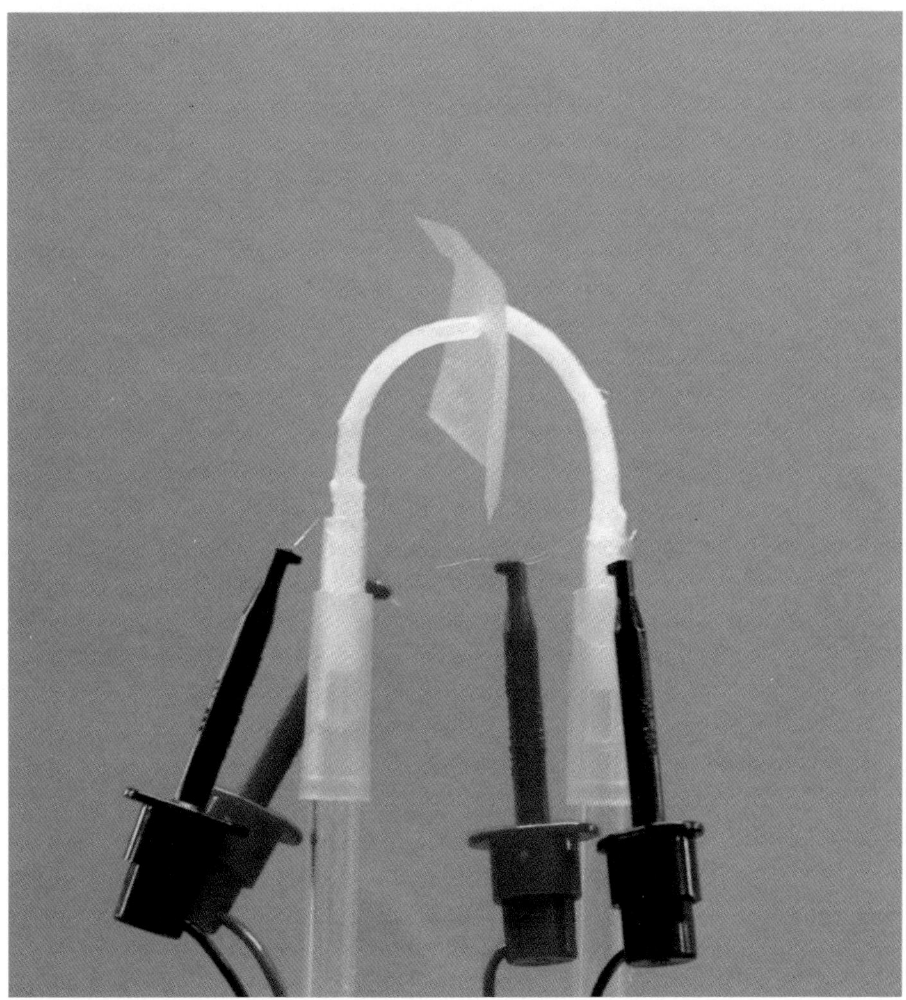

**Plate 10** *Polymer gels that change shape in electric fields have been used to create robot "fingers" that flex in response to electrical signals. (Photograph courtesy of K. Kajiwara, Kyoto Institute of Technology.)*

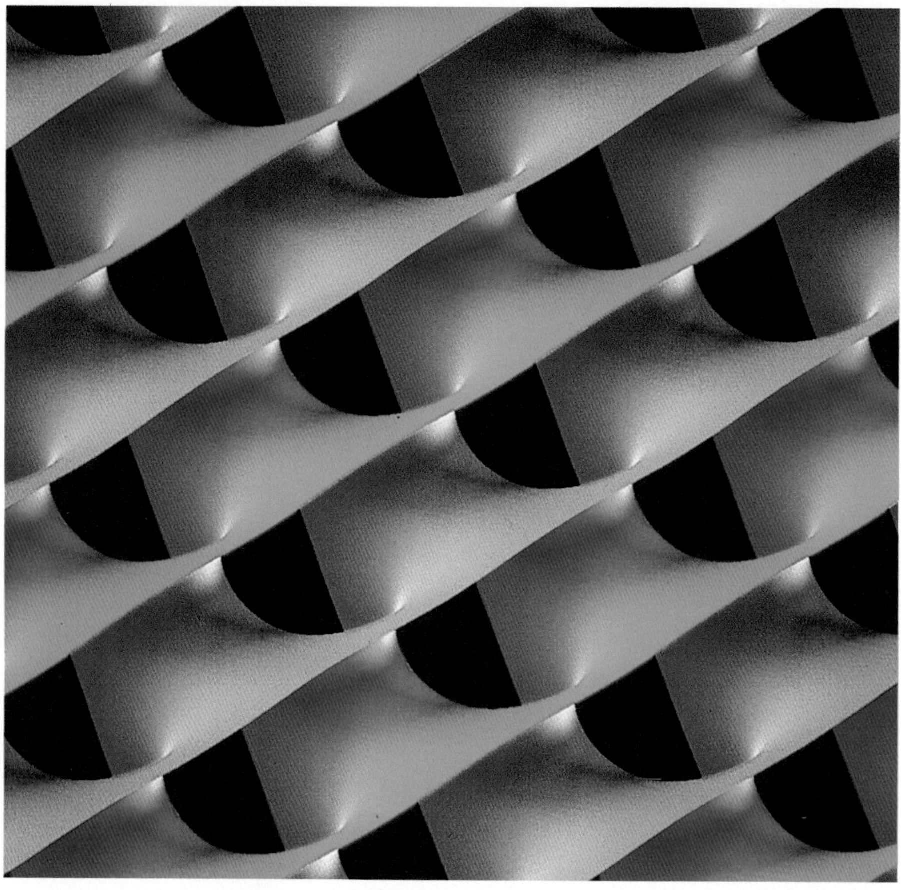

**Plate 11** *A subtle balance between the energy cost of surface area, curvature and stretching in surfactant films can sometimes give rise to ordered, periodic structures in microemulsions stabilized by a layer of surfactants. Shown here is a surface in which the area of the interface between two interpenetrating networks, like those in bicontinuous microemulsions, is as small as can be for a certain, fixed volume of the two phases. The structure is called Scherk's first minimal surface, and has been observed in a so-called diblock copolymer, a polymer comprising two chains joined end to end which do not mix with each other. The two chains separate into two interpenetrating networks, like those of oil and water in bicontinuous microemulsions, between which lies the minimal-surface boundary. This particular structure has not been observed for water/oil/surfactant mixtures, but other periodic minimal surfaces are seen in such systems, with the surfactants lying at the boundary. Periodic minimal surfaces are also formed by surfactant bilayers in water alone, driven by the need to minimize curvature. (Picture courtesy of David Hoffman, Geometry, Analysis, Numerics & Graphics Laboratory, University of Massachusetts at Amherst; generated by Jim Hoffman.)*

a

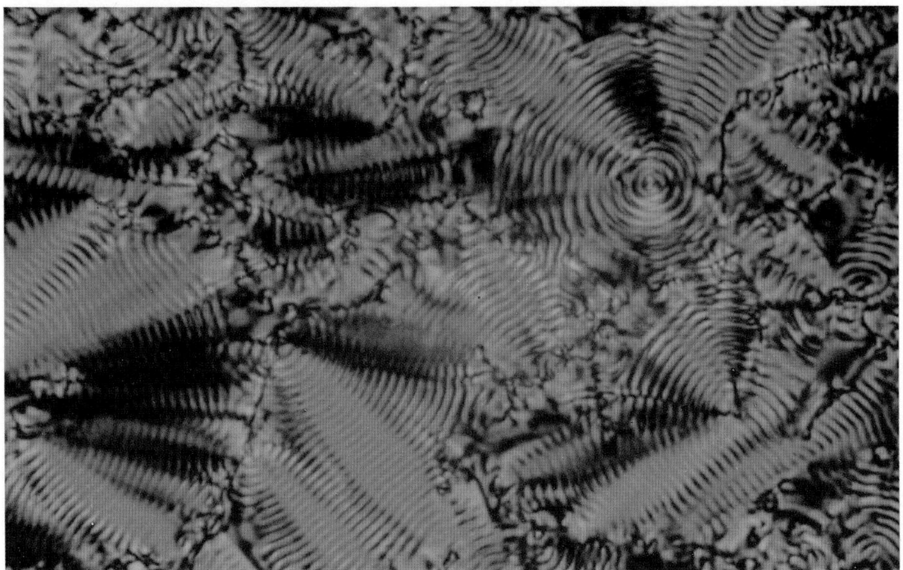

b

**Plate 12** *The birefringence of liquid crystals produces a variety of kaleidoscopic "textures." The liquid crystal in (a) is of a class known as ferroelectric, which is used widely in liquid-crystal displays. The image in (b) shows one of the textures generated by a so-called smectic A phase (see Figure 7.26). (Photographs courtesy of John Goodby, University of Hull.)*

**Plate 13** *A hydrothermal vent at a mid-ocean ridge, which pours out hot water rich in minerals and compounds such as methane and ammonium ions. (Photograph courtesy of Kyung Ryul Kim, Seoul National University.)*

a

b

c

**Plate 14** *Pattern formation in the Hele-Shaw cell, in which a liquid (here the clear liquid) is injected into a more dense fluid (colored) under pressure. Depending on the injection pressure and other factors, different bubble shapes can be observed that are reminiscent of those seen in electrodeposition. Shown here are DLA-like (a), dense-branching (b) and dendritic (c) growth. For the latter, some directional preference must be introduced, for example by scoring the face of one of the two plates with a lattice of grooves. The square lattice in this example generates a "snowflake" with fourfold symmetry. (Photographs courtesy of Eshel Ben-Jacob, Tel Aviv University.)*

**Plate 15** *Chemical waves in the oscillatory Belousov–Zhabotinsky reaction. Incomplete mixing causes local variations in the concentrations of the reactants, which serve as the sources of outward-radiating target- and spiral-shaped wave fronts. (Photograph courtesy of Stefan C. Müller, Max-Planck-Institut für Molekulare Physiologie, Dortmund.)*

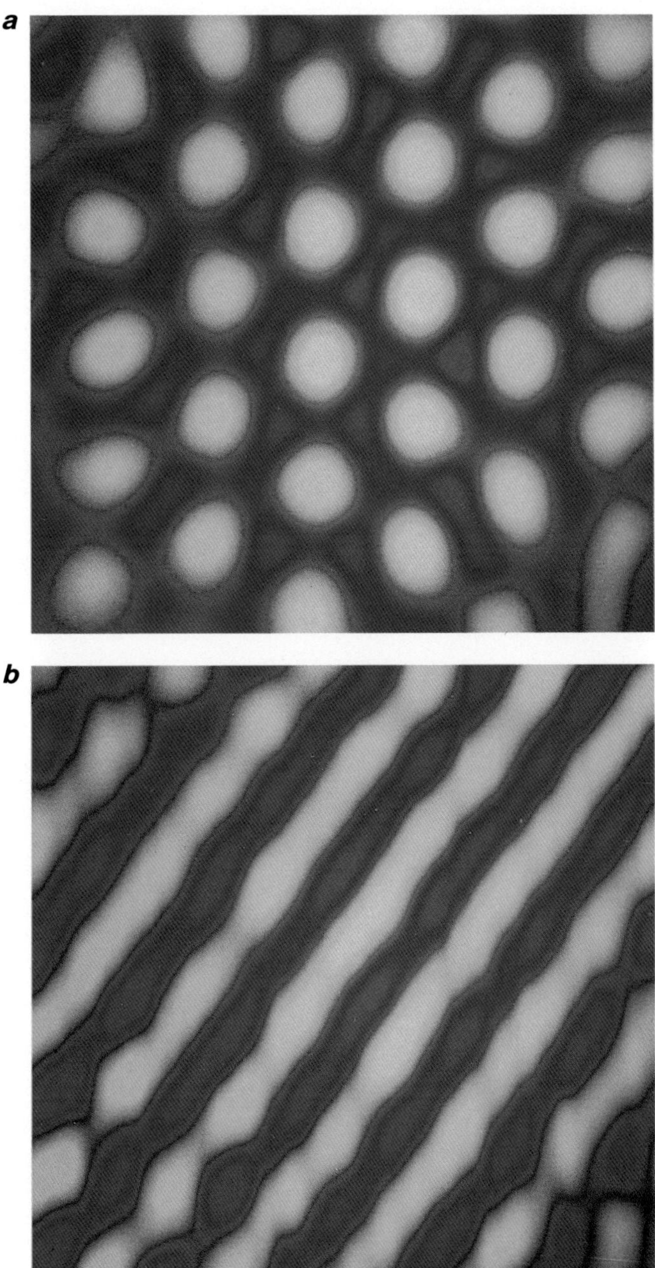

**Plate 16** *Two-dimensional arrays of stationary Turing structures can be generated in the CIMA reaction under appropriate conditions of temperature and reactant concentration. A hexagonal pattern at one temperature (a) changes to a stripe pattern at another (b). (Photographs courtesy of Harry Swinney, University of Texas at Austin.)*

**Plate 17** *Bubbles of ancient air within Antarctic ice provide samples of the atmosphere from past times. The deeper the ice is buried, the longer ago the air was captured. Analysis of the chemical composition of bubbles in cores drilled into the ice thus provide a record of the chemical history of the atmosphere. (Photograph courtesy of B. Stauffer, Universität Bern.)*

**Plate 18** *Polar stratospheric clouds are formed when the air temperature over the poles falls low enough for ice particles to condense. Some PSCs, like that photographed here off the coast of Norway, are composed of water ice; others contain a mixture of frozen water and nitric acid. The cloud particles provide catalytic surfaces for some of the important reactions involved in ozone destruction. (Photograph by NASA, kindly supplied by O. B. Toon, NASA Ames Research Center, California.)*

the chance coming-together of organic molecules on the early Earth; rather, the capabilities of the first replicating chemical systems must have been developed and fine-tuned in a molecular struggle for survival. This is something that I shall discuss in more detail in Chapter 8. The considerable excitement that has greeted Julius Rebek's work is inspired in part by the possibilities that it raises for exploring the kind of chemical processes that led to the appearance of life on our planet.

# 6

# Metals from Molecules

## Electronics goes organic

*Here's metal more attractive.*

*Hamlet* Act III Scene ii

A graduate student working for the Japanese chemist Hideki Shirakawa at the Tokyo Institute of Technology had made the kind of mistake to which inexperience makes one prone. His instruction had been to synthesize a polymer called polyacetylene, which is formed by linking acetylene ($C_2H_2$) molecules into long chains via a catalytic reaction. The normal product of this process is a dark powder; but to the student's dismay, what had materialized instead was a thin silvery film, like baking foil, coating the glass walls of the apparatus. When peeled away from the surface, the film was stretchable, like plastic wrapping. Eventually the hapless student identified his error: he had added one thousand times more catalyst than was specified in the recipe! Usually the result of this sort of mistake is a useless chemical gunk and a morning wasted in trying to clean it from the apparatus; at worst, it can mean the ruin of a great deal of expensive equipment. But in this case, the blunder helped to launch a new field of research.

Hydrocarbon polymers such as polythene (polyethylene) are generally excellent electrical insulators; indeed, this property, in combination with their cheapness, chemical stability and flexibility, makes them most useful as insulating coatings for electrical cables. But the silvery material produced by accident in Shirakawa's laboratory in the

1970s was a plastic that looked like a metal. Could it be that this strange plastic film would, like metals, conduct electricity?

In fact, this form of polyacetylene is not a very good electrical conductor at all. True, it does a whole lot better than an insulator like polyethylene, but it is not a patch on a regular metal like copper. Nevertheless, when Shirakawa made the silvery polymer more widely known, a few researchers considered it a sufficiently peculiar material to warrant further study. In 1976 two US-based chemists, Alan Heeger and Alan MacDiarmid, collaborated with Shirakawa in experiments that involved adding iodine to the plastic films. They found that, when "doped" with iodine, the material's color changed to golden, and the electrical conductivity increased phenomenally – by about a thousand million times.

Many polymers are now known to become electrically conducting when doped with other compounds. Some are as good a conductor as copper. Several of these conducting polymers consist, like polyacetylene, of chain-like hydrocarbon-based molecules; others contain additional or different elements, such as sulfur, nitrogen or phosphorus.

Other "organic metals" have been synthesized that are based not on huge polymer chains but on smaller, discrete organic molecules; as we shall see, however, the molecular structures that these materials adopt in the solid state bear important similarities to those of conducting polymers, and their metallic properties have similar origins. At very low temperatures, some of these organic metals can (like most conventional metals) become superconductors – materials that have no electrical resistance at all – while a few other organic compounds show magnetic properties similar to those of metals such as iron and nickel. Conducting polymers can be incorporated into electronic devices – polymer batteries are already available commercially, and devices such as diodes and light-emitting diodes (LEDs) are now starting to appear. For certain applications, it is likely that heavy, expensive metal cables may soon be replaced by cheap, lightweight wires of conducting plastic.

Organic conductors, superconductors and magnets represent the major focus of a young and burgeoning field of research called molecular electronics, which is concerned with devising and synthesizing chemical compounds with novel and potentially useful electronic properties. These materials represent a new breed of synthetic "designer metals," in comparison with which the traditional copper wire looks crude and cumbersome. While it remains to be seen whether conducting polymers will spawn a revolution in microelectronics to rival that generated by the silicon chip a few decades earlier, there can be no doubt that some very exciting prospects have sprung forth from a mistake in a Tokyo laboratory.

## Current understanding

### *What makes a metal?*

There is a joke about the man who tied a knot in a dangling electricity cable to stop the electricity from flowing out; but I have to confess to some sympathy for the poor fellow, who was after all only following his intuition about things that *flow*. To explain to him his misconception, we'd at least have to delve into a little circuit theory and the idea of electrical resistance. If we wanted to be more rigorous, we'd first have to be clear about what it really is that flows along an electrical cable, and what is peculiar about a metal that makes this flow possible.

A material may be a conductor of electricity if it contains mobile, electrically charged particles. Pure water is able to conduct weakly because it contains small concentrations of charged ions such as $H_3O^+$ and $OH^-$. Molten ionic salts such as sodium chloride will also conduct a current, and there exist some crystalline solids, such as silver iodide, in which a few of the ions are mobile enough to make the material reasonably conducting. But by far the most common carrier of electrical charge in conducting solids is the electron. (Note that, while it is common to speak of the flow of current, this is technically incorrect: it is charge that flows, and the current corresponds to the flow of charge.)

Of course, both a metal and an insulating material, such as wood or rubber, contain electrons. Yet in a conductor some of the electrons are free to move throughout the bulk material, whereas in an insulator they are not. Most solid metals have a crystalline structure comprising stacks of metal atoms packed together in a regular, periodic way. An insulator with a much simpler molecular structure than wood or rubber is diamond, which is also a crystal, consisting of a periodic array of carbon atoms. Why does crystalline copper conduct while crystalline carbon does not? The simplistic answer is that in diamond, neighboring carbon atoms form *localized* covalent bonds in which the pairs of electrons are trapped, whereas in copper the bonding is communal and *delocalized*: every atom donates its bonding electrons to a general "sea" that pervades the solid, through which individual electrons can drift freely. This picture reflects the way in which the two elements tend to behave in chemical compounds: carbon forms covalent bonds whereas metals tend simply to cast off electrons to form positive ions.

We are used to the fact that an electrical signal passes through a wire almost instantaneously, but this does not mean that the electrons themselves travel this fast; in fact, they drift rather langorously through the lattice of ions at speeds of typically less than a millimeter per second. Yet it is one of the perhaps surprising results of the quantum theory of bonding in metals that if the stacking of ions in the metal crystal is perfectly periodic, the moving electrons do not "see" them at all and can pass through the metal unchecked. But real crystals are never so perfect. For a start, the ions inevitably have some thermal energy and so vibrate about their equilibrium positions, temporarily distorting the regularity. There will also inevitably be faults or "defects" in the stacking arrangement, and perhaps impurity atoms of other elements. There is some chance that the electrons will collide with vibrating ions or defects, which will scatter them from their path and disrupt the flow of charge. These interactions are the source of electrical resistance. The greater the scattering of electrons (in other words, the greater the density of "scattering centers" in the metal), the higher the resistance. When an electron is scattered it loses some of its energy, which goes towards heating up the metal. This heating can be sufficient, as in the tungsten filament of a light bulb, to make the metal glow.

If the sole criterion for good electrical conductivity is that electrons be free to move between many atoms rather than being constrained within molecule-type localized bonds, we can immediately make some suggestions as to how polyacetylene acquires its metallic properties. Chapter 1 explained how the alternating single and double bonds in the six-membered carbon ring of the benzene molecule lead to the formation of ring-like orbitals above and below the plane of the molecule, in which the electrons are delocalized. By placing benzene molecules in a magnetic field, these electrons can actually be made to circulate around the ring in a "ring current." But polyacetylene is nothing other than a series of benzene rings cut open and linked end to end, for it consists of long carbon chains with alternating single and double bonds (Figure 6.1). This sequence allows the pi orbitals of the double bonds to overlap in a continuous fashion, giving rise to snake-like molecular orbitals that run along the carbon backbone. This type of bonding is called "conjugated." The electrons in the conjugated orbitals should be able to move along them like trains on a track; they are, in other words, delocalized along the polymer chain. According to this simple picture, a voltage placed across the two ends of the chain should turn it into a tiny wire down which the electrons can flow.

These electrons are, however, free to move in just one direction – along the polymer chain – whereas in a metal one direction is as good as any other. A consequence of this is that if the polymer is stretched (or specially prepared in some other way) so that all the chains line up, it will conduct electricity much more readily in the direction of chain alignment than in others (indeed, it can become virtually insulating in directions perpendicular to the chains). Another consequence is that the conductivity of polymers is very sensitive to the presence of defects in the molecular structure such as breaks in the chains. Defects in metals don't have a pronounced effect unless they are present in extremely large numbers, because it is relatively easy for electrons to find a way around them. In a linear polymer chain, however, that is not possible – if the "track" is fractured, the electron can't pass.

This picture of electrons travelling in delocalized, conjugated bonds along the polymer backbones is good enough to get us over the possible surprise of finding electrical conductivity in plastic materials. But it cannot be the whole story. If it was, we should expect polyacetylene itself to be a good conductor, whereas Heeger and MacDiarmid were able to obtain good conductivity only after doping the material

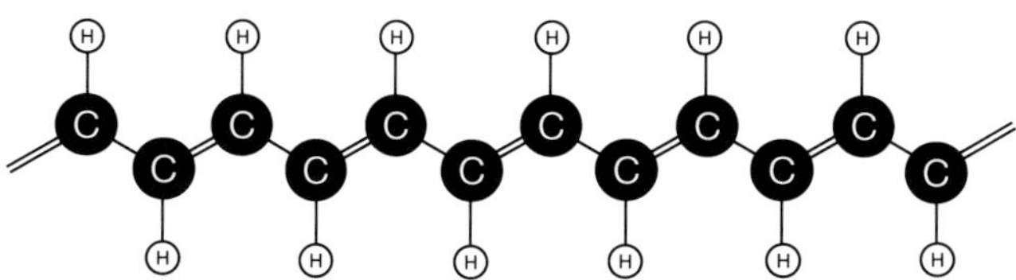


**Figure 6.1** *Polyacetylene is a hydrocarbon polymer in which single and double bonds alternate along the chain. The pi orbitals of adjacent double bonds overlap, forming continuous conjugated orbitals along which electrons can pass.*

with iodine. And then there is the question of how the electrons get from chain to chain (since the polymer molecules will not be long enough to span the whole of a typical sample). More generally, it is not hard to see that arguments based on localized versus delocalized bonding do not constitute a foolproof criterion for deciding whether a substance will conduct or not. Take diamond, for instance. It is the paradigm of a solid held together by strong, localized bonds, yet the addition of small amounts of dopants such as boron or phosphorus can induce conductivities sufficient to make the material a semiconductor, like silicon. Surely all those bonds do not suddenly become delocalized because of the presence of a few foreign atoms? To understand these observations requires a more careful consideration of what bonding is really like in solids.

## *From bonds to bands*

Electrons in a single atom are constrained to spend their time in orbitals close to the nucleus. In molecules those electrons involved in bonding gain a measure of freedom, being able to roam between two or more nuclei. We saw in Chapter 1 that one can view the process of covalent bond formation as the overlapping of atomic orbitals to produce molecular orbitals – bonding molecular orbitals, in which the energy of an electron is lowered relative to that in the isolated atoms, and antibonding orbitals, in which an electron's energy is raised. We can then picture a solid – any solid, be it a metal, a semiconductor or an insulator – as nothing more than a vast, extended molecule containing billions upon billions of atoms, bound together by the overlap of neighboring atomic orbitals. If we consider building such a solid one atom at a time, we can see what a molecular-orbital description implies for the bonding properties of bulk materials.

Let's stick with a simple example – a one-dimensional solid consisting of a single row of atoms (Figure 6.2). Consider constructing this solid from scratch. First we join together two atoms via a single bond. This involves the overlapping of one atomic orbital on each atom, so it produces one bonding orbital and one antibonding. Each time a further atom is added to the row, one more molecular orbital is added to the set. Before the row has grown very long, the orbitals have energies so close together that one can hardly distinguish individual energy levels at all. By the time the row runs to millions of atoms, the energy levels are indeed no longer discrete, but merge into a continuous band of allowed electron energies (Figure 6.2). More generally, bonds in a three-dimensional solid are formed by the overlap of several orbitals on each atom, creating a set of energy bands. Electrons in these bands can possess energies lying anywhere between the upper and lower limits of the band, but not outside of this range. There are therefore gaps of forbidden energies separating the bands.

In more familiar three-dimensional solids the shapes of the extended orbitals corresponding to these bands are not easy to describe. "Orbitals" is in fact hardly an appropriate term any more, since they effectively encompass every atom within the solid; maybe it is better to imagine them as a three-dimensional network of passageways between atoms, through which the electrons can travel. In some parts of these "orbitals" the passageways may get very narrow, perhaps confining the electrons to specific layers

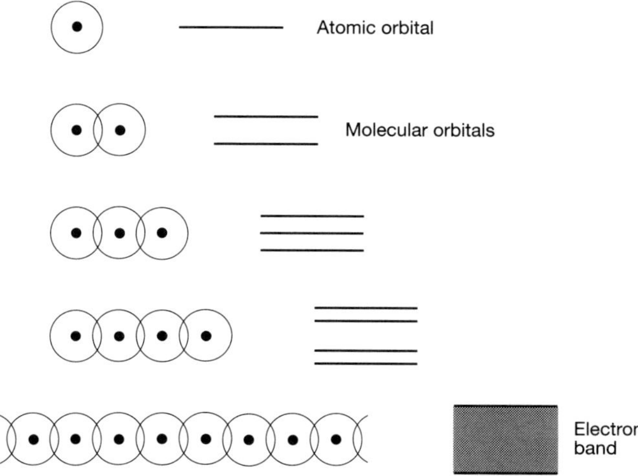


**Figure 6.2** *As atoms are assembled one by one to form a one-dimensional solid, the molecular orbitals become so closely spaced that they merge into continuous electronic bands.*

or rows of atoms, or even forcing them to remain within cavities around individual atoms as if they were still in atomic orbitals. But these pictures should not be taken too literally; physicists have to use special kinds of visualization to depict these structures accurately. They speak – and so shall I – of electrons "filling bands," by which they mean filling the vast, continuous "orbitals" that correspond to these energy bands. Henceforth I intend the word "band" to be synonymous with an extended electron orbital in a solid.

In an insulator such as diamond, a semiconductor like silicon or germanium, and a metal such as copper, the electron bands extend throughout the crystalline lattice. Yet some of the electrons can travel more or less freely through the metal but not through the former two materials. Why is this so? The crucial factor is how many electrons the band contains. The bonding and antibonding molecular orbitals formed by the overlap of one orbital on each of two atoms have an electron capacity equal to that of the constituent atomic orbitals, i.e. four. In the same way, the capacity of bands created by the overlap of atomic orbitals in a solid is the same as the total capacity of all the individual orbitals that constitute them. For example, the overlap of the four second-shell (2s and 2p) orbitals on each atom in diamond gives rise to two bands, equivalent to the sets of bonding and antibonding molecular orbitals, separated by an energy gap. Together these bands can accommodate a maximum of eight electrons per atom: the lower band has room for four electrons per atom, and so does the upper.

The degree to which a band is filled determines the mobility of its electrons. Picture the electrons in the band as billiard balls on a billiard table. Under normal circumstances, the balls can roll around freely on the table top (Figure 6.3*a*). This corresponds to the situation in a partly filled electron band: the electrons are mobile and so the solid can

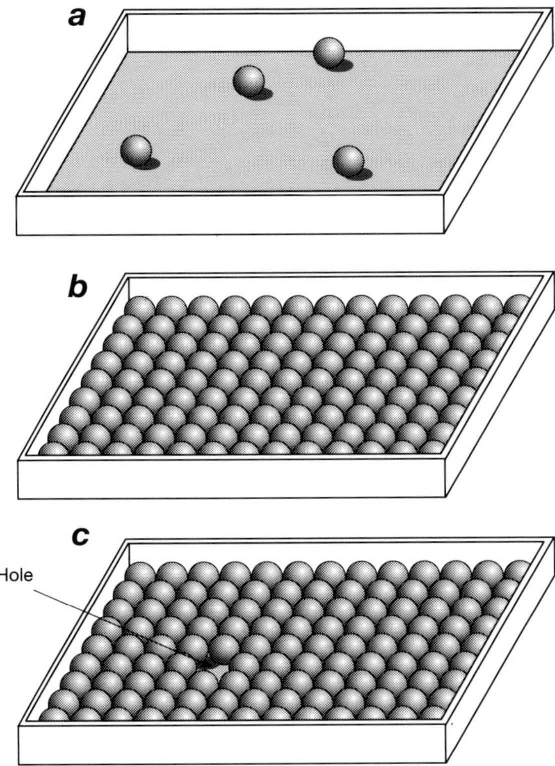


**Figure 6.3** *The mobility of electrons in electronic bands is analogous to that of billiard balls on a table. When there are just a few balls, they are free to roam (a); as the table gets increasingly full, the mobility decreases, and when it is packed completely, the balls are "frozen" into place (b). If a ball is removed, the hole that is left behind can move from place to place by a reshuffling of the balls; in effect, it acts as if it were an "antiball" on an otherwise empty table (c).*

conduct electricity. By definition, therefore, metals have at least one partially filled band (Figure 6.4*a*).

Adding more electrons to a band is equivalent to placing more balls on the billiard table. As we do this, it becomes increasingly unlikely that a given ball will be able to travel any appreciable distance over the table without encountering another and being scattered; in other words, it becomes ever harder for the balls to move around unhindered. Finally we reach the stage at which the entire surface of the table is covered with balls (Figure 6.3*b*). The maximum capacity is reached when the balls are packed together in a regular array. Then it becomes impossible for any of the balls to move; they are rendered immobile by the proximity of their neighbors. So it is for electrons: even if an electron band is continuous throughout the solid, the electrons cannot move thought it once the band is full, so there is then no conductivity. In an insulator such

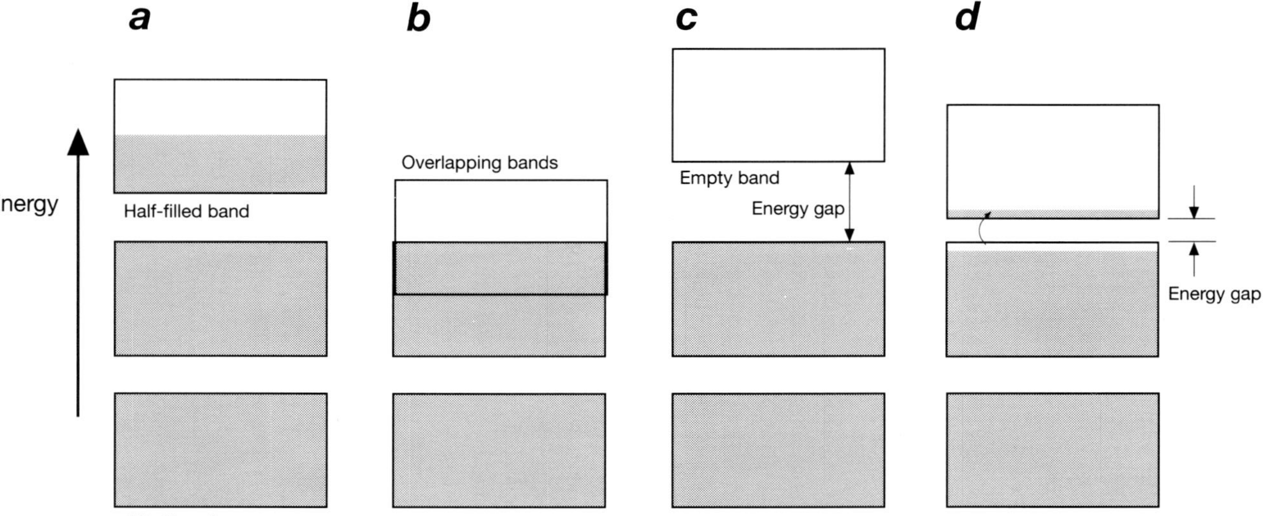


**Figure 6.4** *In a metal, the electronic band of highest energy is incompletely filled by electrons, and so these electrons are mobile (a). Some elements, such as the alkaline earth metals, are rendered metallic by the overlap of a filled and an empty band (b). In an insulator such as diamond, the uppermost occupied band is full, and the next empty band is separated by an energy gap (c). In a semiconductor such as silicon, the highest occupied band is again full, but the energy gap between this and the lowest unoccupied band is small enough for some of the electrons to jump up to it by acquiring thermal energy. These "conduction" electrons are then free to act as mobile carriers of electrical charge.*

as diamond, all the bands are completely filled up to a certain energy (Figure 6.4*c*). Completely empty bands lie at still higher energies, but they are not accessible to the electrons because of the forbidden energy gap separating the bands.

In the alkaline earth metals, the highest occupied bands are made up from the atoms' fully filled outer s orbitals, and one might therefore expect the uppermost band also to be fully filled, making the solids insulators. Yet this band turns out to be only partly filled, containing mobile electrons. What has happened is that the upper bands have become so broad that the gap between the filled s band and the empty p band, lying higher in energy, has disappeared: the two bands merge into one (Figure 6.4*b*). This merging of bands comes to the rescue of several metals, including copper, silver and gold, that might otherwise end up as insulators.

## *Almost metal*

In between the extremes of a conductor and an insulator there lies that curious yet immensely useful class of material, the semiconductor. These substances have conductivities far below those of metals but nevertheless thousands of times greater than insulators. Their technological utility derives from the way in which their conductivity can be changed and controlled by modifying the chemical composition. Semiconductors are generally materials that should, according to the criteria above, be insulators, yet which somehow find a means to obtain a few free, mobile electrons.

The classic example, silicon, has uppermost bands that are constructed by the overlap of the 3s and three 3p orbitals on each atom. Silicon has the same number of electrons per atom, and thus the same solid-state electronic configuration, as diamond: a completely filled upper band separated from the next (empty) band by an energy gap. But silicon has a considerably smaller gap between the two bands than does diamond (Figure 6.4*d*). The difference in energies is of the same kind of magnitude as the thermal energies that the electrons possess at room temperature, so electrons near the top of the filled band can occasionally pick up enough thermal energy to make the jump up to the unoccupied band, where they then have the run of the crystal. These electrons leave behind "holes" in the occupied band, into which the remaining electrons in this band can move. To return to the analogy of a billiard table covered with balls: when one ball is removed (Figure 6.3*c*), another can slide into the hole, in turn leaving a hole behind *it*. In this way the holes can be shuffled about from place to place, traveling across the surface as if they were themselves phantom balls on an otherwise empty table. The holes can be eliminated only by replacing the extracted balls. So when electrons in a semiconductor are excited up into the empty band, there is a double boost for the conductivity: not only are there now mobile electrons in the upper band, but also mobile holes in the previously filled band. Rather than trying to describe the motions of all the shuffling electrons around a hole, physicists find it easier to treat the holes as if they were a kind of charge-carrying particle themselves, a sort of "inverse electron" – like an electron but with a single positive charge rather than a negative one.

The conductivity of a semiconductor is determined by the number of electrons that can make the jump across the band gap. Raising the temperature provides the electrons with more thermal energy to boost them across the gap, so a semiconductor's conductivity increases as it gets warmer. This contrasts with the behavior of metals: for them, the conduction electrons require no thermal stimulation to become mobile, and the effect of raising the temperature merely makes the atoms in the crystal vibrate more vigorously. These vibrations degrade the conductivity, since they make the atoms seem bigger and therefore more likely to scatter moving electrons. Consequently the conductivity of metals decreases as the temperature is raised. In practice it is this temperature-dependence of the conductivity, rather than its absolute magnitude, that distinguishes a semiconductor from a metal.

The conductivity of semiconducting materials can be enhanced by doping – by adding to them foreign "dopant" atoms. Their effect is to increase the number of electrons (or holes) available to carry a current, either by adding electrons to the empty band (called the conduction band) or by making it easier for them to jump out of the filled (valence) band, thereby adding holes to this band.

The conductivity of silicon can be enhanced by doping it with boron or phosphorus atoms. These are incorporated into the crystal lattice in sites that would normally be occupied by silicon atoms. But a boron atom has one electron fewer in its outer shell

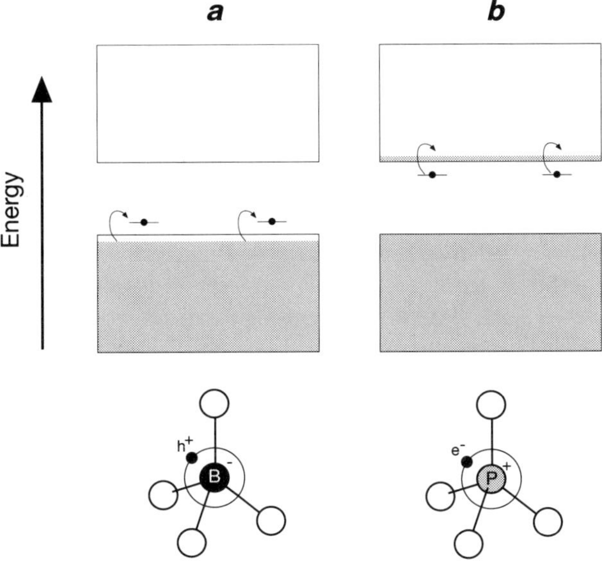


**Figure 6.5** *The conductivity of silicon can be increased by doping it with foreign atoms such as boron or phosphorus. A boron atom in the silicon lattice creates a hole ($h^+$) in the valence band, which can act as a positive charge carrier (a). A phosphorus atom, meanwhile, can place an electron ($e^-$) in the empty conduction band, which acts as a negative charge carrier (b). These dopant atoms can be visualized as hydrogen-like atoms in which a single charge carrier orbits a "nucleus" of opposite charge; these atoms can be ionized relatively easily to free the charge carrier.*

than silicon, and it therefore produces a deficiency of electrons in the valence band: in other words, a hole appears in the valence band in the vicinity of the boron atom. At low temperatures, this hole stays stuck to the boron atom, but it takes only a relatively small amount of energy to make the hole mobile. In effect, the boron dopant atoms add empty energy levels that lie just above the filled valence band. A small thermal kick will suffice to allow the electrons to jump up to these levels (since they are much closer than is the conduction band), leaving an incompletely filled valence band (Figure 6.5*a*). Because the charge carriers in this case are positively charged holes, this type of doping is called p-type.

Phosphorus atoms have one electron more than silicon in their outer shell, so a phosphorus atom in the silicon crystal has an electron left over once it has formed all the bonds that would be required of a silicon atom in the same position. Again this electron remains bound to the phosphorus at very low temperatures, but a small amount of thermal excitation will make it a mobile charge carrier. The effect of the phosphorus dopant atoms on the band structure is to add electron-containing energy levels at energies just below that of the conduction band, so these electrons can easily jump up into that band (Figure 6.5*b*). As negatively charged electrons are responsible for the current in this case, the doping is called n-type.

## Forging synthetic metals

### *Plastic bands*

We could describe the electronic structure of polymers in terms of bands formed by the overlap of orbitals on individual atoms. More usefully, however, the bands can be viewed as being comprised of overlapping *molecular* orbitals on the polymer molecules. That is, first the atomic orbitals overlap to form discrete molecular orbitals on each polymer molecule, and then the molecular orbitals of adjacent molecules overlap to form continuous bands. In polyacetylene, the uppermost bands are comprised of the conjugated pi bonds that run along the polymer backbones. These bonding orbitals are

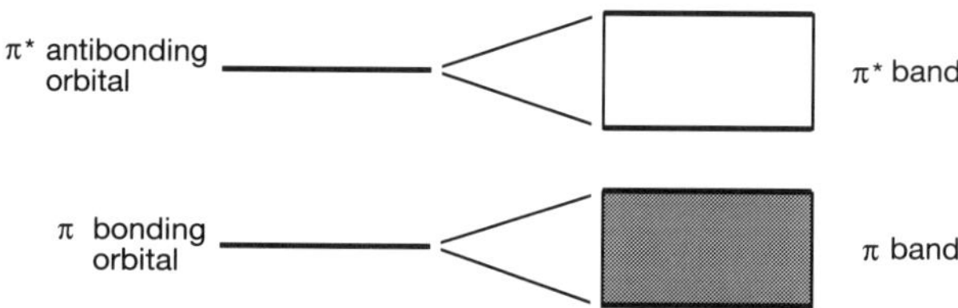


**Figure 6.6** *In polyacetylene, the electronic bands are comprised of overlapping molecular pi (π) orbitals. The highest occupied band is formed from the bonding molecular orbitals, and is filled; but the energy gap between this and the next, empty band formed from antibonding (π*) orbitals is small enough to make the material a (poor) semiconductor.*

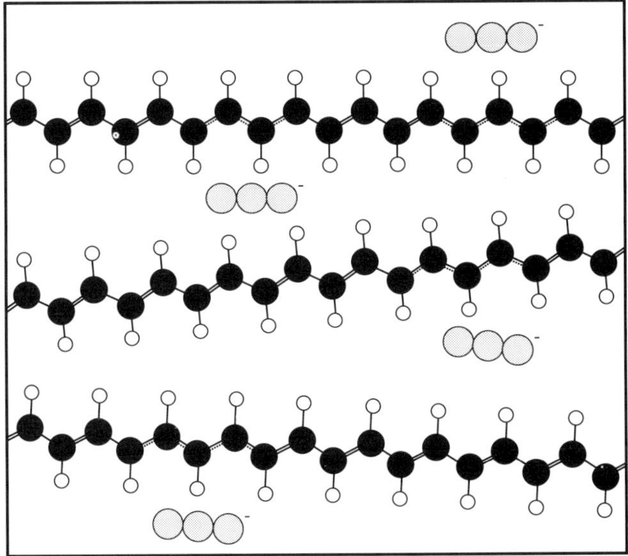

**Figure 6.7** *Dopants in polyacetylene sit between the hydrocarbon chains. They either inject electrons into, or extract them from, the delocalized pi orbitals, creating localized regions of charge ("excitons") on the polymer backbone. In these regions, the distinction between single and double bonds on the polymer backbone becomes blurred. When the density of excitons is great enough, they overlap to form partially filled bands, and the material becomes conducting.*

filled with electrons, while at higher energies are entirely empty antibonding orbitals. In the solid, therefore, the result is a fully filled valence band, formed from overlapping bonding orbitals, and a completely empty conduction band derived from antibonding orbitals (Figure 6.6). While in principle this would make polyacetylene an insulator like polyethylene, in practice the band gap is small enough that the plastic turns out to be a semiconductor.

Doping is, as we have seen, the key to getting good conductivities out of these conjugated polymers. But the dopants, unlike those used for silicon, are not foreign atoms that displace "native" ones from the crystal lattice; instead, they are atoms, molecules or ions that lodge in the gaps between polymer molecules. While the ultimate effect of these dopants is again to add charge carriers to either the conduction or the valence band, the mechanism by which this happens in polymers is a little more subtle. Dopants such as iodine are p-type: they remove electrons from the filled valence band, forming $I_3^-$ ions and leaving "islands" of positive charge on the polymer chains (Figure 6.7). When the concentration of dopant species is sufficiently high, these islands start to overlap on neighboring chains to form new energy bands in the gap between the conduction and valence bands. The process is much the same as that in p-doped silicon, except that it is accompanied by a slight distortion of the polymer chains around the dopant (due to the appearance of the islands of charge) and by the formation of entire bands, rather than individual energy levels, in the gap. The conductivity of the polymer can also be increased by n-type doping – this involves adding atoms of elements such as sodium, which donate electrons into the conduction band with much the same consequences.

The conductivity of doped polymers is not yet completely understood. There is still no detailed description, for example, of how charge carriers pass between chains. It is likely that some "hopping" mechanism exists for transferring electrons from one chain to the next, but other possibilities exist too.

## *Conductivity without metals*

Polyacetylene is one of the most versatile, cheap and efficient conducting polymers available, and for these reasons one of the best studied. The serendipitous discovery of its conducting properties heralded the beginning of work on carbon-based plastic molecular conductors and thereby helped to make molecular electronics a topic of real technological significance. But the idea that molecular, and in particular polymeric, solids might have useful electrical properties dates back much further; indeed, nonmetallic compounds with relatively high conductivities were known before the start of this century.

In 1842 the German chemist W. Knop synthesized a compound containing molecular units in which four cyanide ions, arrayed at the corners of a square, surround a central platinum atom (Figure 6.8). This unit is known as a tetracyanoplatinate (TCP) group, and bears a double negative charge. It therefore forms crystalline salts with positive metal ions such as potassium (for example, $K_2Pt(CN)_4$). But whereas most metal salts have a mineral-like appearance, often being translucent or colored, Knop's compound was distinguished by a golden bronze, metallic luster.

A very different material was prepared in 1910 by the Englishman Frank Playfair Burt. His was an unusual polymer containing just sulfur and nitrogen, which again had a suggestive metallic appearance. It is relatively rare to find polymers that do not contain

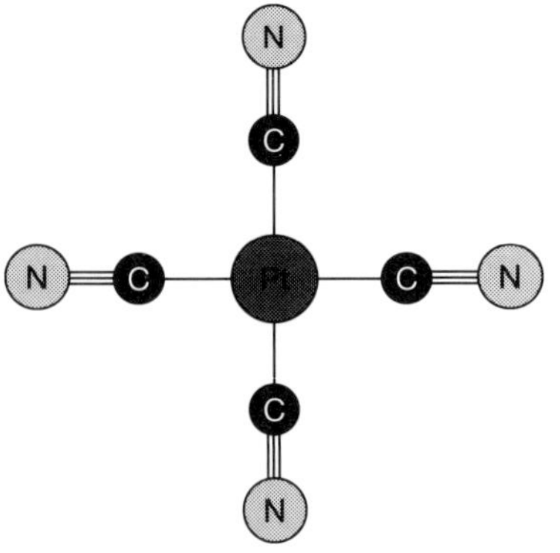


**Figure 6.8** *The structure of the tetracyanoplatinate ion.*

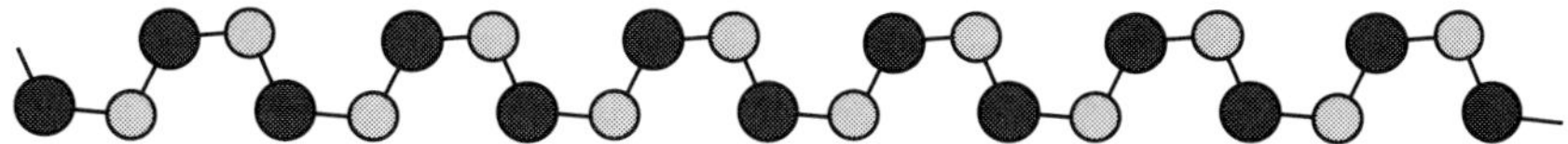

**Figure 6.9** *The conducting material $(SN)_x$ contains kinked polymer chains of alternating sulfur (dark gray) and nitrogen (light gray) atoms.*

carbon in their backbone, since few elements share carbon's inclination to form long chains. Burt's polymer, however, was a remarkably simple compound consisting of kinked chains of alternating sulfur and nitrogen atoms (Figure 6.9). In the notation of polymer chemistry, we can represent this compound as $(SN)_x$ ($x$ denotes that the SN unit repeats a large and unspecified number of times).

The electrical properties of neither of these materials were investigated closely, however, until the 1970s. They both proved to be respectable conductors, certainly more so than most other molecular materials. It is now recognized that $(SN)_x$ is one of those materials in which a completely filled and a completely empty electron band overlap, so that the elecrons acquire a degree of mobility. The conductivity of the TCP salts has a different origin. These solids contain isolated $Pt(CN)_4$ units rather than polymeric chains. The crystal structure of such salts, determined in 1964, shows that

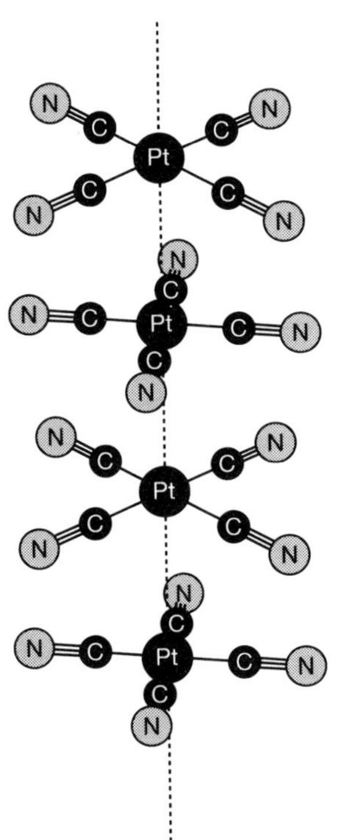


**Figure 6.10** *In tetracyanoplatinate salts the $Pt(CN)_4$ units are stacked on top of each other, so that dumbbell-shaped orbitals on the platinum atoms, protruding above and below the plane of the square units (along the dashed line shown here), can overlap to form a "one-dimensional" band. To make the salts conducting, their composition is adjusted so that this band is only partially occupied.*

the square units are stacked on top of one another like piles of plates (Figure 6.10). While some of the electron orbitals on the platinum atoms point towards the cyanide groups, the metal atoms also possess dumbbell-like d orbitals that protrude perpendicular to the plane of the square. When the TCP units are stacked together, these latter orbitals can overlap to form electron bands (which are "one-dimensional" since the electrons they contain are confined to the linear chain of overlapping orbitals). Conducting TCP salts are created by adjusting the composition so that the ratio of negatively charged $Pt(CN)_4$ groups to the counterbalancing positively charged ions (generally potassium ions) is not a simple integer – then the one-dimensional electron band formed from the stacked $Pt(CN)_4$ groups is only partially filled. For example, the potassium salt $K_{1.75}$ $Pt(CN)_4$ is a reasonably good conductor. Because of the one-dimensional nature of the conduction band, the conductivity of TCP salts is anisotropic (different in different directions), like that of polymers. A technique developed at the Xerox Corporation in Webster, New York, allows $K_{1.75}Pt(CN)_4$ to be grown in the form of highly pure and almost perfect crystals, which have conductivities many thousands of times higher than those grown by conventional means. This method involves passing a current through a solution containing potassium and TCP ions, whereupon needle-like crystals grow at the positive electrode (Plate 6). The stacks of $Pt(CN)_4$ groups lie parallel to the long axis of the needles.

In 1973 Alan Heeger and his colleagues at the University of Pennsylvania prepared an ionic salt consisting of two organic compounds – both containing only carbon, hydrogen, sulfur and nitrogen – which at a temperature of about minus 220 degrees Celsius has a conductivity approaching that of copper at room temperature. One of the compounds in this salt rejoices in the imposing name of 7,7,8,8-tetracyano-*p*-quinodimethane, mercifully abbreviated as TCNQ. The other is tetrathiofulvalene, or TTF (Figure 6.11). TCNQ is an avid acceptor of electrons, and therefore forms ionic salts with species that are good electron donors, such as metals. As TTF is an electron donor, readily giving up an electron to form a stable, positively charged ion, TTF and TCNQ are natural partners: the TTF molecules happily provide the electrons that the TCNQ molecules crave.

The molecules in crystalline TTF–TCNQ are stacked in such a way that the pi orbitals on successive molecules can overlap to form bands. Both molecules are flat, and each of the two forms separate stacks. But unlike the TCP salts, in which the square TCP ions lie at right-angles to the axis of the stack, the molecules in the TTF and TCNQ stacks are tilted, since this arrangement allows them to pack together more efficiently (Figure 6.12). Pi orbitals sticking out of the plane of the molecules are nevertheless still able to overlap with those of the molecules above and below. If each of the TTF molecules gave up a *whole* electron to each TCNQ molecule, the valence band formed from the TTF molecules would be entirely emptied, and the conduction band of the TCNQs completely filled. The compound would then be an insulator (like most ionic compounds), or at best a semiconductor. But in TTF–TCNQ, just three-fifths of an electron, on average, is transferred per molecule, with the result that the TTF band is emptied only partially and the TCNQ band is only partly filled, leaving the charge

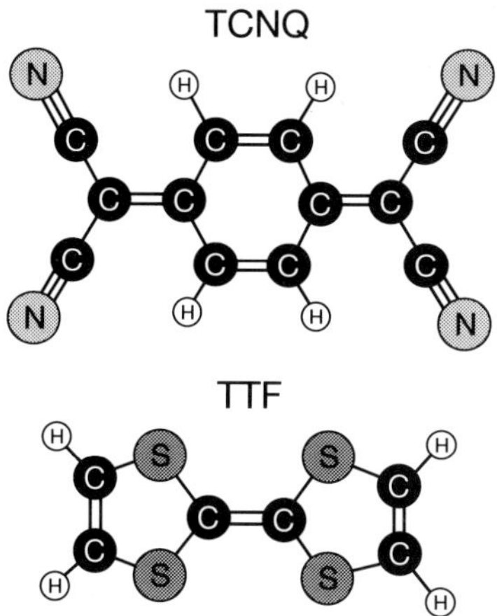


**Figure 6.11** *The molecular structures of the organic molecules TCNQ, an electron acceptor, and TTF, an electron donor.*

carriers mobile. If the idea of three-fifths of an electron seems peculiar given that electrons are indivisible particles, remember that this is just an average; it means, if you like, that only three TTF molecules in five donate electrons to the TCNQ conduction band.

The TTF–TCNQ salt exemplifies a general class of molecular conductors, called charge-transfer compounds because their conductivity depends on the transfer of

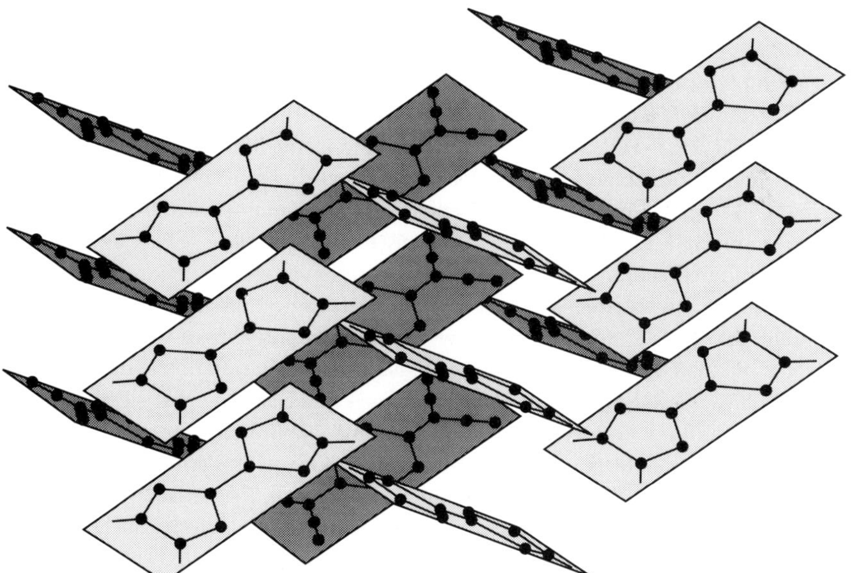

**Figure 6.12** *In the charge-transfer salt TTF–TCNQ, the flat molecules are stacked in a "herring-bone" array, in which pi orbitals on adjacent molecules can overlap to form continuous one-dimensional bands. Transfer of electrons from TTF to TCNQ partially empties the valence band of the former and partially fills the conduction band of the latter.*

charge from an electronic band of donor molecules into that of acceptor molecules. These compounds demonstrate that conducting molecular materials do not have to be polymers; rather, the requirement here is that electron orbitals on proximal molecules be able to overlap to form extended bands, which are then partly filled or emptied by charge transfer. Because the conductivity arises from interactions between at least two different kinds of molecule, there is a lot of scope for tinkering with the electronic properties of charge-transfer compounds by interchanging different donors or acceptors, or by modifying their chemical composition.

### *Conducting polymers in action*

Since the discovery of conducting polyacetylene, conjugated carbon-based polymers have received the greatest attention from those who wish to put molecular electronics to practical use. In part this is because these polymers are easy and cheap to prepare, can be resistant to chemical degradation, and have appealing mechanical properties (such as toughness and flexibility). Another attraction is that their chemistry is so versatile: one can often fine-tune a polymer's properties by making slight changes in their chemical constituents or structure. Many carbon-based polymers have now been developed that are good conductors when doped, amongst them polyparaphenylene, polypyrrole, polythiophene and polyaniline (Figure 6.13). Some of these materials have now found their way into the kinds of electronic applications that have conventionally relied upon metals and semiconductors, as well as having given rise to applications that are entirely novel.

An example of the former is the polymer battery. In the early 1980s, MacDiarmid and Heeger developed a rechargeable battery in which the electrodes were made from doped polyacetylene. When charge flows through conventional batteries, the metal electrodes partially dissolve, releasing metal ions into the battery's electrolyte; recharging involves the reverse process of depositing metal ions back onto the electrode surfaces. In principle this means that a discharge–recharge cycle leaves the electrodes unchanged, but in practice they suffer from deterioration after many cycles of dissolution and redeposition. In polymer-electrode batteries, on the other hand, the ions that are responsible for charge storage and flow do not constitute part of the electrodes themselves but instead remain all the time in solution, so deterioration should not occur. Equally significant is the fact that metal electrodes – especially the lead electrodes found in lead–acid batteries – are very heavy. In some applications, such as electrically powered vehicles, weight is of crucial importance: much of the power generated by a heavy battery will be used up in propelling the extra bulk that it adds to the vehicle. Polymer batteries containing electrodes fabricated from light atoms such as carbon, hydrogen and nitrogen can provide a high power-to-weight ratio. Furthermore, the components of these batteries are generally nontoxic, unlike those in conventional lead–acid or nickel–cadmium cells. Some of the polymer batteries now available commercially have longer shelf-lives and produce higher power outputs than their metal-based counterparts.

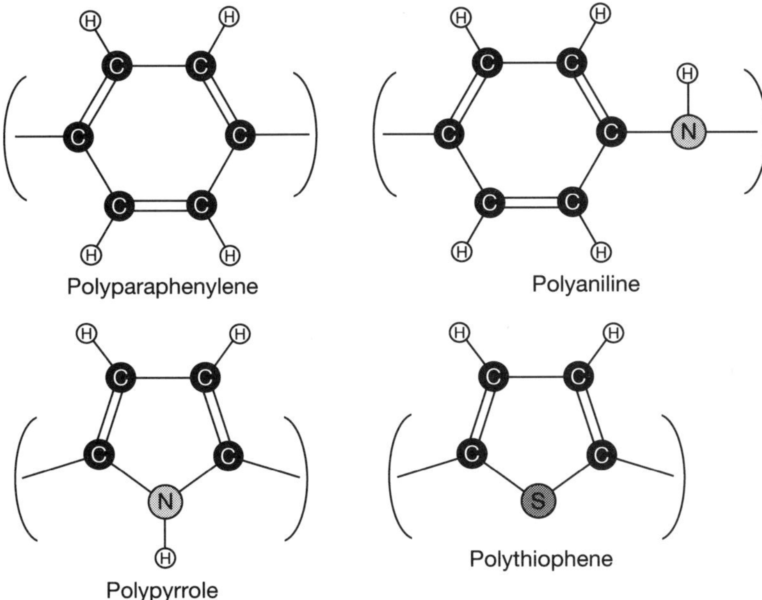


**Figure 6.13** *The repeat units of some other conducting organic polymers.*

Amongst the microelectronic devices that have been fabricated either partially or wholly from conducting polymers, some of the most eye-catching are light-emitting diodes (LEDs). The first of these was prepared in 1988 by Richard Friend and colleagues at the University of Cambridge. Friend's group had succeeded earlier in making standard diodes and transistors, ubiquitous devices in microelectronics, based on polyacetylene. The polymer LED sprung from their observation that another conducting hydrocarbon polymer, poly(paraphenylene vinylene) or PPV, could be made to emit light when excited electronically. By "wiring up" a PPV film and applying a voltage across it, Friend and colleagues could inject electrons into the conduction band of the polymer, and at the same time remove electrons from the valence band, creating holes. These electrons and holes can move along the polymer backbone, but when one encounters the other they stick close together because of their mutual electrostatic attraction, forming a bound pair. In a process called recombination, the electron eventually falls down into the hole in the valence band, annihilating the pair of charge carriers. In doing so, the electron loses energy, which is radiated away in the form of light. When many injected electrons and holes recombine in this way, the PPV sample glows with a yellow radiance (Plate 7).

By tinkering with the chemical structure of the PPV chains, the Cambridge group have been able to generate LEDs that emit light of different colors. The color depends on the amount of energy released when the injected charge carriers recombine, which in turn depends on the width of the gap between the valence and conduction bands. Polymers with slightly different compositions may have slightly different band gaps, and

LEDs made from modified versions of PPV and other polymers can therefore glow right across the spectrum – red, orange, yellow, green and blue.

In addition to these conventional electronic devices, conducting polymers are finding their way into novel applications. Because the doping procedure is so simple, usually involving nothing more than exposure of the polymer film to a vapor of the dopant atoms, conducting polymers can be exploited in sensitive chemical sensors which detect the substances that act as dopants. Such sensors contain a film of the undoped polymer, whose conductivity is monitored continuously. When the film is exposed to air containing the dopant atoms, these become incorporated into the film and its conductivity increases in proportion to the amount of dopant present. The conductivity thus serves as a measure of the level of dopants in the air.

Some conducting polymers, such as polyaniline and polythiophene, change color when doped. The latter, for instance, is deep blue in the undoped state but red when doped. Films of these materials might be used in "electrochromic" displays, which change color when a voltage is applied across them. The color can also be sensitive to changes in temperature – thin films of polyacetylene are red at low temperatures but turn blue as the temperature increases (Plate 8). This property, called thermochromism, could turn out to be useful for the development of new kinds of thermometer. Much work remains to be done, however, before the electrochromic and thermochromic properties of conducting polymers can be exploited in a practical way.

Perhaps the most intriguing of the potential applications yet to be realized are in medicine and physiology. One suggestion is that the tough, flexible nontoxic polymers could be used as artificial nerves. The signal-carrying axons of the nervous system are rather like tiny wires that bear currents from the biochemical sensors at one end to the spinal column and thence to the brain. Could damaged nerves one day be replaced by polymeric ones? Polypyrrole is a candidate for this application, as it is nontoxic and can be rendered conducting by using as a dopant the natural anti-blood-clotting agent heparin.

## The path of no resistance

### *Under pressure*

In 1979 the French-based researchers Michel Ribault, Klaus Bechgaard and Denis Jerome carried out an elaborate experiment on a charge-transfer compound called tetramethyltetraselenafulvalene hexafluorophosphate. The molecule responsible for the first half of this impressive name, abbreviated to TMTSF, is a variant of TTF. Like TTF, TMTSF is a good electron donor. In the salt studied by Ribault and colleagues, TMTSF donates electrons to hexafluorophosphate ($PF_6$) units, of which there is one for every two TMTSFs. The compound can be represented by the formula $TMTSF_2PF_6$.

Crystals of $TMTSF_2PF_6$ have a metallic appearance and are good electrical conductors. But unlike similar compounds such as TTF–TCNQ, which becomes a poorer conductor (and ultimately a semiconductor) as the temperature is reduced, this salt remains a good conductor even at just 20 degrees above absolute zero. The French team aimed, however, to probe the properties at conditions even more extreme than this. First they squeezed the crystals to pressures of 12,000 times normal atmospheric pressure. The crystals were then cooled very slowly to less than 1 degree above absolute zero. As the temperature fell to this value, the resistance of the salt began suddenly to plummet – in other words, it became a much better conductor. And at 0.9 degrees above absolute zero, all electrical resistance vanished. This means that a current could be passed through the material without any of the electrical power being lost as heat: the material was a perfect conductor. A substance that possesses this property is said to be a superconductor. Before this experiment the only superconducting materials known were metals and metallic alloys; and so the salt $TMTSF_2PF_6$ became the first molecular superconductor.

The high pressures used by Ribault and colleagues are essential for this effect to be observed in $TMTSF_2PF_6$. At atmospheric pressure, the salt becomes insulating at 12 degrees above absolute zero (minus 261 degrees Celsius) and stays that way on further cooling. But if perchlorate ($ClO_4$) units are substituted in place of $PF_6$ as the electron acceptors, the salt becomes a superconductor at 1.2 degrees above absolute zero without the need for any squeezing at all. (Rather than expressing these low temperatures as negative values on the Celsius scale, scientists use a scale for which the zero is at absolute zero. One degree of this temperature scale is equivalent to one degree Celsius, and the units are called degrees Kelvin (after the physicist Lord Kelvin), denoted K. So minus 261 degrees Celsius corresponds to 12 K.)

## *Superconductivity in metals*

One might reasonably ask what prompted the researchers to search for superconductivity in this exotic material under such extreme conditions. To answer this question we must first place the work on molecular superconductors within the context of what one might call mainstream superconductor research, a field initiated by some baffling experiments conducted by the Danish physicist Heike Kamerlingh Onnes in 1911.

Kamerlingh Onnes was interested in the conductivities of metals at very low temperatures, because it was believed that their resistance should become very small as absolute zero was approached, finally falling to zero at zero degrees Kelvin (a limit that is unattainable in practice). We saw earlier how the vibrations of the atomic lattice in metals cause electron scattering and thereby contribute to electrical resistance. At absolute zero, the atoms would become frozen into immobility and this scattering should then be eliminated. Defects in the lattice due to impurities or misaligned atoms would remain even at absolute zero, however, and could still act as scatterers. But if one could grow perfect crystals, the conduction electrons could then travel unhindered at absolute zero. The expectation was therefore that superconductivity should be possible, at least in principle, under these coldest of cold conditions. While he could neither reach absolute zero nor grow defect-free crystals, Kamerlingh Onnes hoped at least to see metals approach a superconducting state as the temperature got lower.

Yet when he monitored the resistance of mercury metal cooled with liquid helium, he found to his amazement that its resistance vanished well before absolute zero. The transition to a superconducting state took place more or less at the boiling point of helium, 4.2 K. This was in spite of the fact that the mercury atoms were unquestionably still vibrating in the crystal lattice, and that no doubt defects were present too. Somehow, the electrons appeared to have found a way to avoid or ignore these. It was discovered soon afterwards that other metals also show this behavior: tin, for example, is a superconductor at 3.7 K, and lead undergoes the transition to a superconducting state at a temperature as "high" as 7.2 K.

Low-temperature experiments over the succeeding years showed that most metals could be made superconducting when cooled. For pure metals the superconducting transition temperature is always fairly close to absolute zero, but alloys – mixtures of two or more elements – perform rather better. A vanadium–silicon alloy, for instance, becomes superconducting below 17 K, while for niobium–tin the transition temperature is 18 K and for niobium–germanium it is 23.2 K. The transition in this latter alloy, observed in 1973, stood for many years thereafter as the record holder for the highest transition temperature. Although researchers had been gradually pushing up these temperatures by experimenting with new mixtures, they faced ever-diminishing returns. Superconducting transition temperatures seemed to have come up against the ceiling. This meant that, of the many practical uses to which superconductors could conceivably be put, few seemed likely ever to be realized because of the need for bulky, expensive liquid-helium cooling systems in order to attain the superconducting state. So the outlook for superconductor applications throughout the 1970s and early 80s looked decidedly bleak.

### *The ceramic revolution*

In 1986, all that changed almost overnight. The physicists Georg Bednorz and Alex Müller, working at the IBM laboratories in Zürich, reported that they had made a compound with a superconducting transition at 35 K. This impressive leap of about 12 degrees above the previous record generated considerable excitement amongst physicists. Quite apart from its new record temperature, the Swiss scientists' compound was interesting because it was unlike any superconductor known previously. It was not a metal alloy but an oxide of the metals lanthanum, barium and copper. A material of this sort, containing both metals and nonmetals, is called a ceramic. These compounds tend to be hard and brittle, in contrast to the tough, ductile characteristics of metals and their alloys.

Bednorz and Müller's breakthrough won them the Nobel prize for physics in 1987, but by that time their ceramic superconductor had been overshadowed by far more dramatic developments. Even if the room-temperature superconductor might never be found, physicists agreed that they would be content with a more modest goal. A material that remained superconducting above the boiling point of liquid nitrogen, 77 K, would be a tremendous boon, as it would allow relatively inexpensive liquid-nitrogen coolant systems to replace the liquid-helium systems that had always been necessary previously. Such a material would therefore permit physicists to contemplate some of the applications of superconductors that were just not viable with helium cooling. Lanthanum barium copper oxide had a transition temperature of less than half the liquid-nitrogen barrier, yet it was no more than a year after its discovery that the barrier was broken, and this by no less than 16 degrees. Paul Chu and coworkers from the University of Texas at Houston reported in 1987 that another, related ceramic material, yttrium barium copper oxide, had a superconducting transition at 93 K.

It would be hard to overemphasize the fever that these findings generated at the time in the scientific community. Researchers would stay up all night working in their laboratories, experimenting like medieval alchemists with arcane mixtures of substances in an attempt to find the magic brew that would put them at the head of the race, with all the attendant fame and perhaps fortune that it might bring. These studies culminated with the ceramic devised in 1988 by researchers from the NEC Laboratories in Tokyo: thallium barium copper oxide, which has a transition temperature of 125 K (minus 148 degrees Celsius). But five years later, the record for a superconducting transition has been pushed up by no more than 8 degrees: a mercury-containing copper oxide discovered in 1993 by researches at the Eidgenossische Technische Hochschule in Zürich superconducts at 133 K. After its mad burst of activity, the superconductor research community seems to have found a new ceiling.

Superconductor devotees have now had to knuckle down to the daunting task of trying to put the new materials to good use. One obstacle to applications is the brittle nature of the copper oxide ceramics, which makes them more difficult to work with than metals. But the major problem is that of getting the materials to carry the large electric currents that are required in most applications. Superconductors are converted back to a nonsuperconducting state once the current passing through them exceeds a certain critical value. This critical current is smaller for the copper oxide materials than for conventional metallic superconductors, a hitch that now seems likely to become the main factor limiting the uses to which the copper oxide superconductors can be put.

Some of these applications exploit the superconductor's ability to carry an electrical current without losing power through resistive heating effects. When electricity is transmitted along the copper cables of power lines over many miles, a significant fraction of the power generated is dissipated before it reaches its intended destination. But if one were to replace the copper cables with superconducting ones, no power would be lost at all. (Strictly speaking this is true only for direct-current (d.c), not alternating-current (a.c) supplies.)

Superconductors are also valuable to the computer and microelectronics industries. The absence of heating effects enables smaller, more densely packed circuit boards to be fabricated without the risk of their melting; and switch-type microelectronic devices based on superconducting materials can respond to signals much more rapidly than those based on semiconductors.

Besides their current-carrying properties, superconductors have other unusual characteristics which could be exploited in applications. In 1913, the German scientists K.

W. Meissner and R. Ochsenfeld found that superconductors are repelled by magnetic fields. If a pellet of the ceramic superconductor yttrium barium copper oxide is placed on a magnet and cooled below the superconducting transition temperature using liquid nitrogen, this "Meissner effect" causes the pellet to rise into the air and remain levitating above the magnet (Figure 6.14). One of the uses that has long been proposed for these levitation forces is for the construction of railway trains that float on air rather than running on rails. Without the friction due to physical contact between the vehicle and the rails, tremendous speeds should be possible at little cost in energy. The Meissner effect might also be exploited for friction-free mechanical bearings. The sensitivity of superconductors to magnetic fields has been put to use for some years now in devices called superconducting quantum interference devices (SQUIDs), which are used to detect very small magnetic fields such as those produced by the tiny electric currents that flow between neurons in the brain. SQUID-based sensors can provide maps of the brain's electrical activity, which may guide neuroscientists towards a better understanding of how this mysterious organ works. SQUIDs based on yttrium barium copper oxide, operating at liquid-nitrogen temperatures, are now marketed commercially.

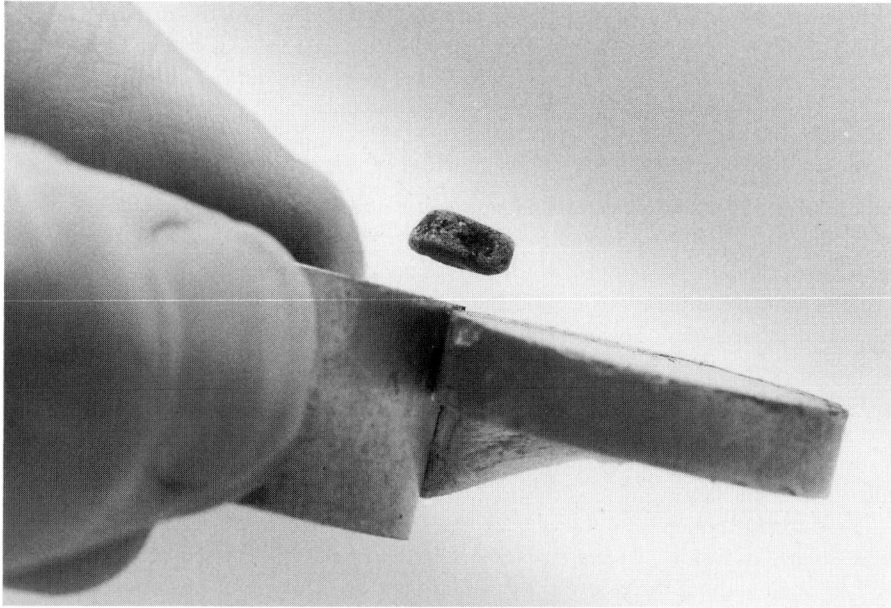

**Figure 6.14** *Because superconducting materials expel magnetic fields, they can be made to levitate above a magnet. Here a pellet of the high-temperature superconductor yttrium barium copper oxide cooled to liquid nitrogen temperature (which is below the temperature of the superconducting transition) levitates over a hand-held magnet. (Picture courtesy of Colin Gough, University of Birmingham.)*

*Take your partners*

The discovery of the copper oxide "high-temperature" superconductors has revitalized interest in these and other applications. But the possibility of superconductors with high transition temperatures had been postulated long before the breakthrough of Bednorz and Müller. In the 1960s, William Little of Stanford University suggested that one-dimensional molecular conductors might prove to be superconducting at unprecedented temperatures.

To understand why Little picked these materials as candidates for high-temperature superconductors, we need to appreciate how superconductivity arises in the old-fashioned low-temperature materials. The explanation, put forward in 1957 by John Bardeen, Leon Cooper and Robert Schrieffer and known now as BCS theory, seems paradoxical at first glance because it states that in superconductors, electrons (which normally repel each other because of their like electrical charges) experience a mutual *attraction*. This attractive force results in the formation of electron pairs, called Cooper pairs, and the supercurrent is then carried through the crystal lattice by the motion of these pairs.

How can two particles with the same charge attract each other? While there is no doubt that electrons in a superconducting metal do experience the electrostatic repulsion that one would expect, this force can be overwhelmed by an opposite tendency that arises via the intermediation of the positive metal ions. An electron traveling through the crystal exerts an attractive force on these ions, pulling them together as it passes between them. But whereas the electron is a very light, nimble particle, the ions are much more heavy and ponderous. The ions therefore remain for some time in relatively close proximity even after the electron has sped past, before relaxing slowly back to their original positions. The effect is for the electron to set up in its wake a ripple in the crystal lattice (Figure 6.15). While the metal ions remain close together they create a region of abnormally concentrated positive charge, to which a second electron can be attracted. In other words, the first electron leaves a kind of ephemeral trail of enhanced positive charge, which causes a second electron to be dragged along behind it just as if there were a genuine attraction between the two electrons.

There are two important points to note about this effect. First, the two electrons in the Cooper pair stay a relatively large distance apart, since the first electron has long since departed by the time the second gets drawn towards the ripple. The "size" of the Cooper pair can be as much as ten thousand times the distance between neighboring ions in the lattice. The second point is that the attraction is fairly weak, and will be disrupted by anything that disturbs the drawing together of the metal ions. Random motions caused by thermal vibrations of the ions have this disruptive effect, so electron pairing is possible only when these vibrations are very small – that is, at low temperatures.

The supercurrent is more conveniently visualized as the motion of the combined Cooper pair rather than that of the two individual electrons (since they need not necessarily be moving in the same direction). It is useful, therefore, to think of the

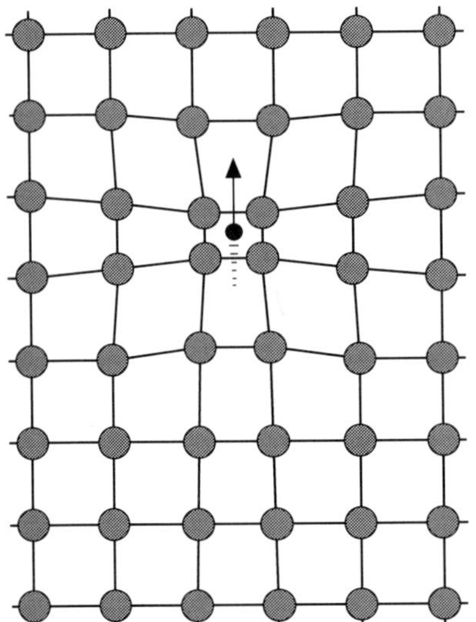

**Figure 6.15** *According to the BCS mechanism for superconductivity, electrons in superconducting metals become paired via their interaction with the lattice of metal ions. A passing electron pulls ions in the lattice together, creating a transient region of concentrated positive charge. A second electron can be pulled along in this wake, since the ions relax relatively slowly to their original positions.*

Cooper pair as if it were itself a kind of composite particle, called a quasiparticle, with a charge and mass equal to those of two electrons. When described in this way, however, the Cooper-pair quasiparticle turns out to possess properties that are very different to those of its constituent electrons, and it is these properties that hold the key to superconductivity.

Electrons belong to a class of elementary particles known as fermions, no two of which are allowed to be in the same quantum-mechanical state. Cooper pairs, meanwhile, are members of the class known as bosons, which also includes protons, neutrons and the "particles" of light, photons. Unlike fermions, it is quite permissible (according to the rules of quantum mechanics) for any number of bosons to occupy the same quantum state. Thus, while electrons are forced into states of higher energy when the lower-energy states are filled up, the Cooper-pair quasiparticles will all jump gleefully into the state of lowest energy – the paired electrons all occupy a single quantum state. It is as if, at the superconducting transition temperature, the energy *band* containing the conduction electrons suddenly collapses into a single energy *level* with enough room for them all.

Once in this "condensed" state, the Cooper pairs cannot easily be scattered. To do so means changing their energy. For electrons in a continuous energy band this is easy; but to alter the energy of a Cooper pair, it has to be kicked right up to the next "collapsed" energy level, which generally takes more energy than a mere vibrating ion has got to offer. The Cooper pairs therefore move through the lattice like one vast, unstoppable crowd, ignoring all scatterers. There is no resistance to the flow of charge.

BCS theory is very successful in explaining superconductivity in metals. It cannot, however, accommodate the new ceramic "high-temperature" superconductors; indeed, any transition temperature higher than about 30 K poses severe problems for the conventional BCS theory. It is generally agreed that all manifestations of superconductivity must involve the formation of Cooper-pair bosons and their condensation into a single quantum state, but exactly how they are formed in the high-temperature materials is not yet known.

## *The molecular superconductors*

William Little's proposal was that a different pairing mechanism might operate in one-dimensional molecular superconductors which could enable them to surmount the BCS limit on transition temperatures. Building on the idea that Cooper pairs are formed in a metal when a moving electron leaves a temporary trail of positive charge, Little suggested that the same effect could be engineered for electrons passing along the backbone of a conjugated polymer. He considered a polymer chain to which were attached side groups with rather "sloppy," easily polarizable electron clouds – dye molecules, for example, which usually have delocalized electron orbitals. An electron passing along the chain will repel the side-group electrons, creating a region of positive charge (Figure 6.16). A second electron can then be drawn towards this region just as the distorted lattice in a superconducting metal attracts the second member of a Cooper pair.

The crucial difference between Little's mechanism and that in a metal crystal is that, in these one-dimensional systems, electron pairing does not involve the motion of atoms at all – only the much lighter electrons need move. This, suggested Little, should make pair formation much easier, so that perhaps it might occur at higher temperatures. Little's rough calculations predicted that superconductivity induced in this way might be possible at room temperature. In fact, he even went so far as to suggest that superconductivity at temperatures as high as 2000 degrees Celsius was not out of the

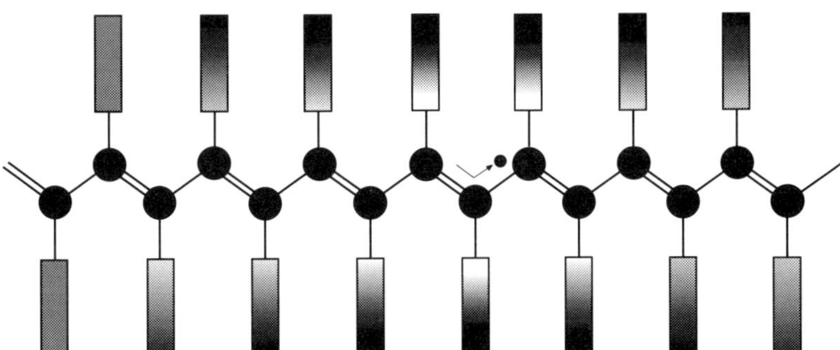

**Figure 6.16** *In William Little's proposed model for superconductivity in conjugated polymers, the electron-pairing mechanism is mediated by "polarizable" groups attached to the polymer backbone, which have delocalized, sloppy electron clouds. An electron passing down the backbone will repel the electrons in the side groups, creating a region of positive charge which then attracts a second electron. This pairing mechanism involves the motion only of light electrons, not of heavy ions.*

question, provided that the molecular conductors were themselves stable at such temperatures (which was most unlikely).

Although this proposal was highly controversial, the potential payoffs were sufficient to stimulate the search for molecular superconductors. Sadly, no compound has yet been found that behaves in a way even remotely similar to Little's suggestion. Nevertheless, when Alan Heeger and colleagues succeeded in preparing TTF–TCNQ in 1973, its one-dimensional stacked structure invited speculation about whether Little's mechanism, or something like it, might make the compound a superconductor under appropriate conditions. Experiments soon proved otherwise: as mentioned earlier, the compound loses its high conductivity as it is cooled, eventually becoming a semiconductor. An explanation for this is to be found in the work of the British physicist Rudolf Peierls, who showed in 1954 that at low temperatures a regular array of molecules with a linear, chain-like structure can lower its energy by distorting so as to make the distances between successive molecules alternate between long and short. This rearrangement, called the Peierls instability, changes the character of the uppermost, partly filled electron band, splitting it into a filled and an empty band separated by an energy gap. The material is then a semiconductor.

The Peierls instability, which occurs at 53 K for TTF–TCNQ, appeared to pose profound difficulties for attempts to make superconductors based on linear polymer chains or stacks of molecules. It was thought, however, that squeezing these solids might help to suppress the distortion. For TTF–TCNQ, squeezing was to no avail; in fact, it actually encouraged the instability to occur. But Jerome and Bechgaard, together with the Dane Jan Andersen, had more success with a closely related compound in which TTF is replaced by TMTSF and TCNQ by a slight modified molecule, 2,5-dimethyl-TCNQ (DMTCNQ). The charge-transfer compound formed by these two molecules, TMTSF–DMTCNQ, behaves much like TTF–TCNQ at atmospheric pressure, conducting at room temperature but becoming an insulator at 41 K. If compressed, however, the transition to an insulator does not take place; instead, TMTSF–DMTCNQ remains a conductor at least down to the boiling point of liquid helium (4.2 K). The researchers felt that they were on the right track to the molecular superconductor.

Their studies showed that the conductivity depended most critically on the presence of the electron donor TMTSF, and they therefore set about making salts in which other electron acceptors replaced DMTCNQ. When, amongst these, they tried hexafluorophosphate ions, they found that they had hit upon the magic mix.

Many variants on this theme have now been discovered that show better superconducting properties. The best of these are based on the molecule bis(ethylenedithio)-tetrathiafulvalene (BEDT-TTF or simply ET), another relative of TTF (Figure 6.17). In 1988 G. Saito and colleagues from the NEC Research Laboratories in Tsukuba, Japan, showed that the charge-transfer salt of BEDT-TTF and the electron acceptor copper thiocyanate, $Cu(NCS)_2$, has a superconducting transition temperature of 10 K, considerably higher than that of most metals. The record breaker is currently a closely related compound prepared by Jack Williams and coworkers at Argonne National Laboratories in Illinois, which superconducts at about 13 K. Whether the super-

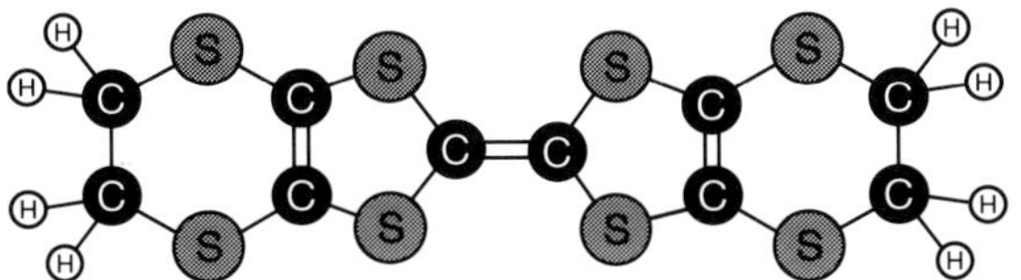


**Figure 6.17** *The structure of the electron acceptor BEDT-TTF, which has provided the "warmest" organic superconductors to date.*

conductivity originates from the interaction between the conduction electrons and atomic motions in the solid, as in conventional BCS theory, or whether some other mechanism is at play in these linear structures, is still not clear.

## Superconducting soccer balls

The family of molecular conductors has recently acquired a new member so spectacular that it has all but eclipsed the painstaking work on linear molecular systems that has gone before. The sixty-atom carbon soccer ball buckminsterfullerene ($C_{60}$), described in Chapter 1, has, it seems, no end of talents. Although the pure $C_{60}$ solid ("fullerite") is a poor conductor, early in 1991 a team from AT&T Bell Laboratories in New Jersey, led by Robert Haddon and Arthur Hebard, found that doping fullerite with alkali metal atoms – lithium, sodium, potassium, rubidium and cesium – causes the material to develop a respectable conductivity. Doping is achieved simply by exposing the solid to a vapor of the metal. In $C_{60}$ crystals the molecules lie close enough for their orbitals to overlap, like those in TTF–TCNQ, forming a filled valence band and an empty conduction band that can accommodate electrons from donors such as the alkali metals. One difference between doped $C_{60}$ and a doped polymer, however, is that in the former the conductivity is isotropic (that is, the same in every direction) because the electrons are not confined to linear chains.

Initially doping $C_{60}$ brings about a rise in conductivity, but once the doping level exceeds about three alkali metal atoms per $C_{60}$ the conductivity starts to fall again. This is consistent with the idea that the metal atoms are donating electrons into the $C_{60}$ conduction band: at high dopant concentrations the band starts to become filled, and when there are six metal atoms per $C_{60}$ the band is completely full and the compound is insulating.

As AT&T Bell Laboratories is one of the homes of superconductor research in the United States, it came as no surprise to find what Haddon and colleagues tried next. Even so, they could scarcely believe their luck when, at only a little below 30 K, the electrical resistance of a doped sample containing three potassium atoms per $C_{60}$ ($K_3C_{60}$) started to drop. And at about 18 K, it plummeted to zero – the doped fullerite (Plate 9) was indeed a superconductor, and with a transition temperature a good six or seven degrees above that of the best molecular superconductor then known.

**Table 6.1** The superconducting fullerene compounds

| Compound | Transition temperature (K) |
|---|---|
| $K_3C_{60}$ | 19 |
| $Rb_3C_{60}$ | 29 |
| $K_2RbC_{60}$ | 23 |
| $K_2CsC_{60}$ | 24 |
| $Rb_2KC_{60}$ | 27 |
| $Rb_2CsC_{60}$ | 31 |
| $RbCs_2C_{60}$ | 33 |
| $Na_2KC_{60}$ | 2.5 |
| $Na_2RbC_{60}$ | 2.5 |
| $Na_2CsC_{60}$ | 12 |
| $Li_2CsC_{60}$ | 12 |
| $Ca_5C_{60}$ | 8.4 |
| $Ba_6C_{60}$ | 7 |
| $(NH_3)_4Na_2CsC_{60}$ | 30 |

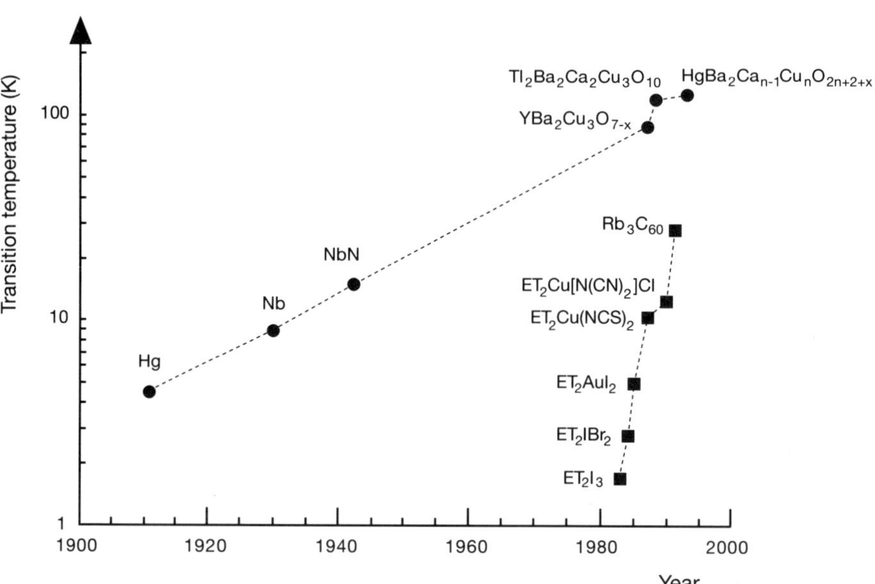


**Figure 6.18** *Will molecular superconductors overtake the copper oxide high-temperature superconductors by the end of the decade? If the trends in this plot continue, they will. (After Jack Williams, Argonne National Laboratory, Illinois.)*

There was more to come. When instead of potassium the Bell Labs team used the alkali metal rubidium as the dopant, the doped $C_{60}$ underwent a superconducting transition at a temperature of no less than 30 K, a value exceeded only by the copper oxide superconductors. A whole family of these superconductors soon began to appear, all with three metal atoms for each $C_{60}$ molecule; in some members of the family, more than one kind of alkali metal is present (Table 6.1). The present record is set by $RbCs_2C_{60}$, with a transition temperature of 33 K; temperatures higher still have been reported but so far not substantiated. A new family sprang up in 1992 with the creation of a $C_{60}$ superconductor in which an alkaline earth metal – calcium – was the dopant. This compound, $Ca_5C_{60}$, has a transition temperature of 8.4 K, and has since been joined by the barium-doped material $Ba_6C_{60}$, which superconducts at 7 K.

It is hard to overstate the extraordinary nature of these results. Before the discovery of fullerene superconductors, researchers had become accustomed to the belief that only the copper oxide materials were ever likely to bring us within spitting distance of a room-temperature superconductor. Molecular superconductors had looked fated to end up as a curiosity that never quite achieved their potential. Now some are convinced that $C_{60}$ will take molecular superconductivity to greater heights than the copper oxide materials in the near future. Jack Williams has contrasted the recent performance of conventional, copper oxide and molecular organic superconductors (Figure 6.18): if the trends are maintained, the molecular superconductors will overtake all comers before the end of the century!

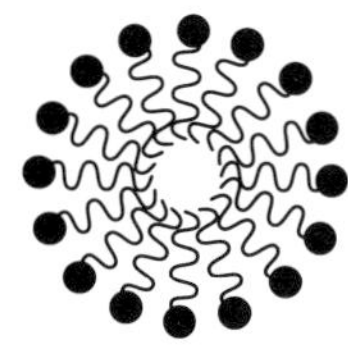

# 7

# A Soft and Sticky World

## The self-organizing magic of colloid chemistry

*I soon realized that manufacturing paints is a strange profession.*
Primo Levi

I attempted to suggest in the introduction to this book that there are interesting things to say about paint. It probably seemed unlikely. Unless, that is, you are contemplating a spot of interior decorating, in which case you may be in a position to appreciate the appeal of a paint that does not spatter hair, skirting boards and carpets with stray drops. Nondrip paint is surely a curious substance. In the tin or sitting on the brush, it is almost like a solid: you can cut through it with a knife and leave two flat faces. Yet once you start to spread it, it flows like a liquid. Isn't that a rather fancy trick? How is it done?

Equally puzzling perhaps (but also equally familiar) is a substance that *thickens* when disturbed. Just try mixing custard powder with enough water to make a thick paste. If stirred slowly it flows freely enough, but give it a rapid whisk of the spoon and it becomes stiff. Ease off the stirring . . . and again it is a sluggish liquid. Think also of children's "magic" putty, which is like dough in the hands but shatters when cast against a wall.

This property of undergoing pronounced changes in "thickness" (technically, in viscosity) in response to mechanical disturbances is called thixotropy. It seems to have been recognized even in the Middle Ages. Amongst the sacred medieval relics of the

Roman Catholic church in Italy are fourteenth-century vials said to contain the blood of saints: nothing more than a solidified brown mass, as you might expect, but one that becomes liquid when handled (which is to say, shaken gently) during religious rites. Needless to say, the priests do not regard this behavior as akin to that exhibited by anything so mundane as nondrip paint; rather, it is taken to be a miraculous phenomenon. (Italian chemists have shown, however, that substances with just the same appearance and properties can be made from compounds easily obtainable in the fourteenth century, including iron oxides from the slopes of Vesuvius.)

What are the general principles that guide chemical technologists in their attempts to produce substances with specific properties like these? Such behavior is not the result of chemical reactions – there is no making and breaking of strong covalent bonds during these processes, and no atoms are changing hands. Rather, we are dealing with something akin to the supramolecular interactions encountered in Chapter 5, but on a larger scale, in the sense that the structures being assembled consist not of a couple or a handful of molecules but vast numbers of them. These substances are examples of *colloids* – molecular assemblies in which the size of the structures ranges from a nanometer (the size of a medium-sized molecule such as $C_{60}$) to a micrometer (comparable to the size of a bacterium). Thus defined, colloids encompass a vast range of substances: paints, greases, toothpaste, bitumen, liquid crystals, soap bubbles and foams. They are generally "soft" forms of matter, easily deformable, able to flow. And in the sense that the human body is comprised of cells that are molecular assemblies of micrometer dimensions, we too are colloidal structures.

Colloid science is an eminently applied discipline – it is the concern of a great number of industrial scientists, and its relevance extends beyond paints to food science, cosmetics, lubrication and agriculture. Some of the principles on which today's understanding of colloids is based have been known for a very long time; others have emerged from research in exotic and often seemingly unrelated areas of physics, chemistry and biology. The ancient Egyptians showed a knowledge of the topic in their ability to create stable colloidal suspensions of soot in gum arabic, which we now know as Indian ink. The Romans and Babylonians too had an appreciation of the virtues of colloidal materials, as evidenced by their use of bitumen as a watertight sealant for boats and buildings. The field is too broad for a book, let alone a chapter; but by describing here just a few selected topics in colloid chemistry I hope to be able to provide a hint of this breadth.

## The incredible shrinking gels

### *Solid or liquid?*

Let me tell you first how the trick is done. Many nondrip paints contain long, chain-like polymer molecules dispersed (along with microscopic pigment particles) in an oil- or water-based solvent. The polymer chains contain chemical groups at several points along their length that are insoluble in the solvent (ionic groups, for example, are insoluble in oily solvents). To minimize their exposure to the solvent, these groups will cluster together so as to become surrounded by others of like kind. They can therefore be thought of as "sticky patches," which adhere to those on neighboring molecules through interactions that are relatively weak and can be easily broken. But because each polymer molecule contains a great number of sticky patches, their cumulative effect is to bind the molecules together into an interlinked network that is reasonably rigid: a framework that holds the liquid solvent within it. When part of the network is sheared (deformed by an applied pressure, such as is exerted by a paintbrush), the weak links will become unstuck and the polymer molecules, solvent and pigment will flow freely. As soon as the shear is removed, the sticky patches will again be able to adhere to their neighbors (Figure 7.1*a*).

The same approach can be used to make a substance that has the opposite property of increasing its viscosity when sheared. This time the polymer molecules are designed so that their shape, and the strength of the interactions between the patches, encourages patches on the *same* molecule to stick together. When this happens, there is no formation of an interlinked network; instead, the chains curl up on themselves like tangled balls of sticky tape. Each polymer molecule is therefore able to slide over the others without sticking to them. But when the fluid is sheared, the interactions between the sticky patches are disrupted as the shearing stretches out the curled-up molecules into long chains. The polymer molecules then become aware that their neighbors too have sticky patches, and bind together to form the rigidifying network (Figure 7.1*b*).

Nondrip paints exemplify a class of substances known as gels. The term probably has a familiar ring to it, conjuring up images of gelatin and jello. These materials occupy a sloppy middle ground somewhere between solids and liquids. It is hard to see how one could reasonably call them liquid, given that they can be sliced or molded into shapes that they will retain (if only temporarily), and yet one might have a hard time convincing oneself that they can be classed as true solids. A gel is actually a sort of composite substance – it contains a liquid, but this is bound within a network of polymer molecules which displays a certain degree of rigidity, like a solid. The liquid fills the gaps in the polymer framework and prevents it from collapsing into a tangled mass. Just how rigid the gel is depends on the degree of cross-linking between the polymers – gels can range from highly viscous fluids to fairly sturdy solids.

Nature has found many uses for gels. They are just the ticket when the situation demands a combination of liquid-like and solid-like properties: liquid-like in the sense that molecules need to be able to move around in the system (body fluids, for example, need to be mobile within tissues), but solid-like insofar as the substance must be able to support some weight. Gels can be found in the eye, and also in skeletal joints, where they act as lubricants. Many natural substances can be made to form gels: gelatin, for example, consists of a network of the natural protein collagen "filled up" with water. Collagen is found in tendons, skin, bone and in the eye's cornea. In its natural state the chain-like protein molecules twist around each other into helical fibers; but if the fibers are heated the protein helices become untangled, freeing individual chains. When

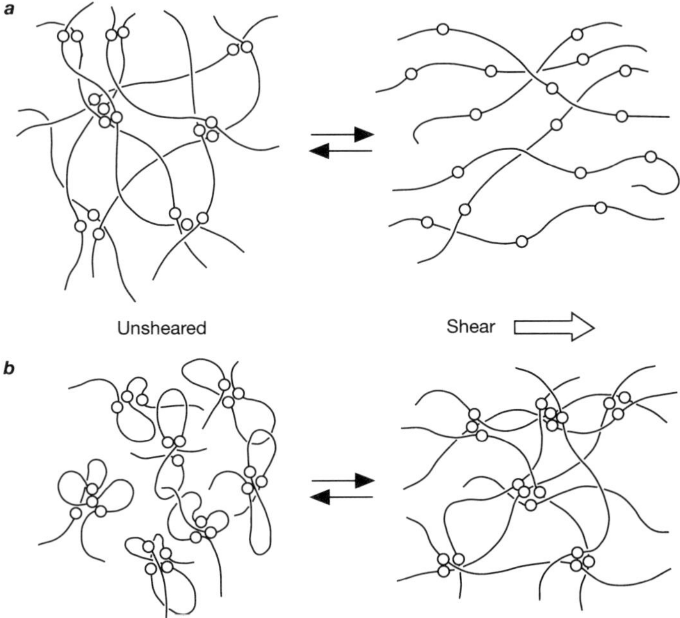


**Figure 7.1** *Thixotropic polymers change their viscosity when stirred. The polymer chains may contain, for example, ionized groups (white circles) which will cluster together in certain organic solvents, thereby acting as "sticky patches." When the clustering is predominantly between groups on different chains, the polymers form an interlinked, somewhat rigid network which can be broken apart by shear forces (a). When the clustering is between groups on the same chain, the individual chains curl up on themselves; but they can be cross-linked when a shear force straightens them out (b).*

the system is then cooled again, the molecules coil around each other once more, but at random, so that rather than forming discrete fibers they interlink into a three-dimensional network – gelatin (Figure 7.2).

The volume occupied by a gel network depends on the amount of liquid solvent trapped within it. The polymer network is flexible, so that as it takes up more solvent it swells like a sponge. When "dried," the gel shrinks. But there are factors other than the amount of encapsulated solvent that can influence the gel's volume. The polymer molecules experience forces of mutual attraction and repulsion, and how far apart they sit (and thus how much volume the gel occupies) is determined by a delicate balance between these opposing forces. Because the balance is very sensitive to ambient conditions such as temperature, the nature of the solvent or the acidity, it is possible to vary the volume of a gel by altering these conditions.

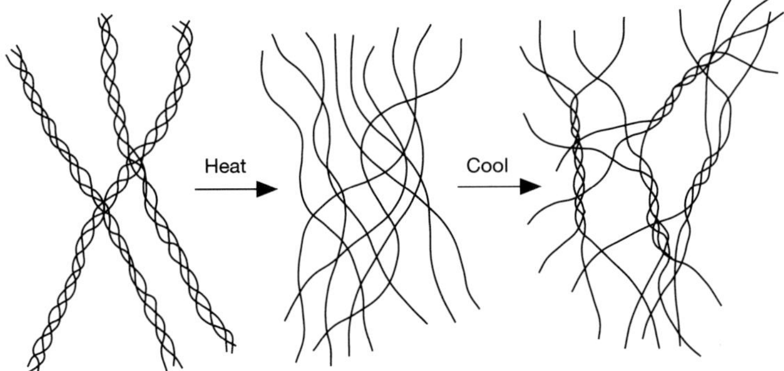


**Figure 7.2** *Formation of gelatin from the protein collagen. The natural state consists of strands in which three protein chains are twisted around each other in a triple helix. On heating, the strands become separated. Subsequent cooling allows the chains to regroup into triple helices, but as this occurs at random, different parts of a single collagen chain may twist around several others, creating a cross-linked gel.*

## *Volume control*

At the Massachusetts Institute of Technology, Toyoichi Tanaka has been devising gels that swell or shrink in response to a whole range of stimuli. Tanaka's gels are based on polymers called polyacrylamides, consisting of a hydrocarbon backbone along which are located amide groups (—$CONH_2$) at regular intervals. Unlike gels such as gelatin, these polyacrylamide networks are held together by strong covalent bonds between the polymer molecules. The links are made at just a few points along the chains, so that the network remains highly flexible and able to expand or contract.

Tanaka modifies this robust network by soaking it in an alkaline solution. This treatment converts some of the amide groups to carboxylic acid groups (—COOH), a process called hydrolysis. These acid groups can shed a hydrogen ion ($H^+$) to form a negatively charged carboxylate group (—$COO^-$). The gel ends up with negatively charged regions dispersed throughout the polymer network, and it is these that are primarily responsible for the swelling and shrinking properties (Figure 7.3). When placed in a mixed solvent of acetone and water, the volume of the hydrolyzed polyacrylamide (PAA) gels depends on the relative proportions of the two components of the solvent: as the amount of acetone increases, the gel shrinks. If the gel contains only a small number of carboxylate groups, this shrinking occurs gradually with increasing acetone content. But for a gel with many carboxylate groups, a curious thing happens: the gel scarcely shrinks at all to begin with, and then at some critical proportion of acetone it collapses suddenly (Figure 7.4). This shrinking can be very pronounced – for a gel soaked in alkaline solution for 60 days (so that many of its amide groups have been hydrolyzed to carboxylates), the volume decreases by 350 times at the jump. The abrupt shrinking transition can also be induced by changing the temperature rather than the composition of the solvent, or by changing the solvent's acidity. A great many gels,

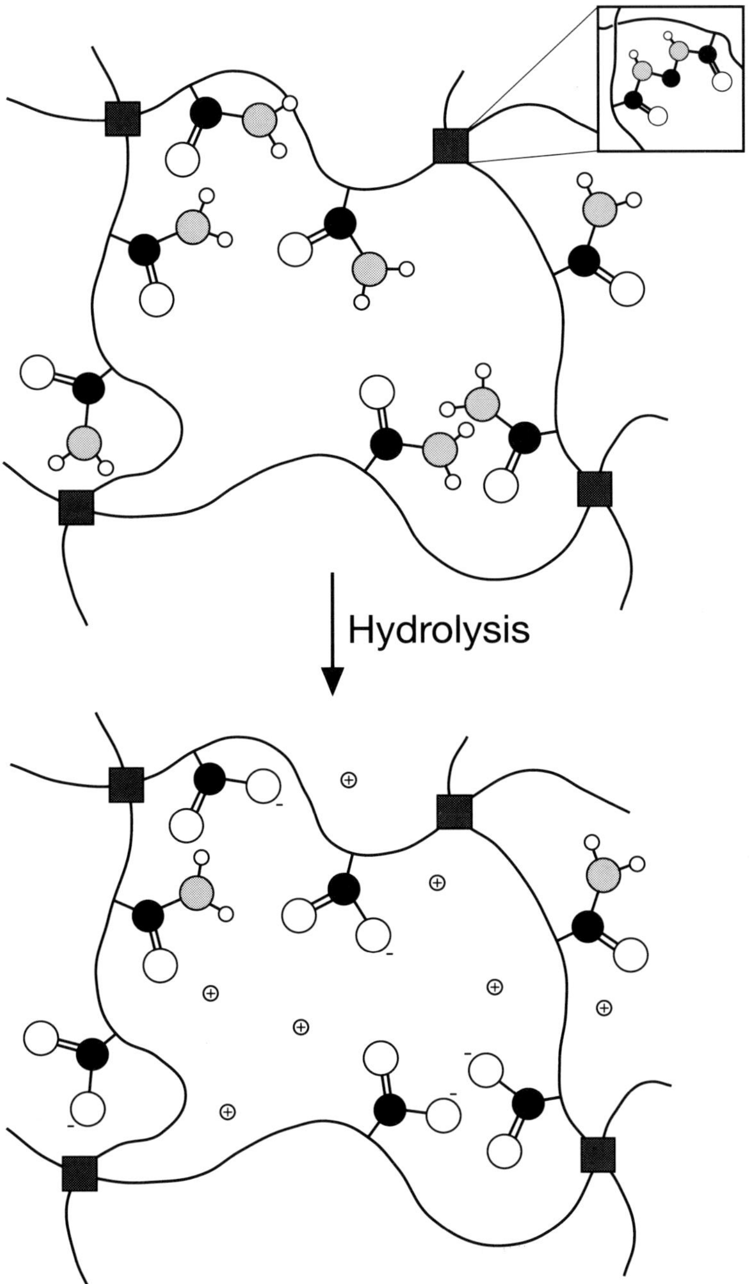


**Figure 7.3** *The polyacrylamide gels studied by Toyoichi Tanaka contain polymer chains linked by covalent bonds (squares). When exposed to alkalis, some of the amide (—CONH$_2$) groups are hydrolyzed to acid groups (—COOH), which then lose a hydrogen ion to form ionic carboxylate groups. Hydrolysis is relatively slow, taking place gradually over several days, so that the proportion of carboxylate groups can be controlled by varying the hydrolysis period. The swelling and shrinking properties of the gel are very sensitive to the degree of hydrolysis.*

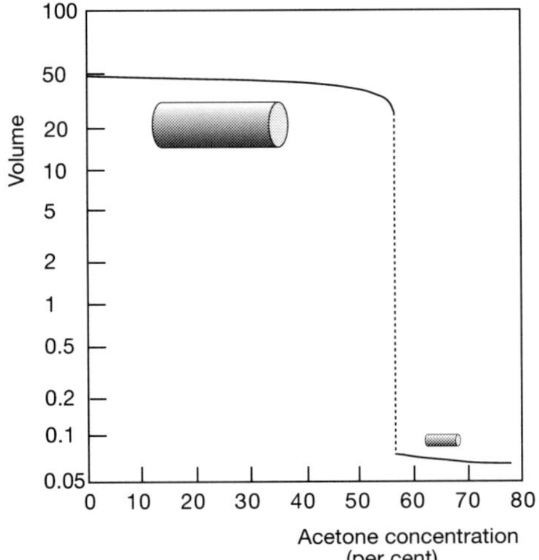


**Figure 7.4** *As the composition of the water/acetone solvent in a PAA gel is changed, the gel swells or shrinks. For an extensively hydrolyzed gel, an abrupt change in volume – a volume transition – may occur at a certain critical composition of the solvent. (After Toyoichi Tanaka, Massachusetts Institute of Technology.)*

both natural and artificial, can be made to undergo swelling or shrinking "volume transitions" of this sort.

The changes in volume are brought about by the competing effects of three forces. First is the elastic force in the polymer network. The polymer molecules have an innate tendency to curl up, which makes the network akin to a system of interconnected springs: it can be stretched and deformed, but such deformations are resisted by forces of elasticity that act to return the molecular chains to their initial configuration (Figure 7.5*a*). Squeezing below the equilibrium volume is also opposed by elastic forces, because the jiggling thermal motions of the building blocks of the polymer chains will tend to keep them apart.

The two other forces involve the solvent which permeates the gel. (In fact the solvent plays a role in elastic forces too, since it is the interactions between the polymer and solvent molecules that determine, in part, how much they coil up.) Depending on their chemical nature, the polymer chains may prefer to be surrounded preferentially either by other chains or by solvent molecules. When there is more than one kind of solvent molecule (as in the water/acetone mixture) the chains may also exhibit a preference for one over the other. The polyacrylamide chains in Tanaka's gels interact more favorably with water molecules than with acetone (that is, the attractive forces are greater), but more favorably still with other polyacrylamide chains. The PAA gel therefore possesses an inherent tendency to shrink, squeezing out solvent and allowing the polymer chains to be surrounded by other chains. Since water molecules oppose this tendency more effectively than acetone, the gel will have a larger volume in water than in acetone;

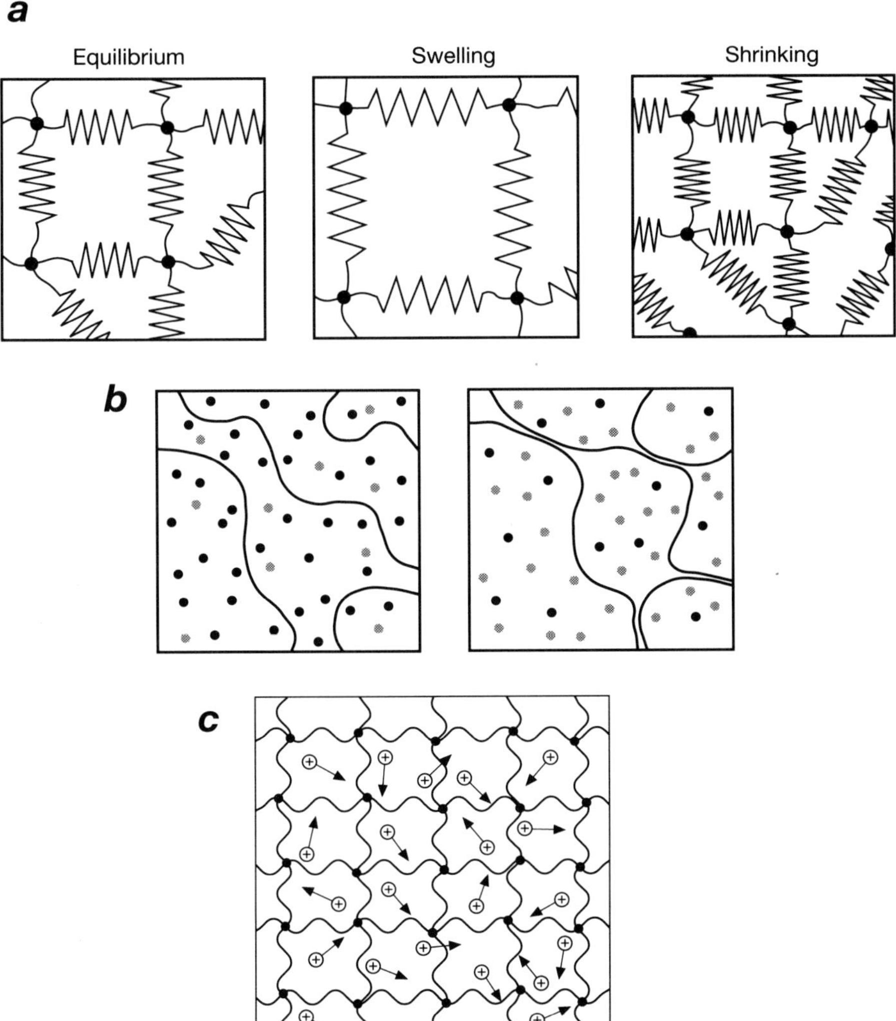


**Figure 7.5** *A gel's volume is determined by the balance between three forces. The polymer chains behave somewhat like a network of springs, subject to elastic forces which act to maintain a certain equilibrium volume, opposing both swelling and shrinking (a). The preference of the chains for one or other of the components of the solvent, or for other chains, makes the volume sensitive to solvent composition (b). Finally, the "osmotic" pressure of hydrogen ions acts to prevent collapse (c).*

shrinking occurs in a mixed water/acetone solvent as the proportion of the latter is increased (Figure 7.5*b*).

Finally, we have seen that the carboxylic acid groups on the hydrolyzed PAA gel can lose a hydrogen ion into solution, so that the network becomes pervaded by hydrogen ions floating in the solvent. The result of interactions between these positively charged ions and the negatively charged carboxylate groups left behind on the polymer network is similar to the effect of gas molecules moving through some sponge-like porous material: there is a kind of pressure, called osmotic pressure, exerted on the network as a consequence of the relatively high concentration of hydrogen ions within it (compared with their concentration in the solvent outside the gel). This has the effect of swelling the network (Figure 7.5*c*). The greater the number of ions in the enmeshed solvent, the greater the osmotic pressure. The carboxylic acid groups lose hydrogen ions more readily in some solvents than in others – water, for instance, encourages the dissociation into ions more effectively than does acetone. (Like the pressure of a gas, the hydrogen-ion osmotic pressure also depends on temperature.)

The interplay between these various forces determines the gel's volume. Changing the composition or the pH (that is, the hydrogen-ion concentration) of the solvent, or changing the temperature, can switch the dominant force from one to another, thereby causing a change in volume. Under some circumstances, these volume changes may be sudden, almost as though one of the forces abruptly gives up the ghost and allows another to take over. Tanaka has demonstrated that a great variety of external influences can be brought to bear on the balance of forces and can thus trigger swelling or shrinking transitions. Electrical fields, for example, exert an effect on the negatively charged carboxylate groups attached to the polymer network, and consequently on the osmotic pressure. By exposing just one region of the gel to an electric field, volume changes can be induced just in that region, producing for example a contracted "neck" in a cylinder of gel (Figure 7.6). One potential use for systems of this sort is as electrically controlled artificial muscles that contract or relax in response to electric currents. Tanaka has also made gels that shrink in response to light by trapping within the polymer network a natural light-absorbing compound called chlorophyllin.

One can imagine many possible uses for materials that can switch or move mechanically in response to light, electric fields, temperature and so forth. Artificial muscles for robotics spring immediately to mind, and several prototypical "polymer hands" have already been demonstrated (Plate 10), along with polymer "fish" that squirm through water in response to an electric field. The fact that polymer gels, unlike metals and semiconductors, can be compatible with biological systems suggests that they might prove useful in designing artificial replacement parts for malfunctioning organs, such as artificial heart valves, and has already prompted their use in other biomedical applications. Swelling caused by changes in acidity can be exploited for delivering drugs to specific locations within the body. Drug molecules entrapped within the network of small polymer gel capsules may be flushed out by the solvent when the gel swells in response to a pH change, such as that encountered in passing from the stomach to the intestines.

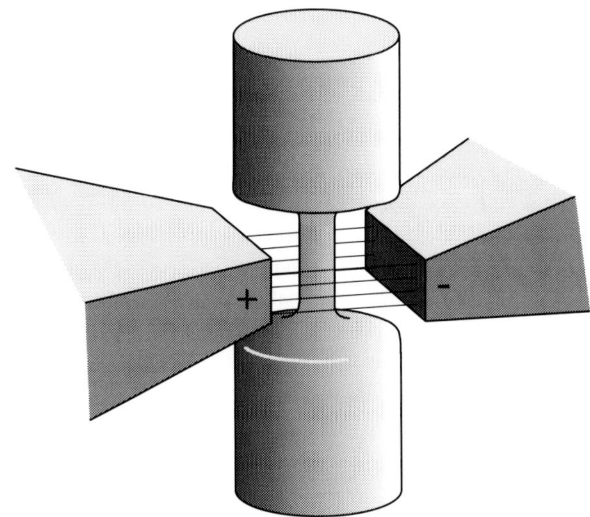


**Figure 7.6** *The influence of an electric field on the ionized groups in the solvent can cause a change in volume, leading to "necking" in a gel cylinder placed in the field.*

These substances represent just one aspect of a burgeoning field of research in materials science – the development of "smart" materials, which alter their properties in response to changes in the surroundings. Electrorheological fluids, for example, can switch from liquid to solid when placed in an electric field, promising a new type of ultrasmooth, wear-free clutch for motor vehicles. Other smart materials may heal their own fractures or change color when they become damaged. Studies of this sort presage a new and exciting concept in materials, according to which changes in ambient conditions provoke not a passive, helpless response, but an "active" change which may, in some cases, produce essentially a different type of material altogether.

## From soaps to cells

### *The chemistry of the kitchen sink*

Advertisements for detergents place great emphasis on the fact that, when it comes to getting clothes clean, those greasy or oily stains are amongst the most stubborn. It is no mystery why this is so: oils and fats are not soluble in water, and so will remain clinging to fabric fibers rather than dissolve. Soap molecules remove particles of grease from fabrics, crockery and kitchen surfaces by giving them a water-soluble coat. One part of a soap molecule is soluble in oil or fat, and therefore becomes embedded in the surface of a grease globule; the remaining portion, which is water-soluble, is left protruding from the surface. The soap molecule therefore has a double nature: part of it likes water, and part likes oil.

Molecules of this sort are called amphiphiles; amphi- derives from the Greek for "both", and -phile from the word for "loving." The amphiphilic molecules in soaps are often called "surfactants," alluding to the fact that they are "surface-active" molecules that do their job at the interface between two different (and in general incompatible) substances. As a general rule, like dissolves in like. Oils and fats contain hydrocarbon chains, and so does the oil-soluble part of a surfactant molecule. The water-soluble part is generally a negatively charged (anionic) "head" group, such as carboxylate ($COO^-$) or sulfonate ($SO_3^-$) (Figure 7.7). Most commercial soaps are carboxylate surfactants. For overall charge neutrality, the negative charge of the head group must be balanced by a positive ion, and in soaps this is usually the sodium ion, $Na^+$. A soap therefore typically has the chemical formula [$CH_3$—$(CH_2)_n$—$CO_2^-Na^+$], with $n$ lying in the region of 10 to 18.

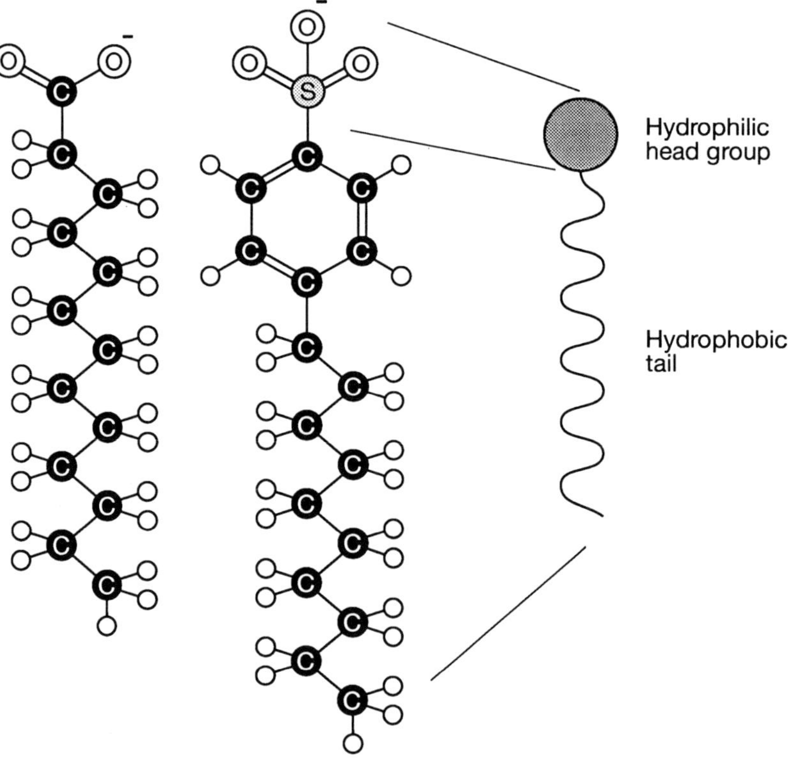


**Figure 7.7** *Surfactant molecules contain a charged "head" group which is soluble in water and a hydrocarbon tail which is soluble in oil.*

The water-loving head groups are called "hydrophilic," and the oil-loving (and thus water-fearing) tails "hydrophobic." Surfactant molecules will dissolve in water, but they prefer if possible to shelter their hydrophobic tails from the water molecules. Burying their tails in oil globules is one way of doing so; but there are many other ways in which surfactants achieve this end, and these give rise to an extremely rich variety of molecular structures which now provide one of the primary focuses of colloid chemistry.

## *The pull of the crowd*

When present in solution in small quantities, surfactant molecules will tend to gather at the water surface. Here they can keep their hydrophobic tails out of contact with the water by lying head-down with the tails poking out into the air. At the surface of pure water, the $H_2O$ molecules cannot experience as many attractive, stabilizing interactions with neighbors as do molecules in the bulk of the liquid, and so the surface molecules have a high energy relative to the bulk. Therefore, the presence of the surface carries an energy cost; the larger the surface area, the greater the energy cost. We generally speak of this surface "excess energy" as the surface tension, since its effect is to "pull" the liquid into a compact form, keeping the surface area as small as possible. This is why water droplets in mist are spherical and why, when they sit on a plastic or oily surface, the droplets form lens-like beads rather than spreading under the pull of gravity (Figure 7.8). But a layer of surfactants at the water surface has the effect of lowering the energy cost of the surface (in other words, lowering the surface tension), because the surface layer now comprises the hydrophobic tails of the surfactants, which didn't want to be in the water *anyway*. Adding a small amount of soap to a water droplet sitting on a surface thus enables it to spread.

Surfactants at the surface of water essentially form a membrane between the liquid and the air, allowing extremely thin liquid films to be stabilized in bubbles and foams. A hypothetical bubble of pure water would simply collapse into a droplet of minimal surface area; but by decreasing the surface tension, surfactants reduce the energy cost of large surface areas. A foam is simply a large number of bubbles packed together. Foams are of considerable commercial and industrial interest, being useful in processes as diverse as fire-fighting and mineral extraction. Because they can be rather robust despite their very low density, foams provide a semirigid yet extremely light blanket that will float on top of burning oil and thereby exclude the air that the fire needs to keep burning.

If the amount of surfactant in solution is increased, there comes a point at which it can no longer all accumulate at the surface. The surfactant molecules must then find other ways of shielding their hydrophobic tails from water. One such way is for the molecules to aggregate into clusters in which the tails point inwards, with the head groups forming a water-soluble shell (Figure 7.9). These structures, called micelles, are just like those formed when the surfactants surround a globule of grease, except that there is generally nothing inside the micelle but the grease-loving tails themselves. The formation of micelles, which occurs when the amount of surfactant exceeds the "critical micelle concentration," can be detected by passing a beam of light through the

**Figure 7.8** *A water droplet on a hydrophobic surface. Surface tension acts to pull the droplet into a lens, but the addition of surfactant reduces the surface tension and allows the droplet to spread. (Photograph courtesy of Isao Noda, The Proctor & Gamble Co., Cincinnati.)*

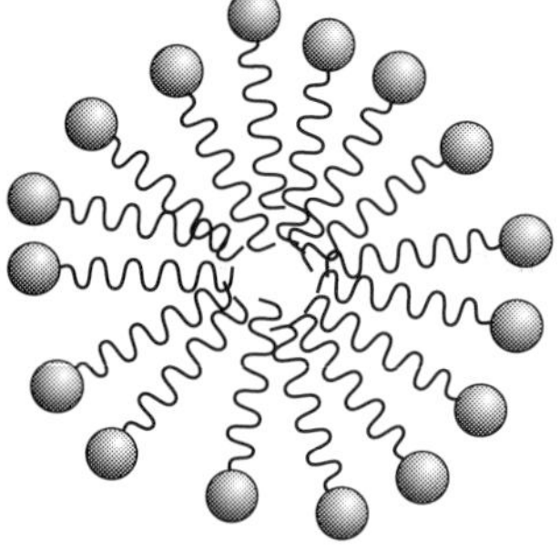

**Figure 7.9** *Surfactants in water may shield their hydrophobic tails by forming micelles, clusters in which the tails are buried in the interior.*

solution, whereupon one can see the path of the beam clearly illuminated. This effect was discovered in the nineteenth century by the British physicist John Tyndall, and is due to the scattering of light by the micelles. Because they have dimensions similar to the wavelength of visible light, many colloidal systems scatter light strongly, and the Tyndall effect is a characteristic signature of their presence.

One might imagine that it would take considerable ingenuity to arrange for a large number of amphiphilic molecules to come together as a micelle. But the unfavorable interactions between water and the hydrophobic tails provides all the driving force that is necessary to enable the molecules to organize themselves. This is a process of supramolecular self-assembly of the kind encountered in Chapter 5, but on a larger scale, perhaps involving tens, hundreds or thousands of molecules: such structures are said to be self-organizing. Organization is used here in a rather loose sense, however: micelles are somewhat disorderly structures in which the surfactant molecules are packed together imperfectly. Individual molecules can leave the cluster quite easily, and new ones can be incorporated.

In oily solvents such as liquid hydrocarbons (paraffins), surfactants will form inside-out or "reverse" micelles. Here the surfactants try to protect the hydrophilic heads from the solvent by aggregating with the heads inward and the tails poking out (Figure 7.10). Further variation is to be found in the form of cylindrical micelles, in which the molecules gather into rod-like assemblies (Figure 7.10). When in close proximity, cylindrical micelles may line up like stacks of logs, forming structures similar to liquid crystals (described later in this chapter).

The inside of a small micelle is filled up with hydrophobic tails; but larger micelles contain water-free cavities which can enclose water-insoluble substances. Surfactants can therefore stabilize a dispersion in water of a liquid which will not otherwise mix, preventing the two from separating out into distinct layers. Vigorous shaking of a simple oil-and-water mixture will disperse one of these two phases in the form of tiny droplets within the other; the mixture turns cloudy because of the strong light scattering from the tiny colloidal droplets. But as French dressing demonstrates, the two phases will settle out again when left to stand (in French dressing, the "water" phase is vinegar, essentially a solution of acetic acid). If a surfactant is added to the mixture, however, it will stabilize the dispersed droplets by coating them with a layer that is soluble in the other phase. If you can bear the waste, you might try adding a little washing-up liquid to French dressing to verify this. The resulting stable dispersion is an example of an emulsion – a colloidal dispersion of one liquid in another. To stabilize an emulsion (that is, to prevent it from separating out) is an important practical challenge for industrial colloid science, particularly in the food and paint industries. Emulsions occur also in nature, the most familiar being milk. This is a dispersion of fats and proteins in water, and very strong scattering of light by the colloidal fatty particles is what gives milk its opaque whiteness – if separated out, the various components would be transparent.

The interiors of reverse micelles are now being exploited by some chemists as miniature reaction vessels for chemical reactions. Researchers have found this approach

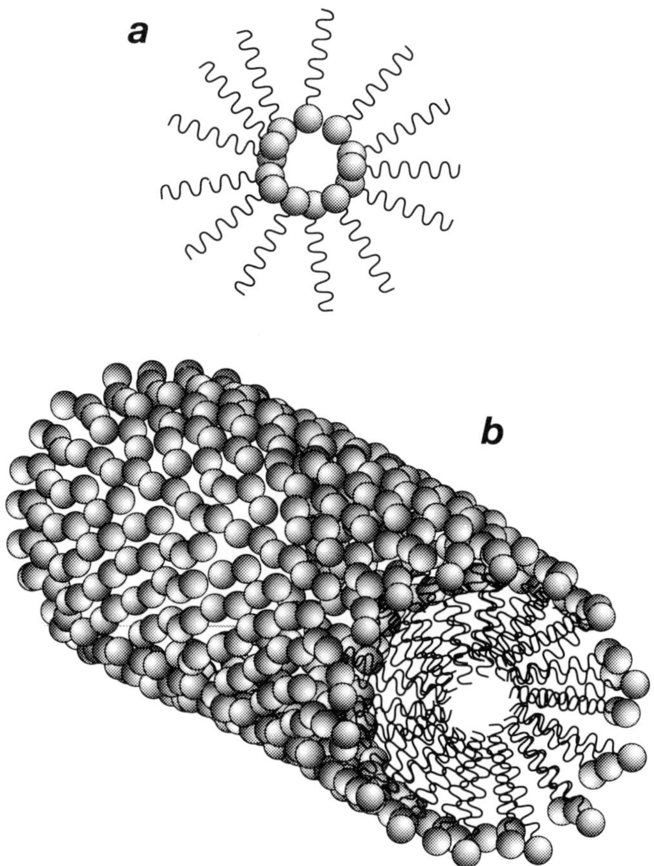


**Figure 7.10** *A reverse micelle (a) formed by surfactants in an oily solvent; and a cylindrical micelle (b).*

particularly valuable for synthesizing colloidal-scale solid particles, which may have useful catalytic or electronic properties. When added to an oily solvent containing reverse micelles, an aqueous solution of the (generally ionic) components of the solid particles will find its way to their water-loving interior. In this encapsulated water pool the concentration of the ions may increase until the solid precipitates out, and the reverse micelle acts as a kind of mold to determine the size and the shape of the precipitated particle. Because it is possible to arrange for the reverse micelles all to have a similar size, this technique permits the formation of roughly spherical crystals of almost uniform diameter. Researchers at AT&T Bell Laboratories in New Jersey and at Sandia National Laboratories in Albuquerque have used the reverse-micelle method to prepare small particles of the semiconducting material cadmium selenide, which they hope will show a novel type of luminescent behavior.

*Micelles that come to life*

In Chapter 5 we saw how synthetic molecules can be designed that makes copies of themselves in a way that recalls the self-replication of DNA, the central molecular component of all known life. The creators of these chemical systems hope that they might provide insights into the way that life developed on Earth, a topic discussed in detail in the next chapter. A very different kind of replicating chemical system has been devised by Italian chemist Pier Luigi Luisi and colleagues at the Eidgenossische Technische Hochschule (ETH) in Zürich. The ETH team has explored ways in which micelles can be made to enhance the rate of their own assembly, just like the molecular "template" replicators of Julius Rebek (page 178). Within the same highly restricted sense pertaining to the latter, these self-replicating micelles can be considered to show some of the characteristics of living organisms.

Luisi's idea is a simple one. Given the known capacity of micelles (and their inside-out relatives) to act as tiny vessels within which chemical reactions can take place, might it be possible to carry out inside a micelle the very reaction that produces its amphiphilic components? If this reaction takes place more readily within the micelle than in the solution outside, the micelle will speed up the rate at which further micelles are formed; it will, in other words, be autocatalytic.

It turns out that this kind of behavior can be coaxed out of a wide variety of micelle-forming amphiphiles. The first autocatalytic system identified by Luisi and colleagues involved the soap sodium octanoate ($CH_3—(CH_2)_6—CO_2^-Na^+$) in a mixed solvent of nine parts isooctane to one part octanol. Isooctane is a water-insoluble hydrocarbon, so the surfactant forms reverse micelles in this solvent. (The role of the octanol is subtle, as it is somewhat soluble in both the hydrocarbon and in water.) When a small amount of water is added to this system, it is encapsulated in pools within the hydrophilic interior of the reverse micelles.

The ETH researchers added to this colloidal dispersion reagents that will form further octanoate surfactant: ethyloctanoate (a type of compound known as an ester) and lithium hydroxide, which hydrolyzes the ester into octanoate ions and ethanol. Lithium hydroxide is rather insoluble in isooctane, so the hydrolysis reaction is not very efficient when these components are added to the isooctane solvent alone. But with water-containing reverse micelles present, the ester and the lithium hydroxide can dissolve in the water pools, and hydrolysis proceeds there readily (Figure 7.11). The octanoate molecules so produced then escape from the reverse-micelle interiors and group together into new reverse micelles. Thus the micellar structures replicate, in a crude sense, when provided with the raw materials.

In their early experiments, Luisi and colleagues needed a few reverse micelles ready-made in the solvent to set the ball rolling; but they later produced replicating micelles starting from nothing but the ester precursors to the surfactants, along with a hydrolyzing agent. Once enough of the ester had been hydrolyzed to surfactant to allow reverse micelles to form, the process suddenly took off as the micelles catalyzed further hydrolysis. Micellar structures then proliferated abruptly, like a living colony finding its feet.

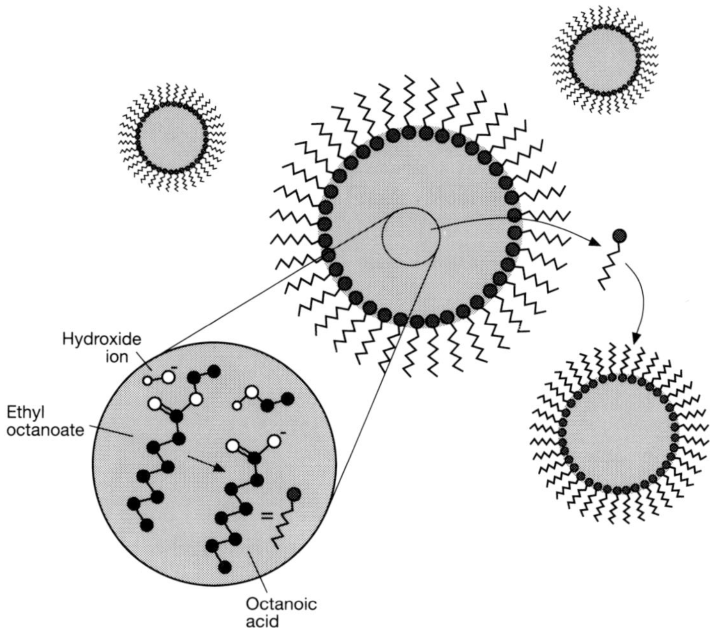


**Figure 7.11** *Reverse micelles that make copies of themselves. In the water pools (gray) within the reverse micelles, ethyl octanoate is rapidly hydrolyzed to form octanoic acid, which then escapes to form more reverse micelles. The reverse micelles thereby act as autocatalysts, speeding up the rate of their own formation.*

Is it conceivable that the earliest proto-organisms on our planet might have been self-replicating micelles? One consideration that renders this a suggestive possibility is the similarity between micelles and cell membranes, both of which are self-organized structures of amphiphilic molecules. But there is much more within a cell than merely water, of course. An empty cell membrane – even one that can replicate – cannot be considered a very good approximation to a living organism. For one thing, it has no means to store genetic information and pass it on to subsequent generations; in other words, it cannot evolve.

## *Towards the model cell*

If the concentration of a surfactant in solution is increased far beyond the critical micelle concentration, new kinds of structure appear that have a greater degree of self-organization. The principal structural motif for these new phases is called a bilayer, in which the surfactant molecules line up side by side to form sheets; to shield the hydrophobic tails from water, two sheets lie back to back with the tails pointing inwards. To avoid exposing hydrocarbon tails at the edges of the sheets, they can curl in on themselves to form closed sacs, called vesicles (Figure 7.12). Cell walls are essentially vesicles comprised of bilayers of natural amphiphiles, most commonly those called

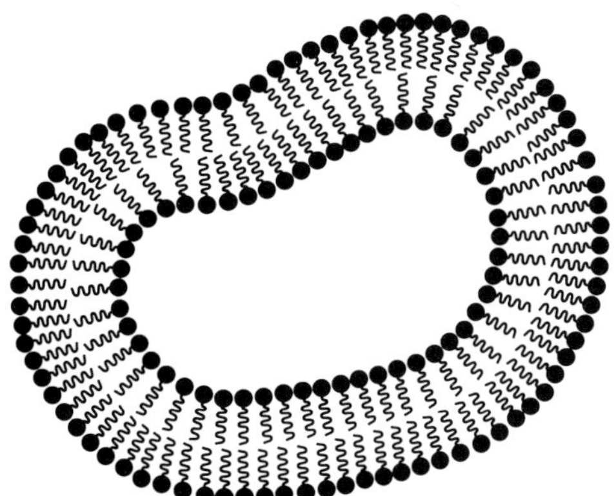

**Figure 7.12** *Cross-section of a vesicle formed from an amphiphilic bilayer. Amphiphiles line up back to back in sheets, which then close up to form enclosed, sac-like structures.*

phospholipids. These molecules have a hydrophilic phosphate head joined to two hydrocarbon tails (Figure 7.13*a*).

The spontaneous self-organization of phospholipids and other amphiphiles into vesicles, first observed in 1961 by Alec Bangham of the Institute of Animal Physiology in Cambridge, enables scientists to study some aspects of cell behavior using "model cells" – vesicles filled with nothing but water. Like micelles, bilayer vesicles are generally rather loosely bound molecular assemblies – their components are not linked by chemical bonds but are held in place by weaker "hydrophobic" forces arising from the aversion of the hydrophobic tails to water. The amphiphiles are relatively free to move laterally through the layer, like people jostling past each other in a closely packed hall. In the bilayer membrane that encompasses a single bacterium, a phospholipid molecule can move from one end of the membrane to the other in about one second.

Not all cell walls are so fluid, however. The membranes of some cells in animals contain cholesterol molecules, which are amphiphiles in which part of the hydrophobic tail is rigid, unlike a flexible hydrocarbon chain (Figure 7.13*b*). Cholesterol acts as a membrane rigidifier, making the membrane stiffer and more robust. The bilayer membranes of red blood cells, meanwhile, are strengthened by a protein skeleton, a web of strands of a protein called spectrin which is attached to the membrane via other proteins embedded within it. The diseases spherocytosis and elliptocytosis, which cause anemia, result from genetic mutations which affect the body's ability to produce the proteins comprising this cell skeleton, giving rise to red blood cells with abnormal shapes.

More generally, the bilayers of cell walls provide a kind of matrix in which "active" protein components are incorporated. These membrane proteins control aspects the cell's behavior, such as the way that it responds to molecules encountered in the solution around it. Molecular recognition processes controlled by membrane proteins play a central role in the biochemistry of the immune system. We saw in Chapter 5 that most cell membranes, such as those of nerve cells, are pierced by "channels" which allow substances (in nerve cells, metal ions such as potassium and calcium) to pass through. Differences in the concentrations of metal ions on each side of nerve cell walls gives rise to the electrical signals that travel through the nervous system.

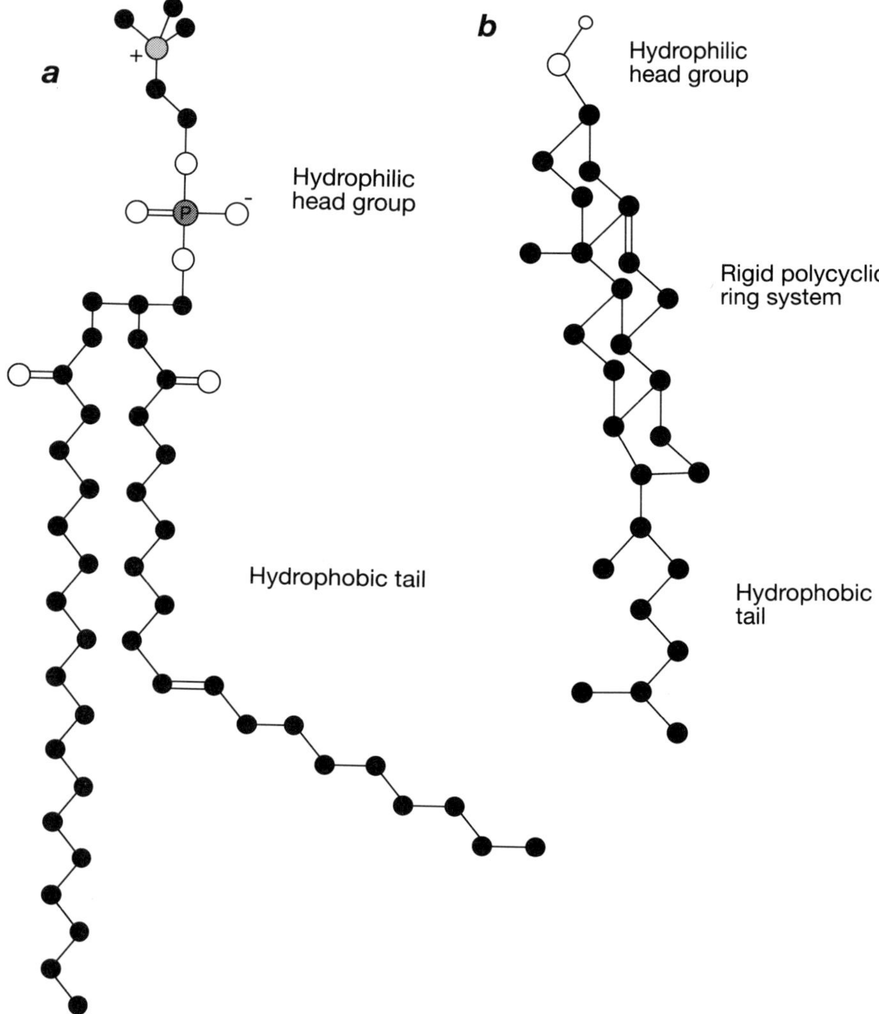


**Figure 7.13** *Cell membranes are composed primarily of phospholipid amphiphiles (a). Cholesterol molecules (b) act as rigidifiers which make the membrane stiffer. (Hydrogens not shown for carbon atoms.)*

## *Molecular delivery bags*

Because of the lack of chemical bonds between the molecular components of vesicles, they can be pierced or burst fairly easily, like soap bubbles. One of their more striking properties, however, is the ability to fuse or divide. When two vesicles touch each other, their membranes can join together at the point of contact to create a single, larger vesicle. Conversely, a vesicle can develop a "bud" – a kind of sprouting pseudopod –

which can close up on itself and become detached to form a smaller vesicle. These processes are extremely important for cell biology, as they provide means by which particles or substances can be incorporated into cells or expelled from them. Foreign particles can enter a cell by being enveloped in a vesicle that buds off from the cell wall (Figure 7.14*a*). This is known as endocytosis. (Cyto- comes from the Greek *kytos*, meaning hollow vessel; the study of cells is called cytology.) Conversely, foreign particles can be removed from inside a cell by enclosing them in a vesicle which then merges with the cell wall, a process called exocytosis (Figure 7.14*b*). Researchers have recognized that artificial phospholipid vesicles might be able to carry substances such as drugs and to inject them into cells via processes such as membrane fusion and endocytosis. These phospholipid vesicles were called liposomes in early research. Today the term liposome is often applied to bilayer vesicles in general.

Liposomes used as agents of drug delivery are prepared outside the body in the presence of the drug that they are supposed to deliver, thus encapsulating the drug molecules. The drug-filled liposomes are then injected into the body, where they prevent the drug from having any physiological effect until the liposomes reach their target cells and deliver their charges.

The way in which a liposome interacts with cell wall depends on the precise chemical nature of the phospholipids in the two membranes. Most liposomes will simply stick to cell walls; the drug molecules in the liposome can then diffuse slowly out of it and into the cell. By varying the composition of the liposome, one can control its readiness to stick to different types of cell and the length of time taken for the drug to diffuse out, allowing a great amount of selectivity in the delivery system.

Alternatively the cell may swallow the liposome whole through endocytosis. Once inside the cell, the liposome will be gradually broken apart, allowing its contents to be released. A further possibility is that exchange of phospholipid molecules takes place between the cell wall and a liposome attached to it. By binding the drug molecules to exchangeable amphiphiles in the liposome wall during the preparation stage, the drug can be introduced into the cell by this process. In rare instances, the liposome and cell walls may actually fuse. These interactions are illustrated in Figure 7.15.

The use of liposomes as drug delivery systems holds much promise. It has already met with some success in the treatment of diseases: liposomal delivery of the drug doxorubicin, for example, which is used in the treatment of malignant tumors and leukemias, reduces the otherwise severe side effects. The use of liposomes to administer the anti-cancer drug anthacycline affords a tenfold increase in the efficiency with which tumors can be targeted selectively. Liposomes are now regarded as a highly promising vehicle for gene delivery, whereby encapsulated segments of DNA and RNA are delivered into cells to replace faulty genes. It is hoped that this approach, called gene therapy, will prove effective for the treatment of a variety of diseases.

But liposomal delivery is not without its difficulties. In particular, the immune system is remarkably good at spotting foreign intruders into the body, even when they resemble cells. Liposomes in the bloodstream are often quickly recognized as such and destroyed by antibodies.

**a**

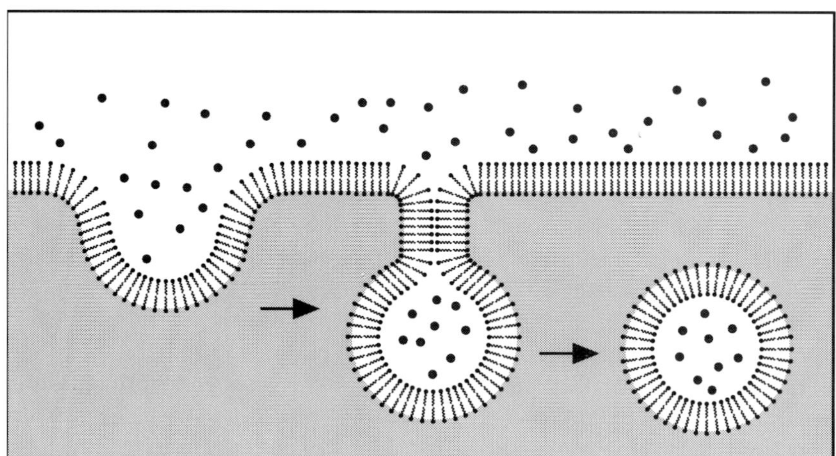

**b**

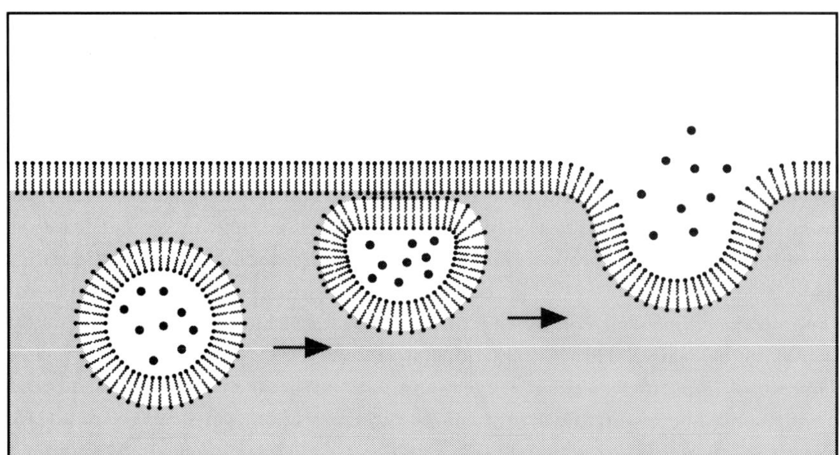

**Figure 7.14** *Cells ingest and expel foreign particles via the processes of endocytosis (a) and exocytosis (b), respectively. The gray region is the cell interior.*

## *A three-sided game*

We have seen how surfactants can be used to stabilize colloidal dispersions of one liquid in another, when they would otherwise separate out. Oil droplets can be dissolved in water by enclosing them within micelles, and water can be similarly dispersed in oil within reverse micelles. These represent just two of a tremendous variety of oil-and-water

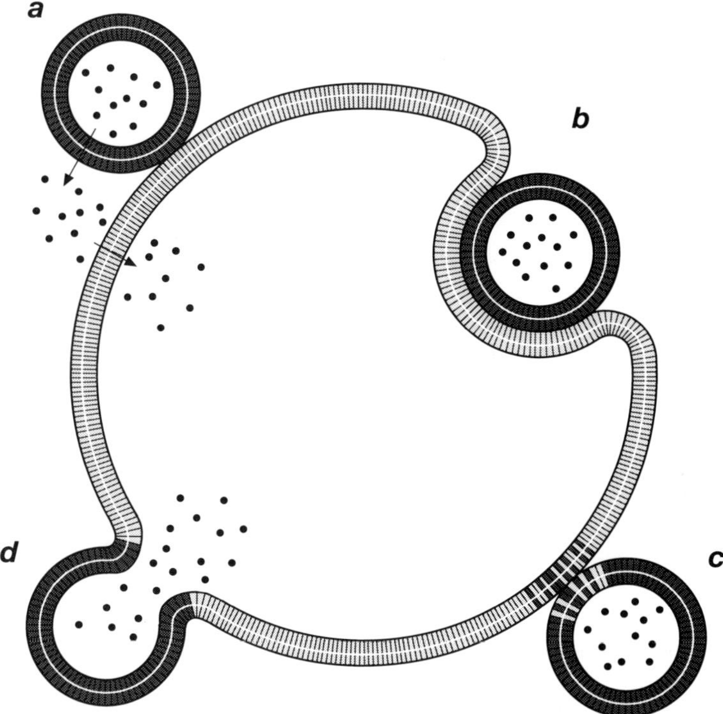


**Figure 7.15** *Mechanisms of drug delivery into cells via liposomes. Many liposomes simply adhere to the cell surface, whereupon the drug molecules that they contain will diffuse through the liposome wall and into the cell (a). Liposomes may be ingested into the cell by endocytosis (b); they are subsequently broken down, releasing their contents. A liposome adhering to the cell wall may exchange lipids (c). Fusion of the liposome with the cell wall (d) is more rare.*

arrangements, or phases, that can be achieved by the addition of a surfactant. Which phase is adopted will depend, in general, on the relative proportions of each of the three components. Dispersions of two insoluble liquids that are stabilized by surfactants into colloidal structures that are intimately mixed at the microscopic scale are called microemulsions.

If we add more oil to the oil-filled micelle phase, the micelles swell to large and squashy proportions. When two oil-filled micelles encounter each other, they may merge. Ultimately, the micelles will link up to form extended, branching globules of surfactant-coated oil throughout the solution. It is then no longer clear that we should regard this as a dispersion of oil in water any more than of water in oil: the system has become two random labyrinths of these components, with surfactants stabilizing the interface between them (Figure 7.16*a*). In this microemulsion the oil network and the water network are more or less continuously connected, forming a so-called *bicontinuous* double-labyrinth structure.

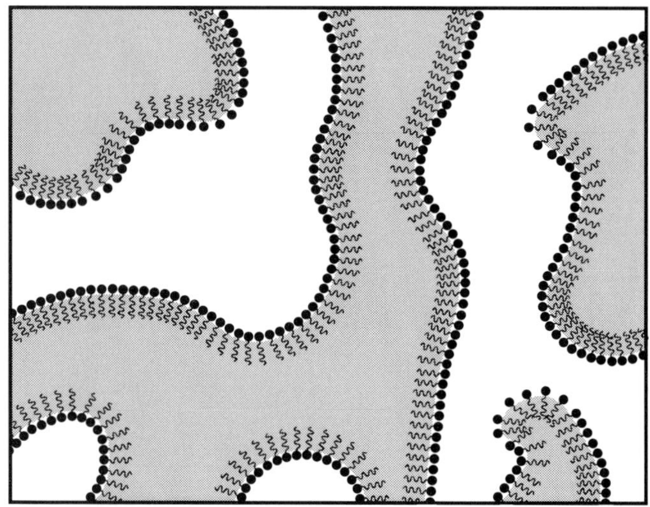



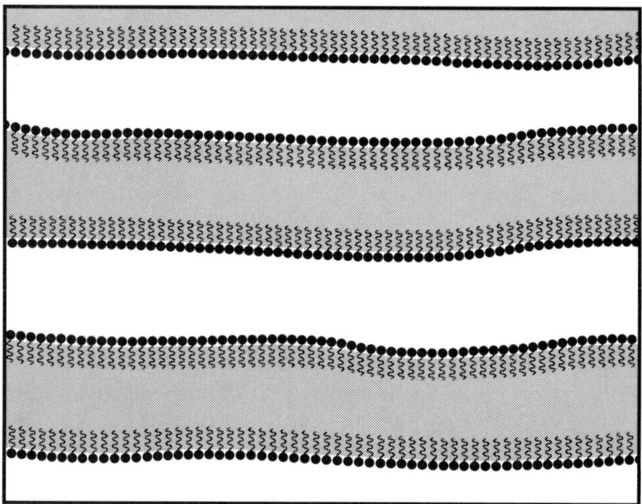

**Figure 7.16** *In microemulsions, oil (gray) and water (white) are mixed at the microscopic scale, stabilized by surfactants at the interfaces. The oil and water phases may be disorderly (a), sometimes forming continuous interpenetrating networks. In lamellar phases, the oil and water lie in flat slabs separated by a layer of surfactants (b).*

The irregular surfactant-coated interface between oil and water is highly curved. In microemulsions, and in amphiphilic structures generally, curvature costs energy because it exposes those parts of the amphiphiles that should be "hidden." Curving of the interface can be avoided in a structure in which flat slabs of oil and water alternate with one another, with a surfactant layer between each slab (Figure 7.16*b*). This is called a lamellar phase.

Perhaps the most remarkable phases of the oil-water-surfactant system are the ordered bicontinuous structures. We saw earlier that interfaces have a free-energy cost; to keep this cost to a minimum (that is, to minimize the area of the interface), bicontinuous microemulsions may adopt structures very different from the random, disorderly networks represented in Figure 7.16*a*. Bicontinuous structures of minimal surface area, called minimal surfaces, have long been known to mathematicians and were studied by the German H. A. Schwartz in the nineteenth century. They are periodic, ordered structures in which a certain structural element repeats again and again, like the unit cell of a crystal (Chapter 4). Minimal surfaces have elegant and mind-boggling shapes (Plate 11): some comprise interlocking periodic arrays of tunnels, for obvious reasons called "plumber's nightmare" structures.

## *Crystals in flatland*

Surfactant molecules at a water surface are generally highly disordered: their tails, protruding from the surface, flap about with abandon, and the position of one molecule takes little account of any other. But if the molecules are more closely packed, each feels and responds to the presence of the others, so that it starts to make sense for the molecules to align themselves in a more ordered fashion. With increasing density of molecules at the surface, the arrangement will eventually become highly ordered, with the molecules stacked side by side in a regular, crystalline manner.

In the early part of this century, the Scottish scientist Irving Langmuir discovered that this kind of well-ordered packing can be induced in a very simple way. He prepared a surfactant layer on water in a rectangular trough, across the top of which (in contact with the water surface) was a movable barrier (Figure 7.17). Compressing the surfactant layer at the water surface by moving the barrier brings about several abrupt changes in the structure of the surface film. These two-dimensional surfactant films at a water surface have become known as Langmuir films.

Langmuir films provide the opportunity to compare phenomena in a two-dimensional "flatland" with the corresponding behavior of our familiar three-dimensional world. In the latter, there are three classical states (or phases) of matter: gas, liquid and solid, which are increasingly dense. Changes (transitions) from one of these phases to another can be induced by altering the substance's density – by compressing it, for example. The same is true in flatland. A few widely separated surfactant molecules on the water surface can be considered to comprise a two-dimensional (2D) gas – the molecules are independent of each other, but are constrained to stay within the flat plane. To increase

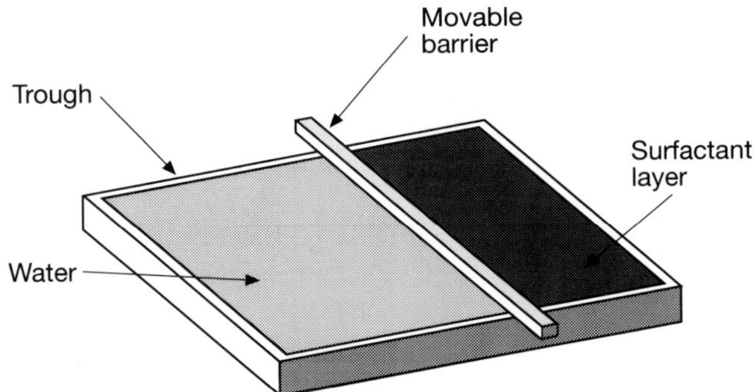


**Figure 7.17** *A Langmuir trough. Surfactant molecules lie at the water surface, and their surface density can be controlled by moving a barrier to alter the area within which they are contained.*

the surface density of the 2D gas, one need simply reduce the area of surface within which they are free to move, which can be achieved using a Langmuir-type barrier.

The structure of Langmuir films can be studied by a technique called fluorescence microscopy. Amongst the surfactant molecules one scatters a small number of "probe" molecules that fluoresce when they are irradiated with light from a laser or an arc lamp. Because the fluorescent molecules are soluble to different degrees in the various 2D phases of the surfactant, regions containing different phases are revealed as patches that differ in brightness when the fluorescent emission is observed with a microscope.

In the 2D gas phase, the probe molecules are widely dispersed, and their fluorescence is too faint to be seen: the phase looks dark. But as the film is compressed by moving the barrier, bright spots suddenly appear – these correspond to regions of a 2D liquid phase, in which the density of molecules (including the fluorescent probe molecules) is much greater. Upon further compression, the bright spots merge to form a continuous bright field, and the dark bubbles of 2D gas shrink and finally vanish.

But as the surface pressure is increased still further, the 2D liquid undergoes a transition to a new type of phase, peculiar to these two-dimensional films. This phase is a kind of intermediate between a liquid (in which the surfactant molecules are disorderly) and a solid (in which they are packed regularly together). It is not a true liquid, because there is some evidence (obtained by X-ray diffraction) of a regularly repeating molecular arrangement, but it is far from perfectly orderly. In this phase the hydrophobic tails poking out above the water are somewhat aligned with one another, but the positions of the molecules themselves remain considerably disordered. This phase is called the liquid-condensed (LC) phase, whereas the preceding, less dense phase is called the liquid-expanded (LE) phase. Because the probe molecules do not dissolve readily in the LC phase, it appears dark in fluorescence microscopy. The formation of the LC phase is therefore marked by the reappearance of dark spots (Figure 7.18). Further squeezing of the film will transform it from the LC phase into a true 2D solid, in which the molecules are packed regularly with their tails aligned (Figure 7.19).

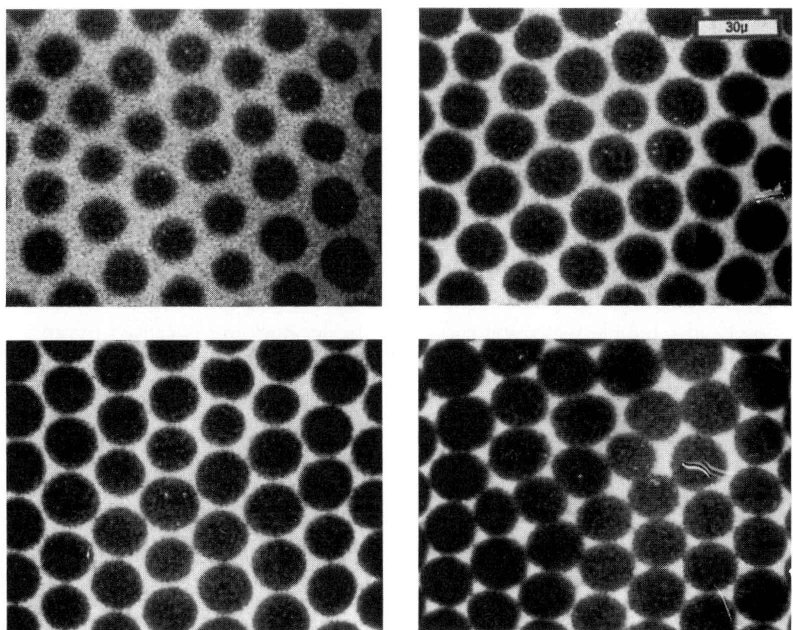


**Figure 7.18** *Growth of the liquid-condensed (LC) phase in the liquid-expanded (LE) phase of a Langmuir film, viewed by fluorescence microscopy. The dark circles are the LC phase; these grow bigger as the surface density of surfactant is increased. (Photograph courtesy of H. Möhwald, Universität Mainz.)*

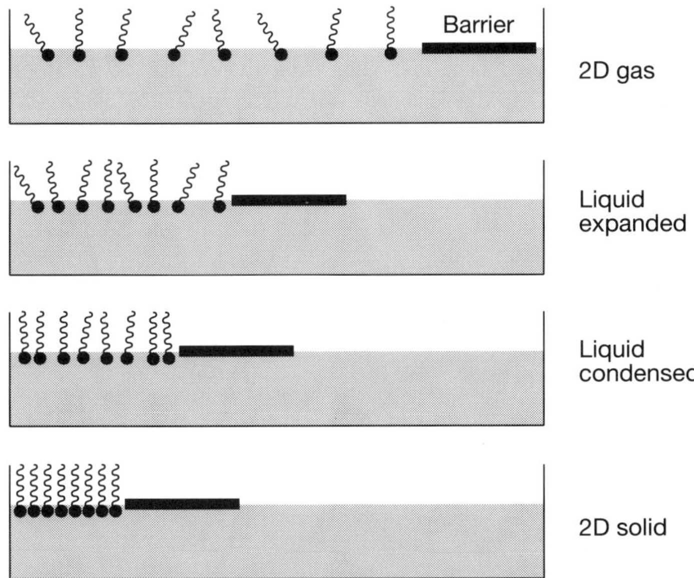


**Figure 7.19** *The structures of the 2D gas, LE, LC and crystalline ("solid") phases of Langmuir films.*

The LE phase can exist only above a certain temperature; below this, the 2D gas condenses directly into the LC phase. Dropping the temperature of a LE film suddenly into this unstable regime creates the psychedelic pattern shown in Figure 7.20, where the gas phase (circles and flower centers) and LC phase (flower "petals") are both forming within the bright LE phase. Another curious structure is produced from bubbles of 2D gas in the LE phase, which can form a kind of 2D foam in which the bubble walls are thin sections of the LE phase (Figure 7.21). Charles Knobler and colleagues from the University of California at Los Angeles found that if the temperature of the foam is raised slowly, the bubble walls will suddenly buckle. Buckling is caused by an increase in the proportion of the LE phase in the bubble walls, which means that they push against each other at their vertices until eventually the network can take the strain no longer.

Another complex pattern that results from phase transitions in Langmuir films is shown in Figure 7.22*a*. This structure, called the stripe phase, can arise when the LC phase grows within the LE phase. The former appears initially as dark circular spots, which repel each other as a consequence of interactions between the surfactant molecules. As the LC domains grow, this repulsion causes the circular domains to deform into worm-like shapes. Equally dramatic is the branching structure shown in Figure 7.22*b*, which is formed by the growth of the 2D solid phase of a phospholipid in a liquid phase. Structures like these, called dendrites, are commonly formed in rapid growth processes – we will encounter some other examples in Chapter 9.

### *Peeling off the layers*

In the 1910s, Irving Langmuir and his student Katharine Blodgett devised a way to transfer compressed, ordered Langmuir films from the surface of water onto a solid surface, such as a glass slide. When a slide is carefully dipped into a film-coated solution, the hydrophilic heads of the surfactants stick to the (hydrophilic) surface of the glass, allowing the film to be pulled out of the water. To prevent the partially adhered film from breaking off and leaving the rest behind, the surfactant layer must be continually

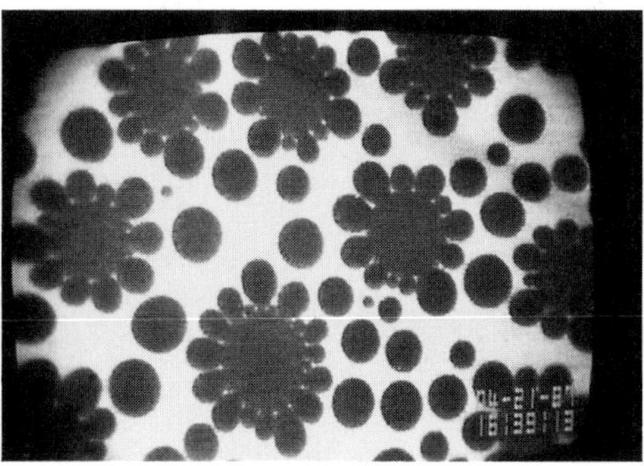

**Figure 7.20** *Patterns formed in the LE phase by rapid cooling (quenching) to a temperature at which only the 2D gas and LC phases are stable. The rapid growth of these two phases can give rise to a variety of striking patterns. (Photograph courtesy of Charles Knobler, University of California at Los Angeles.)*

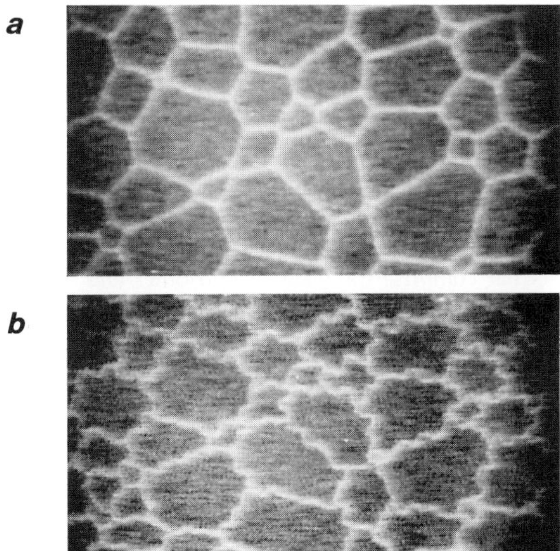


**Figure 7.21** *A foam in flatland. The dark areas are regions of 2D gas, separated by thin walls of the liquid-expanded phase (a). When this foam is heated, the walls buckle (b). (Photographs courtesy of Charles Knobler, University of California at Los Angeles.)*

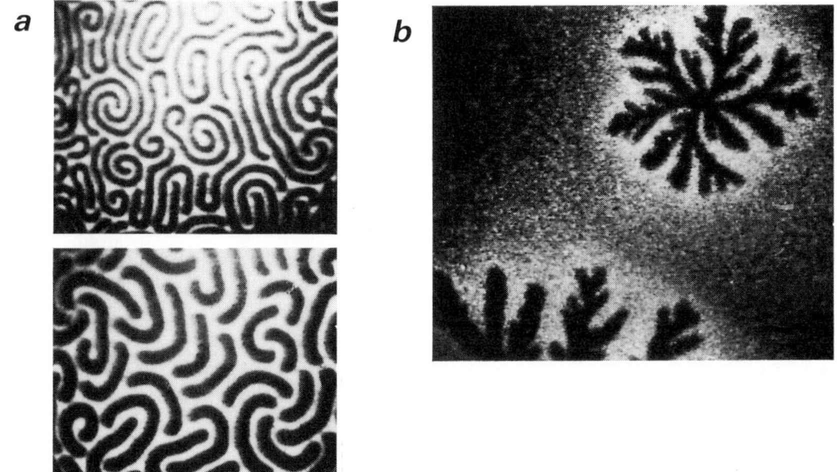


**Figure 7.22** *Some of the more complex patterns formed by Langmuir films. The LC phase growing within the LE phase can give rise to a striped pattern (a), while "dendritic" structures can be formed by the 2D solid phase (b). The structure of these particular dendrites is said to be "fractal"; Chapter 9 will explore what this peculiar property implies. (Photographs courtesy of H. Möhwald, Universität Mainz.)*

pushed up against the glass as it is withdrawn. The result of this process (Figure 7.23) is a layer of aligned surfactant molecules on the surface of the slide, which is called a Langmuir–Blodgett film.

These films need not be restricted to a single layer in thickness: we can send the slide for another dip to pick up a second layer. This time, however, what the Langmuir film sees is not the hydrophilic surface of the glass but the bristling coat of hydrophobic tails from the first layer. The second layer of surfactant molecules therefore sticks tail first (Figure 7.23). The two-layer Langmuir–Blodgett (LB) film so produced is a bilayer structure like those that comprise the walls of liposomes and cells.

This deposition process can be repeated *ad nauseam* to build up whole stacks of

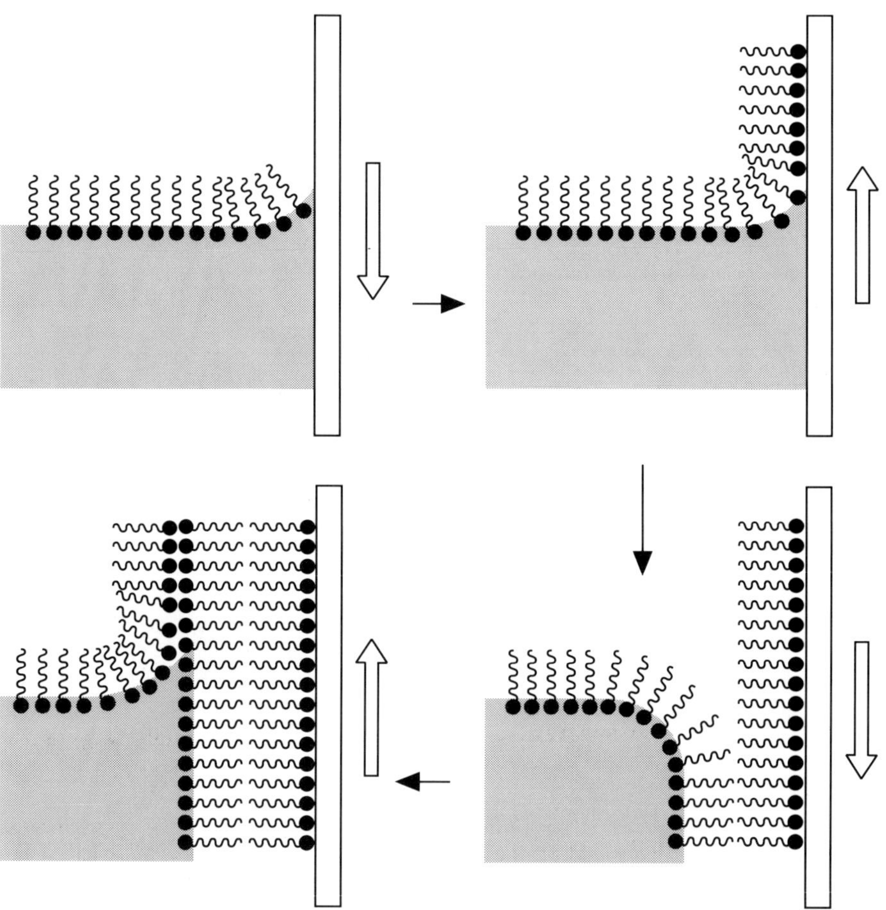

**Figure 7.23** *Langmuir films can be transferred from the water surface onto a solid surface such as a glass slide dipped carefully into the trough. This creates Langmuir–Blodgett films – orderly layers of surfactants – on the slide. Repeated dippings yield many-layered films.*

layers. Different stacking arrangements can be obtained by altering the dipping configuration or the nature of the surfactants. Certain amphiphilic carboxylic acids can be laid down with the hydrophilic heads in every layer pointing upwards when the solution from which they are prepared is alkaline (Figure 7.24); while solely head-down layers are possible for some molecules, such as certain dyes.

Langmuir–Blodgett films languished in obscurity for at least 40 years after Langmuir and Blodgett first described them. But in recent years they have enjoyed a great resurgence of interest, attributable largely to the development of molecular engineering technologies that offer a wide range of potential applications for these ordered molecular assemblies.

The similarity between LB films and biological membranes means that the films can be used as surrogate membranes for holding biological molecules. For example, LB films provide a medium that mimics the natural environment of membrane proteins, and can

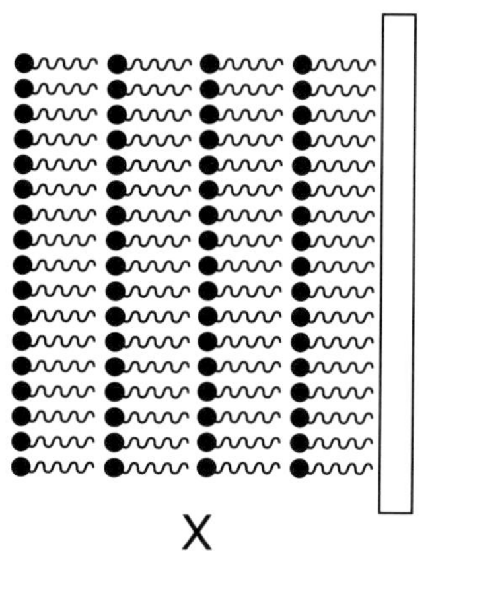


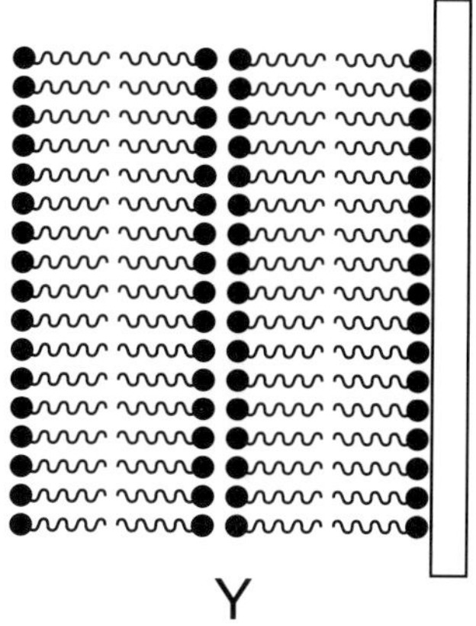


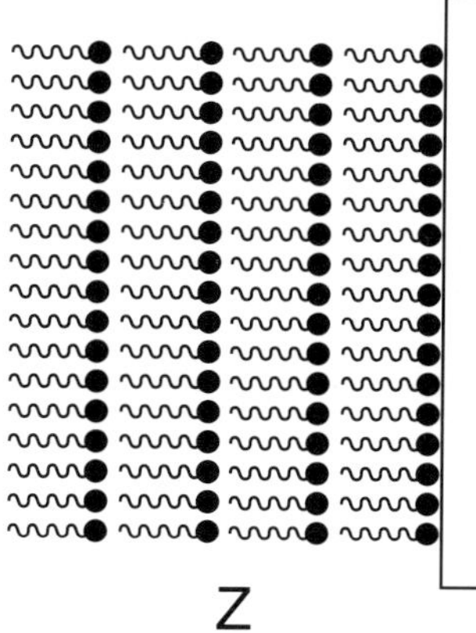


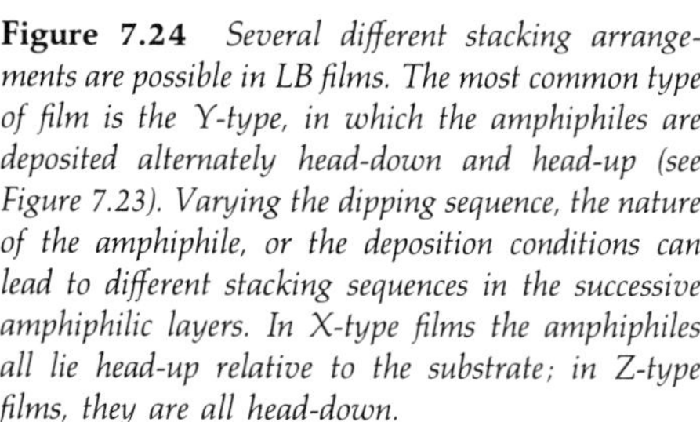

**Figure 7.24** *Several different stacking arrangements are possible in LB films. The most common type of film is the Y-type, in which the amphiphiles are deposited alternately head-down and head-up (see Figure 7.23). Varying the dipping sequence, the nature of the amphiphile, or the deposition conditions can lead to different stacking sequences in the successive amphiphilic layers. In X-type films the amphiphiles all lie head-up relative to the substrate; in Z-type films, they are all head-down.*

therefore be used to create a model system for studying processes that occur at cell surfaces such as those involved in immune responses. Arrays of biological molecules immobilized in LB films might also provide new types of device in their own right. It has been suggested that biochemicals involved in harvesting light energy, such as the protein bacteriorhodopsin, might be incorporated into a solar cell based on natural compounds rather than on the semiconducting materials that have been used traditionally. Enzymes trapped in LB films could be used in biosensors (see Chapter 2). Like natural membranes, LB films can act as selective chemical filters, permeable to some substances but not others.

Some LB films interact with light in ways that can be exploited in optical applications. One would normally expect the light that is transmitted through or reflected from a material to be much the same as the incident light, perhaps with some frequencies removed where the material has absorbed them. But there exist some materials for which the transmitted or reflected light is at frequencies that were not contained within the incident beam at all: the material doubles or even trebles the frequency of the light that passes through. A beam of infrared light shone onto a material of this sort might emerge as blue light. This so-called "nonlinear" optical behavior tends to occur when the incident beam is very intense, as is the case when a laser beam is used. A few inorganic crystals, such as lithium niobate, have frequency-doubling or -trebling abilities. Such materials are extremely valuable for extending the repertoire of colors available from lasers: most lasers (such as the carbon dioxide, argon or helium/neon lasers) emit light of only a single, well-defined frequency, whereas researchers would like to have a whole range of laser frequencies at their disposal. Some LB films turn out to have better nonlinear optical properties than inorganic materials, and moreover these properties can be tailored to specific applications by controlling the thickness or chemical composition of the films. Nonlinear optical behavior can manifest itself in other ways too, some of which – such as the possibility of making optical switches that can be made opaque or transparent by varying the intensity of the incident beam – are enabling researchers to entertain notions of logic circuits and computers based on light rather than electricity. It is possible that LB films will play a major role in such devices.

Indeed, prototypes of optical information-storage devices based on LB films have already been developed. Mostly these films are formed from molecules that can be switched between two stable states by irradiation with light; an ordered array of such molecules then represents a two-dimensional bank of light-operated switches. If, for example, the LB film consists of molecules that can be switched by light between one isomer and another, a finely focused laser beam could be used to encode data as a pattern of the two different isomers. In principle it might be possible to store information at the incredibly high density of one "bit" per molecule. Reading this information back out might then exploit the fact that the two isomers, if chosen carefully, will have different absorption spectra: a read-out beam scans the film and registers the different "colors" of regions containing different isomers. LB films for applications of this sort have been prepared from amphiphiles based on azobenzene molecules, which are photochemically switchable between two isomers (page 166).

Chapter 6 showed how certain molecular materials can form crystals that conduct electricity. As the preparation of LB films is essentially like building up a molecular crystal layer by layer, there has been speculation about whether LB films based on charge-transfer compounds such as TCNQ–TTF (page 200) might be able to provide electrically conducting films. To build LB films from these charge-transfer molecules, hydrocarbon tails are attached to make them amphiphilic. M. Matsumoto and colleagues from the National Chemical Laboratory in Tsukuba, Japan, have combined light-switching and electrical conductivity in LB films based on molecules that incorporate both azobenzene and TCNQ units: the conductivity of these films can be controlled by light (Figure 7.25). Semiconducting and conducting LB films could prove valuable for constructing "sandwich-type" electronic devices, such as metal–insulator–semiconductor (MIS) structures, which electronic engineers presently build from conventional inorganic materials. As yet, however, the conductivities of LB films based on charge-transfer compounds are not especially impressive, partly because it is hard to create films that are perfectly ordered over large distances – defects lower the conductivity.

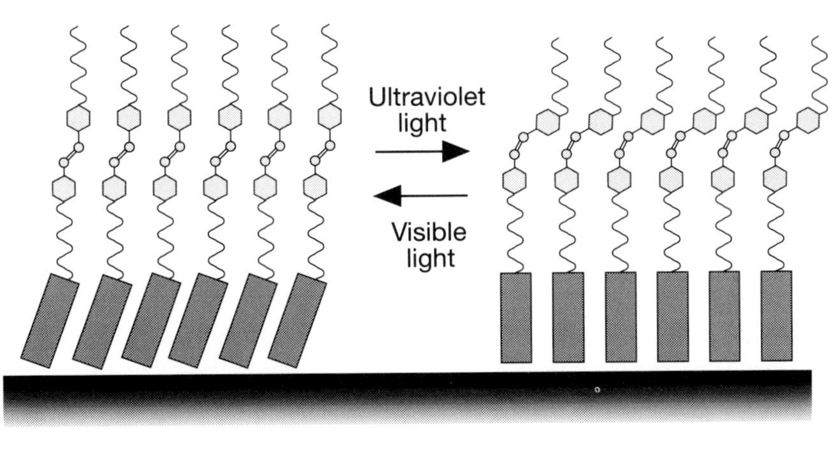


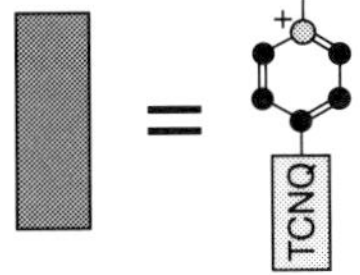


**Figure 7.25** *Optical memories and switches might be created from LB films that alter their properties in response to light. Here light is used to control the electrical conductivity of an LB film. The film contains TCNQ groups to which long hydrocarbon chains are attached, making the molecules amphiphilic. The chains contain azobenzene units, which can be switched between two isomeric forms using visible and ultraviolet light. This photoisomerization alters the way in which the TCNQ groups are stacked, and thereby influences the film's conductivity. This approach to molecular devices has been developed by M. Matsumoto and colleagues at the National Chemical Laboratory in Tsukuba, Japan.*

Some of the applications envisioned for LB films are "passive" ones, in the sense that the film is required to do little more than just act as a kind of surface coating. They might, for example, serve as protective coverings to shield surfaces from corrosion or to provide a mask for etching processes. Just as stencils are used to create patterns on glass by sandblasting, so LB films can mask off certain areas during etching of electronic circuits into semiconducting materials – here X-rays or beams of electrons or ions take the place of a sandblaster. The magnetic-tape industry is hoping for great things from LB films as durable coatings for magnetic tape, both protecting the magnetic particles on the tape from becoming degraded and providing lubrication between the tape and the recording head. Films of compounds such as barium stearate just a few molecular layers thick can greatly reduce the friction as the tape passes the head. Langmuir himself, in fact, showed interest in using LB films to lubricate the bearings in his meters!

## Crystals that flow

### *A paradoxical liquid*

In Langmuir–Blodgett films the amphiphiles have generally been compressed enough in the Langmuir trough, prior to being deposited on the solid surface, to represent truly two-dimensional crystals. But the molecules within the bilayers of vesicles are more mobile and less orderly in their packing arrangement – more like the liquid-condensed phase of Langmuir films. These bilayers are examples of molecular structures that are said to be liquid crystalline. The term "liquid crystal" is now such a familiar one, in these days of liquid crystal displays (LCDs) in watches, clocks and hifi systems, that one can easily overlook the oxymoron it represents. Think about it: by definition a crystal is a substance in which the molecules are highly ordered over long distances, yet a liquid can *flow*, implying that the molecules must be free to move and therefore must surely be packed together in a rather haphazard fashion. How can a substance be both a liquid and a crystal?

This, indeed, was the puzzle facing the discoverers of liquid crystals – Friedrich Reinitzer, an Austrian botanist, and Otto Lehmann, a German physicist. In 1888 Reinitzer synthesized a new organic compound with a most peculiar nature. The compound, cholesteryl benzoate, was derived from the cholesterol molecule, which we encountered earlier as a component of some cell membranes. Reinitzer crystallized the benzoate derivative and duly set about characterizing it according to the standard methodology of organic chemists of the time (which is much the same today): he measured the melting point of the crystals. What perplexed him was that melting of cholesteryl benzoate appeared to be a two-stage process: the crystals became liquid at 145.5 degrees Celsius, but the cloudy liquid then turned clear at 178.5 degrees. There seemed to be two kinds of liquid state, something for which Reinitzer knew of no precedent.

In perplexity the botanist sent a sample of his compound to Lehmann, who was known to be an expert in microscopic studies of crystallization. Lehmann observed that the cloudy form of the liquid exhibited a property called birefringence, which is a characteristic of many crystals that display double refraction (page 86). Double refraction occurs when the effective speed of light in a material differs in different directions. In such materials the plane of polarized light may become rotated as the light passes through the material – this is birefringence.

Lehmann was astonished to find that the first liquid state of cholesteryl benzoate is birefringent. In crystals this property derives from the ordered structure, which means that the properties may differ in different directions. In a normal liquid, on the other hand, the molecules drift about at random, so that on average one direction should look like any other. The only possible explanation for the behavior of Reinitzer's compound seemed to be that it possessed some degree of nonrandom, orderly structure. Here, apparently, was a new, fourth state of matter: neither gas, liquid nor solid, but a "soft" or "liquid" crystal. Under a polarizing microscope, in which polarized light is used to illuminate the sample, the birefringence of liquid crystals gives rise to a variety of beautiful colored patterns (Plate 12).

## *A question of orientation*

In 1924 the German Daniel Vorlander showed that liquid crystalline materials are comprised of molecules with long, rod-like shapes. The realization soon followed that the molecular order in liquid crystals resides in the orientation of these rods. Materials comprised of spherical particles can possess only one kind of order: a regularity in the positions of the particles. But for assemblies of rods, order can also be present in the form of a regularity in the rods' orientations. Regularity in position and in orientation are independent, in that one does not necessarily imply the other. In a crystal formed from rod-like molecules, the simplest kind of periodic array has them packed side by side and all lined up in the same direction – there will be order of both position and orientation (Figure 7.26). Above the melting point, the rods can jostle around so as to lose positional order but may retain some degree of orientational order, remaining pointed in the same direction on average. The proximity of a molecule's neighbors prevents it from tilting too far out of alignment. This is a liquid crystal phase, in which there is no regularity in the positions of their molecules but there is, on average, a well-defined orientational preference.

There are several degrees of disorder through which a liquid crystalline material can pass when its crystals are melted. Just a little above the melting point, the thermal motions of molecules may be insufficient to disrupt the layered structure of the crystal, so that the rods maintain this structure even though within each layer there is no positional order. This liquid crystalline phase does retain a degree of positional order perpendicular to the layers, and is called the smectic phase. Two varieties of smectic phase are common – when the orientation of the rods is perpendicular to the layers, the phase is the smectic A, while if the average orientation is tilted relative to the layers the phase is the smectic C (Figure 7.26). Seven types of smectic phase are currently known.

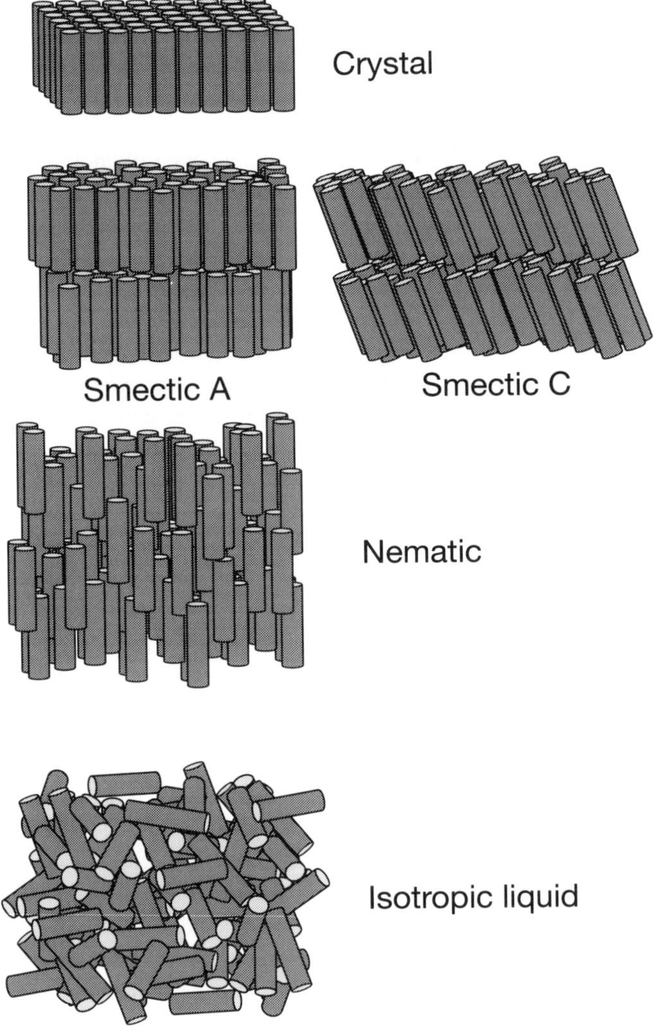


**Figure 7.26** *The structure of solid, liquid and liquid crystal phases formed from rod-like molecules: crystal, smectic A, smectic C, nematic and isotropic liquid. Note that, for simplicity, all of the molecules in the liquid crystal phases are depicted here as having identical tilts, whereas in reality there are appreciable variations around a well-defined average tilt angle.*

More pronounced thermal motions will break down the layers of the smectic phase, removing all vestiges of positional order. Orientational order, however, may remain, in that a preferred direction of the rod-like molecules can still be identified. This phase is called nematic. At greater temperatures still, the molecules will jiggle about so vigorously that even orientational order will be lost – the rods rotate and point in all directions at random. The material is then no longer a liquid crystal but a true (isotropic) liquid. The liquid crystalline phases, occurring between the crystal and the liquid (Figure 7.26), are called mesophases – "middle" phases.

The clear phase of cholesteryl benzoate observed by Reinitzer and Lehmann was the true liquid phase; but the cloudy phase had neither the smectic nor the simple nematic arrangement described above. The structure of this phase is complicated further by the fact that molecules of cholesteryl benzoate are chiral – they have a handedness. For chiral molecules, the nematic phase can lower its free energy by developing a screw-like twist of the direction in which the molecules point. The origin of this twist lies with the molecules' preference to lie at a slight angle to their neighbors, rather than aligning in parallel fashion as in a regular nematic phase. The result of this preference is a helical rotation of the orientation throughout the liquid crystal (Figure 7.27). This twisted phase takes its name from the compound in which it was first observed – cholesteric.

The pitch of the helix in cholesteric phases is generally of the same order of magnitude as the wavelengths of visible light, so that these phases reflect light much as crystals scatter X-rays (Chapter 4). This wavelength-dependent reflection gives rise to iridescent colors; the iridescence of some beetle cuticles and butterfly markings stems from the formation of cholesteric phases by the molecules that comprise the tissues.

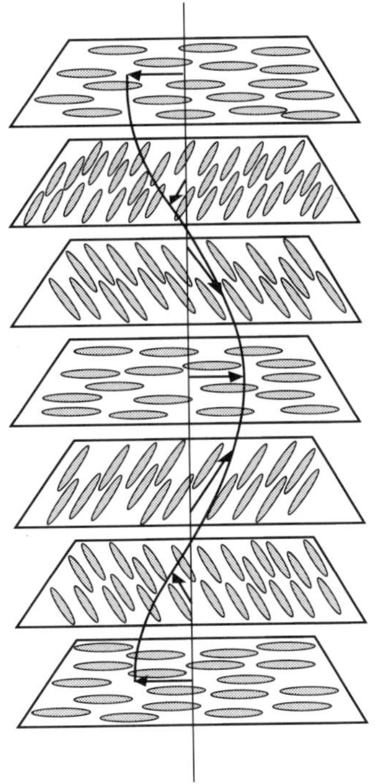

**Figure 7.27** *The cholesteric phase, in which the direction of molecular orientation rotates in a helical fashion. Because they scatter light strongly, cholesteric phases give rise to iridescent colors such as those observed in butterfly wings and insect cuticles.*

Rod-like molecules are not alone in their ability to form orientationally ordered liquid crystalline phases. Indeed, any molecules that are not essentially spherical should in principle be able to show orientational ordering, and in crystals many molecules do. In the fluid state, however, only those that have a pronounced degree of nonsphericity can retain some of this order. For example, the opposite extreme to a molecule that is stretched out into a rod is one that is flattened into a disc. Orientationally ordered arrangements of discs are very familiar – just think of a stack of dinner plates or gramophone records.

The Indian physicist Sivaramakrishna Chandrasekhar of the Raman Research Institute predicted in the 1970s that disc-like molecules might form stacked, "discotic" liquid crystalline phases. In 1977, Chandrasekhar discovered the first discotic phase (Figure 7.28). Researchers have predicted that, if these stacks are formed from molecules that can act as electron acceptors and donors, electrons might be passed down the stacks as if along "molecular wires." An increase in conductivity has indeed been observed when porphyrin molecules with metal ions at their centers form discotic phases.

### *Putting on a display*

Rod-shaped liquid crystals generally possess an electric dipole – an unequal distribution of charge between the two ends – which causes the molecules to align themselves with

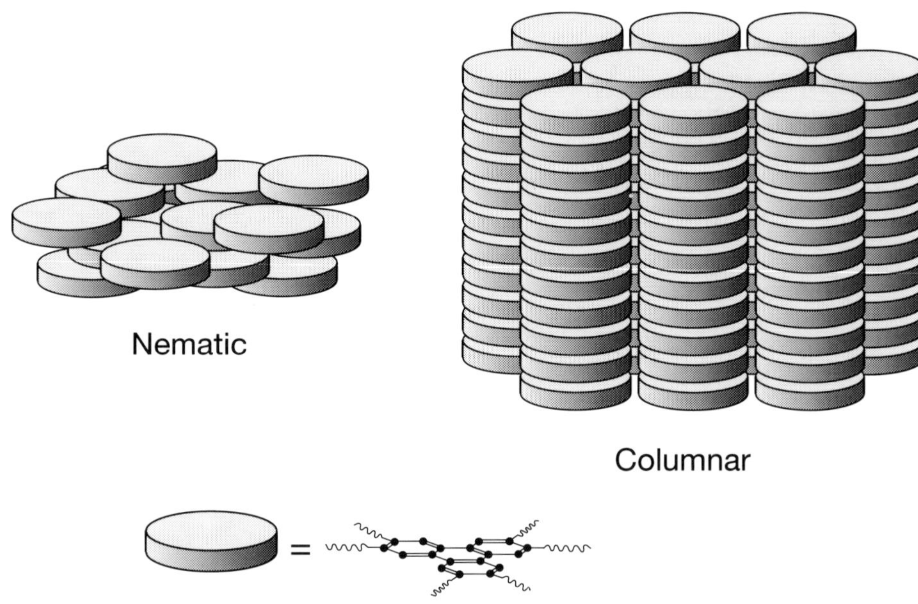


**Figure 7.28** *Discotic liquid crystals. In the liquid crystalline columnar phase the disc-like molecules are stacked like dinner plates, while in the nematic phase they have approximately the same orientation but no positional order.*

the direction of an electric field. Thus the direction of orientation of a liquid crystal phase, called the director, can be switched by an applied electric field. Because of the effect that this orientation has on the refractive and optical properties of the material, it was recognized as early as the 1930s that the electrically switchable nature of liquid crystals might be exploited in electronic displays. Only in the 1960s, however, were compounds developed that were sufficiently stable towards light and heat for these applications to be realized.

Liquid crystal displays make use of birefringence. A liquid crystal placed between crossed polarizing filters can twist the plane of polarization of light passing through the first polarizer so as to enable it to get through the second too. Just the right degree of twist can be introduced by passing the light through a transparent cell containing a nematic phase in which the orientation of the molecules twists smoothly through 90 degrees from one side of the cell to the other (Figure 7.29). This change in orientation was achieved in the first liquid crystal displays by coating the surfaces of the cells with a polymer called polyimide; when prepared by rubbing in a certain direction, the polyimide layer encourages adjacent liquid crystal molecules to line up in the direction of rubbing. A twisted nematic structure therefore results from coating the top and bottom of the cell with polyimide and rubbing them in directions at right-angles to one another. Polarized light can pass through crossed polarizers above and below the cell.

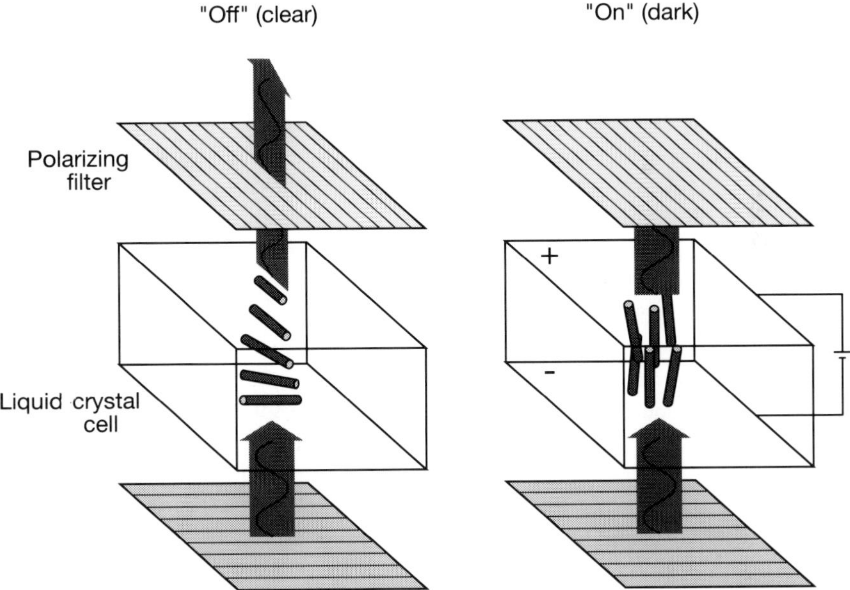


**Figure 7.29** *Nematic liquid crystal displays exploit the ability of the aligned molecules to rotate the plane of polarized light. Aligning agents at the surface of the top and bottom plates of the cell ensure that the molecules twist the light by precisely the right amount to allow it to pass through crossed polarizing filters. But when an electric field is applied across the cell, the molecules line up in a different direction and the light can no longer pass – the cell goes dark.*

To turn the cell black, the light is cut out by switching the director of the nematic phase to a different orientation, using an electric field to realign the molecules. In practice, the pixels of the display are cells coated top and bottom with indium tin oxide, a transparent conducting material. By switching on the electric field between the top and bottom plates, the aligning influence of the rubbed polyimide layers is overcome and the molecules swing right round to lie parallel to the field (Figure 7.29). The plane of polarized light coming through the first polarizer is no longer rotated, so it fails to pass through the second polarizer, and the pixel goes black.

One of the big advantages of liquid crystal displays is their compactness – they can be made as thin as a credit card. But they become less practical as the area of the display increases, since the intricacies of connecting up large numbers of pixels and preventing them from interfering with each other presents a refractory problem. For this reason, the liquid crystal television that you can stick on your wall like a picture has yet to materialize. Progress in this direction was advanced, however, by the development in the 1970s and 1980s of so-called chiral ferroelectric liquid crystals. Displays based on these compounds work on similar principles to the earlier "twisted nematic" devices, but can be switched more rapidly and can be packed more densely without problems of interference between pixels. Screens based on these materials have already reached the size of a magazine page.

## *Liquid color*

For all the neatness of liquid crystal displays, it would be perhaps a little disappointing if this "fourth state of matter" turned out to be useful for nothing more than wristwatches and televisions. But such is not the case. Today, liquid crystals are providing a wide range of exciting new directions in materials science.

Many liquid crystals exhibit nonlinear optical properties like those of LB films, suggesting that they can be used for frequency doubling of lasers. Because their optical properties can be switched electrically, liquid crystals are promising materials for building "optoelectronic" computers, which use light as well as electricity for signal processing.

Polymeric plastics can be imbued with novel and useful properties when the polymeric constituents form liquid crystalline phases. In conventional plastics, the polymer molecules are generally disordered and tangled. But in the liquid crystal state, the molecules may adopt a high degree of alignment, and this order may be retained when the polymer is cooled below the solidification temperature. Plastics made from liquid crystal polymers can exhibit high strengths as a result of this orientation order. Materials of this sort developed at E. I. du Pont de Nemours Company in Delaware, known as aromatic polyamides or "aramids," form fibers that are stronger than steel.

The unusual optoelectronic behavior of certain liquid crystal polymers allows them to be used as a medium for optical memories – patterns can be written into the material with a laser beam and erased again by laser illumination in the presence of an electric field. These compounds are promising candidates for use in erasable compact discs.

The striking colors afforded by liquid crystals have made their study one of the rare scientific enterprises to have leapt boundaries into the realms of art and design. The interest here arises from the fact that the colors of liquid crystals need not be fixed, but can be varied in several ways. This might lend new dimensions to artistic expression, introducing a nonpermanent "kinetic" element to coloration. The hue of iridescent materials changes as the viewing angle is changed, providing the potential for the viewer's motion to become part of a composition. Perhaps more intriguing still are the possibilities suggested by the temperature dependence of the color of cholesteric phases. The pitch of the helical twist in these phases, which determines the wavelength of the light they scatter most strongly, depends on temperature. So these materials are thermochromic; their color changes with temperature. One application that has already been developed is in fabrics that change color in response to body heat. The chemicals company Merck UK now markets thermochromic inks based on liquid crystals which can be screen-printed onto fabrics to yield rainbow-colored, heat-sensitive cloth. Might this perhaps be the way always to be wearing the colors of the season without ever buying a new outfit?

*Part III*

# Chemistry as a Process

# 8

# Chemical Beginnings

## How chemistry came to life

*It is a long way from granite to the oyster . . .*

Ralph Waldo Emerson

The American chemist Julius Rebek has said that "once there was physics and there was chemistry but there was no biology." In other words, unless you are inclined to take a more biblical view, there seems to be no avoiding the conclusion that the former two begat the latter. While evolution from primitive to complex life is the concern of the biologist, the initial appearance of life on Earth sets a puzzle that physical science must solve. In earlier ages the fundamental question was: can life arise from chemistry alone? But now that we understand, in considerable measure, the replicating mechanism of DNA, the manufacture of proteins, the chemical basis of photosynthesis, cell metabolism, immune responses and the host of other molecular processes on which life depends, few scientists would doubt that this question can be met with a resounding affirmative. The issue has instead become that of *how* these exquisitely tuned (bio)chemical systems could have arisen spontaneously on a planet of rock, water and a mixture of simple gases. We don't yet have all the answers, but this chapter will look at how far we have come, and how far we have yet to go, in attempting to formulate a scientific account of the origin of life on Earth.

Some people view such attempts with distaste, even if they do not have a strong predilection for theological alternatives. They may suggest that, if we achieve a purely scientific explanation for life and its origins, we will have somehow stripped our existence of any spiritual content. Curiously, this point of view seems to assume implicitly that such a complete understanding is possible – the argument is that we should not inquire too deeply, that we should knowingly and purposefully leave unexplored some of the mysteries of life *even though* such mysteries would in principle yield to rational investigation. Nothing, it seems to me, could be more damaging to the human spirit than to impose boundaries on the questions that we are permitted to ask or the territories that we may explore, so long as we do so with sensitivity and responsibility. But the attitude in any case seems to me to display a profound failure of imagination. We have nothing to fear from a scientific origin of life. It will not rob us of our dignity, though it may enhance our humility. It will not come even close to explaining why we laugh or cry, nor even will it explain what motivates some of us to probe into such issues in the first place. It will not strip the world of mystery or excitement or romance. It should not even bring crashing to rubble our religious beliefs, if we have indeed invested them with any spiritual content. Life is, in all probability, a chemical process; but living it is perhaps something that is not even fully accessible to scientific enquiry.

Before the nineteenth century these dilemmas did not arise. Scientists then had a much simpler outlook. They considered that there were two types of substance: inorganic, such as that which comprised rocks, and organic, which was ultimately derived from living things. It was a basic tenet that, while one might be able to render organic matter "inert" in the sense of converting it to inorganic forms, the reverse operation was not possible – organic compounds could not be formed from purely inorganic ones. According to the eminent German chemist Baron Justus von Liebig, "one may consider that the reactions of simple bodies and mineral compounds . . . can find no kind of application in the study of the living organism." Of course, this did not address the sticky question of where organic matter came from in the first place, but in those pre-Darwinian times the hand of God was ever ready to lend itself to difficult questions of this sort.

Biology and chemistry were therefore seen as distinct disciplines. At the same time, however, it was recognized that there were clearly similarities between the two: like chemistry, much of biology was concerned with the transformation of matter from one form to another. Living organisms take in organic matter as food and incorporate part of it into the body tissues, excreting the remainder as waste. But scientists saw no reason to suppose that these biological transformations need be bound by the rules of chemistry. Rather, organic substances were considered to be imbued with a "vital force" which distinguished them from inanimate matter. Inorganic or dead materials that entered the body could acquire vital properties and become part of the living organism; and the vital properties of living matter could be "worn away," leading to death. But only living organisms had the power to bestow vital properties – they could not appear spontaneously in inert matter. Indeed, it was widely believed that chemical and vital forces acted *in opposition*: the former acted to degrade organic materials, to render them inanimate, while the latter provided an organism's drive towards growth and reproduction.

The distinction between inorganic and organic chemistry was one that, during the nineteenth century, became ever harder to sustain. Chemical analyses of organic materials showed that they contained carbon and hydrogen, and commonly oxygen and nitrogen, as well as sulfur and phosphorus in some cases. All of these elements were to be found also in inorganic matter such as minerals. Chemists began to discover that it was possible to prepare compounds generally regarded as organic from inorganic starting materials: in 1828 the German Friedrich Wohler prepared urea from ammonium cyanate, a crystalline salt, and several decades later the French chemist Marcellin Berthelot combined carbon and hydrogen to form acetylene. The advent of thermodynamics, meanwhile, brought organic chemistry still closer in line with the inorganic branch by demonstrating that there was no need to distinguish a mysterious vital force from the energy changes that drive inorganic reactions. Respiration, for instance, could be seen as an energy- (or heat-)liberating process in which food is combusted by atmospheric oxygen. There were undeniably differences between biochemical reactions and those of inorganic chemistry – the latter seemed to take place in a more controlled fashion, for example, and often seemed to operate in a concerted, organized manner. But chemists and biologists were ultimately forced to accept that, at the molecular level, there are no boundaries between their disciplines. Matter can pass back and forth from oxygen to inorganic materials, and the chemistry of life is subject to the same principles as the chemistry of gases, salts, minerals and metals.

Implicit in all of this are Julius Rebek's words, and their corollary that, at some point in the past, chemistry gave birth to biology. We need not invoke a mysterious "spark of life" to nucleate living organisms on the planet, nor assume that vital forces were sent from God or gods to permeate matter. Life appeared spontaneously from lifeless matter. Here I will consider what we have deduced about how that happened; Chapter 10 will say a little more about where and when. The question "why?" is not (yet) a scientific one, although no doubt it is the most interesting of all.

## Earthbound beginnings

### *A recipe for primordial soup*

We saw in Chapter 5 that the building blocks of life are proteins. Most of an organism's tissues are protein-based, and certain highly specialized proteins – the enzymes – play a central part in the chemical processes by which life is sustained. For this reason, attempts to understand the chemical origin of life initially focused on the question of how proteins can be made without the involvement of living organisms. But we should not forget that there are many nonprotein molecules that we cannot do without either, not the least of which is DNA. I will consider these in due course.

The basic structural units of proteins are amino acids. Proteins are formed by the linking together of amino acids via peptide bonds. So to address the issue of how proteins formed on the lifeless (prebiotic) Earth, we must first ask how amino acids might have been generated. This is a considerably simpler question, because while the structures of proteins can be enormously complicated, amino acids are relatively small organic molecules which, in living organisms, come in just twenty different varieties.

The fossil record suggests that life appeared on Earth no earlier than 3,500 million years ago; that is, no sooner than about 1,000 million years after the planet itself was formed. Analyses of the chemical composition of ancient rocks led the Russian biologist Alexander Oparin to conclude in the 1920s that the fundamental elemental constituents of amino acids – carbon, hydrogen, oxygen and nitrogen – took different molecular forms on the early Earth from those predominant in today's environment. Nitrogen, for example, today abundant in the atmosphere in the form of the molecule $N_2$, was then far more prevalent as ammonia, $NH_3$. Carbon in the atmosphere would then have been bound up largely as methane, $CH_4$, whereas now it is more abundant as carbon dioxide. Oparin suggested that simple organic molecules could have been generated in a primitive atmosphere of methane, ammonia, hydrogen and water vapor, via reactions sparked off by sources of energy such as lightning or ultraviolet radiation from the Sun.

Oparin's ideas were proposed in much the same form by the British biologist J. B. S. Haldane, while both men remained unaware of the other's work. The putative origin of life from chemical synthesis of organic compounds in the atmosphere, which then accumulated in the oceans to form a "primordial soup," is now often called the Oparin–Haldane scenario.

In the 1950s the American chemist Harold Urey and his student Stanley Miller, working at the University of Chicago, set out to test this hypothesis. Miller proposed that the best way of testing whether organics could really have formed in the primitive atmosphere was to carry out an experiment that simulated the conditions on Earth at that time (as far as these were known). Urey and Miller circulated a mixture of methane, ammonia, water and hydrogen (thought to mimic the composition of the early atmosphere) through a reaction vessel, heating the brew in one part of the cycle and allowing it to cool and condense in another. While in the gaseous state, an electrical discharge was passed through the mixture to simulate lightning (Figure 8.1). After running the experiment for a week, Urey and Miller found a mixture of organic compounds dissolved in the water that condensed in the cool part of the cycle, amongst which were substantial quantities of simple amino acids such as glycine and alanine.

Following this ground-breaking experiment, other researchers obtained similar results using different energy sources, such as beams of electrons or even just plain old heat. The importance of the Miller–Urey experiment lay in demonstrating that organic molecules thought to be critical for life to begin can be generated under crude, nonspecific conditions from simple starting materials; in this respect, its significance in the quest to understand the chemical origin of life is considerable. But there was an initial tendency to overexaggerate the implications of the results, which were sometimes described almost as if to suggest that life itself had been created in a test tube.

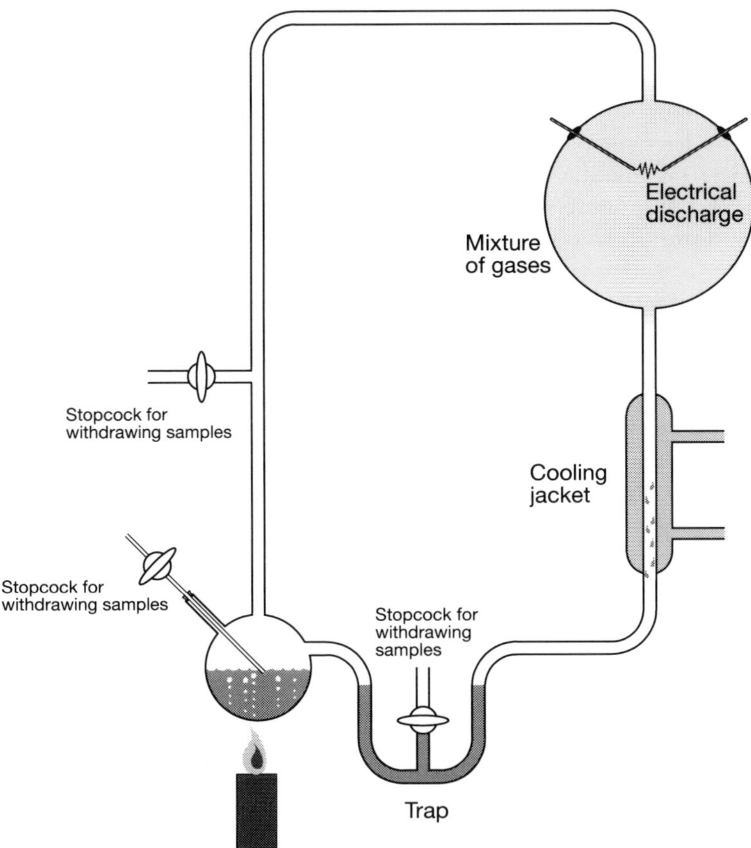


**Figure 8.1** *The apparatus used by Harold Urey and Stanley Miller to mimic the formation of organic compounds in the Earth's early atmosphere. A mixture of hydrogen, ammonia, methane and water is cycled around the system. In one chamber, an electrical discharge induces chemical reactions which form organic compounds, including simple amino acids. These are concentrated and collected by condensing the water (in which these organic compounds dissolve) in another part of the cycle.*

There is now considerable doubt as to whether the mixture of chemicals used in the Urey–Miller experiments really mimics the chemical composition of Earth's early atmosphere. Rather than being made up of predominantly hydrogen-bearing compounds (a mixture that chemists call "reducing"), the atmosphere may have contained significant amounts of oxygen-containing compounds such as carbon and nitrogen oxides. This composition is far less conducive to the formation of amino acids, as the presence of many oxygen-containing compounds means that organic molecules may be "burnt up" during the energetic processes that create them in the first place.

## *Hot springs and fool's gold*

To make organic molecules from inorganic gases requires an energy source to drive the reactions – atmospheric lightning, volcanoes or ultraviolet light are all candidates. But these sources have the potential to destroy delicate organic molecules too. Another, very different proposal for the prebiotic synthesis of amino acids invokes an energy source that lies at great depths beneath the sea, where the products would be shielded from the inclement conditions at the planet's surface.

In the 1970s, automated bathymetric vessels lowered to the floor of the Pacific ocean revealed hot springs that were ejecting water at great temperatures, sometimes exceeding 300 degrees Celsius (the water remains liquid only by virtue of the great pressures at these depths, which raise the boiling point). These springs, called hydrothermal vents, are formed when seawater transported down through cracks and pores in the rocks of the Earth's crust is heated by magma inside the Earth and rises back to the seafloor under pressure. Vents are therefore common in regions where undersea volcanic activity is strong, such as at "mid-ocean ridges." Here molten rock wells up from deep within the Earth's mantle to form new oceanic crust. The hot water issuing from a vent is generally rich in minerals, which are deposited to form tall chimney-like structures (Plate 13). The vent fluid is rendered turbid and opaque, like smoke, by its load of small mineral particles.

John Corliss from NASA's Goddard Space Flight Center in Maryland was amongst the team that discovered hydrothermal vents. He and his fellow researchers found that the vent chimneys were swarming with communities of creatures such as clams, marine worms and bacteria. These bacteria derive their energy and nourishment at least in part from sulfur-containing compounds formed in the volcanic environment and emitted from the vents. They are representatives of a very primitive class, called archaebacteria, which bask happily in hot surroundings and have no need of oxygen – characteristics that would have been necessary in the earliest organisms.

Far from being unpleasantly hot, hydrothermal vents therefore apparently provide a comfortable environment for living creatures. Corliss proposed that they might have been ideal locations for life to begin. The volcanic gases that are mixed in with the hot effluent of the vent are comprised largely of the simple chemical compounds, such as $H_2$, $N_2$, hydrogen sulfide, carbon monoxide and carbon dioxide, from which more complex organics may be formed. Here too was a source energy (the heat of the vent waters) to drive prebiotic chemistry. And there would be nutrients aplenty, in the form of minerals, to sustain the primitive organisms that might develop. (Inorganic nutrients are still an important part of the diet of today's marine microorganisms.) If creatures took up residence *inside* the vents, Corliss suggested that they would be protected from the worst of the natural catastrophes that the young Earth experienced, such as giant meteorite impacts. Corliss's vent hypothesis renders curiously prophetic the words of the Greek Archelaus in the fifth century BC: "When the Earth was first being warmed, in the lower part where the warm and cold were mingled together, many living creatures appeared, deriving their sustenance from the slime."

Some researchers, Stanley Miller amongst them, have little time for the vent hypothesis. Miller points out that temperatures in most vent fluids are so high that they would be more likely to destroy organic molecules than to create them. As a volume of water equivalent to the entire ocean is cycled through vents and mid-ocean ridges every 8 to 10 million years, Miller argues that subsea volcanism might have acted to frustrate the appearance of life, by burning up organic compounds formed elsewhere in the sea, rather than promoting it. It is certainly hard to imagine that the sort of chemistry that might generate to living organisms can carry on amidst temperatures of hundreds of degrees: any amino acids and sugars formed in this environment would survive for scarcely a minute or so before falling apart again in the heat.

Supporters of the vent theory counter this by pointing out that further from the direct outflow, temperatures are milder while the waters are still rich in volcanic gases and nutrients. Geologists from the University of Glasgow in Scotland have reported that "minivents" made of iron pyrites, emitting waters at less than 150 degrees Celsius, can form on the seafloor close to vent systems. They suggest that the chemical environment within these mineral structures would have been ideal for the formation and the joining together of simple organic molecules. The hydrothermal vent scenario seems likely to continue to provoke lively debate, but it cannot be denied that it remains a speculative idea for which the precise details of how it might work are not clear.

One of the protagonists for the vent model is A. G. Cairns-Smith, a chemist from Glasgow who has for many years advanced an alternative theory for the origin of life which borders on the bizarre. He proposes that the first self-replicating "organisms" may not have been made of organic molecules at all, but were instead inorganic crystals. For all the peculiarity of this idea, it is based on principles similar to those that direct the replication of DNA molecules, and has been considered sufficiently intriguing to be afforded serious attention. Cairns-Smith points out that crystals grow via a kind of template-directed process in which the crystal surface acts as a guiding framework for assembly of a new layer, just as a single strand of DNA provides a template on which a new strand can be built. Many crystals can be fractured quite easily along the planes of atoms. The growth and subsequent fracturing of crystals might, Cairns-Smith suggests, be regarded as a kind of replication, like cell growth and division. Moreover, different crystal faces are known generally to grow at different rates, while the speed of growth is also influenced by the presence of irregularities and defects in the perfect crystal structure. This could provide a means by which slightly different crystal shapes might mutate and compete with each other in an evolutionary sense.

Cairns-Smith suggests that clay minerals are particularly good candidates for primitive inorganic replicators. These substances are multidecker sandwiches of negatively charged aluminosilicate sheets stuffed with water and metal ions (Figure 8.2). The metal ions between layers can be exchanged with others quite easily; ion exchange in clays is exploited in some water-purification processes to remove toxic metals from solution. Cairns-Smith proposes that clays might use different sequences of metals between layers as primitive "gene banks" that pass on advantageous traits to new crystals.

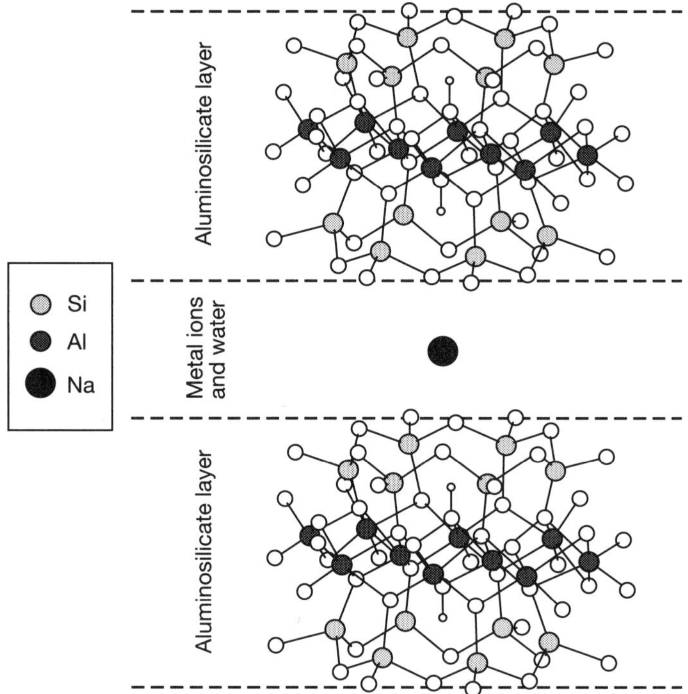


**Figure 8.2** *The crystal structure of sodium montmorillonite is typical of many clays. It consists of negatively charged aluminosilicate layers separated by positively charged sodium ions. In general the spaces between the layers also contain a considerable amount of water.*

There is little doubt that the idea of clay replicators is far-fetched, but Cairns-Smith does not propose that these systems could ever be considered as living organisms (or even "inorganisms"). Rather, he thinks that they might have evolved to a stage at which they were able to facilitate the formation of organic molecules. Initially the organics would have been mere slaves, created because they help the crystals to replicate; but eventually they would have become adept at replication themselves. In support of the idea, we might note that clay minerals are sometimes rather good catalysts for reactions involving organic compounds, and because of their variable layer spacing they can show a degree of catalytic specificity reminiscent of that of zeolites (see Chapter 2). Furthermore, some organic molecules are known to act as growth inhibitors or promoters which engineer preferential growth of certain crystal faces. This kind of control of mineral growth is crucial to the formation of biominerals such as tooth and bone, which is orchestrated very skillfully by proteins.

Minerals lie also at the heart of the recent addition to origin-of-life scenarios put forward by the German Gunter Wächterhäuser, a one-time organic chemist now turned patent lawyer. His theory focuses largely on the energy source that might have driven prebiotic chemistry, rather than being concerned with the details of the molecules so formed. In most biochemical reactions, energy means electrons: the reactions are often assisted by enzymes that are good at supplying electrons to, or extracting them from, the reactants. This is something that inorganic chemistry can do too, and Wächterhäuser singles out in particular iron pyrites – fool's gold, or $FeS_2$ – as a potential electron source; this compound is the very constituent of the "minivents" identified by the Glasgow geologists. Wächterhäuser suggests that iron pyrites might have acted as a kind of battery to drive the metabolism of primitive precellular organisms consisting of layers of organic molecules attached to the positively charged surface of the mineral. Iron and its other principal sulfide FeS could have been sequestered by these "surface metabolists" and transformed into $FeS_2$ with the concomitant release of an electron for use in the "organism's" metabolic processes.

Wächterhäuser is the first to admit that his theory is highly speculative. But it gains a certain degree of respectability from the existence of a rare type of bacteria that do in fact seem to use tiny particles of iron pyrites for metabolism. If Wächterhäuser's "surface metabolists" were able to break free of the mineral surface and close up their organic membranous coat to form a sac with a tiny lump of iron pyrites inside, they could be considered to represent a very crude kind of organism – protocells containing their own batteries.

## Extraterrestrial origins

### *Panspermia – life from the stars*

While there is no shortage of ideas about how to make organic molecules on the early Earth, some researchers see no need – they suggest instead that the building bricks of life may have been delivered ready-made from outer space.

The Swedish chemist Svante Arrhenius proposed in 1908 that the seeds of life could have been sown by deep-frozen spores from another world, blown to Earth by the radiation pressure of the stars. Arrhenius called this the "panspermia" hypothesis. It was a fanciful notion and arguably an unscientific one, since it is hard to see how it might be tested. But this did not deter Francis Crick from resurrecting the idea in his book *Life Itself* in 1981. Crick speculated about the possibility of "directed panspermia," according to which a race of beings from another solar system may have sent spores out into space with the specific intention of seeding inhabitable planets. Here we really do get into the realms of science fiction, and Crick recognized as much: his suggestion was presented as no more than a playful exploration of the panspermia idea (although for a time his wife was forced to wonder whether Crick's Nobel prize had unhinged him).

Set against this backdrop of wild speculation, you may wonder why there should be any reason at all to consider an extraterrestrial origin of life as a serious alternative to the purely Earthbound scenarios outlined above. Yet although it is true enough that we have as yet no firm grounds for believing that life exists beyond our atmosphere (except when we put it there), there is nevertheless no doubt that its raw materials –

organic molecules – can be formed in space. In the spectra of interstellar gas clouds, astronomers can read the signature of simple organics such as methanol, formaldehyde and hydrogen cyanide. The latter in particular is the ideal starting material for the synthesis of a variety of more complex organics, including amino acids. And while the conditions in interstellar space hardly seem likely to encourage prebiotic chemistry, there is compelling evidence that extraterrestrial organic molecules can indeed fall to Earth.

## *The bombardment of Earth*

Much of what we know about extraterrestrial organic chemistry is derived from analyses of the rocky objects that find their way to Earth in the form of meteorites. These lumps of cosmic debris have several origins. Most are from the asteroid belt, the ring of rocky bodies left over from the formation of the planets, which lie strewn between Mars and Jupiter. A few asteroids wander further into the inner Solar System than the orbit of Mars, and on rare occasions a stray asteroid will pass close to, or even collide with, the Earth. In 1908, an explosion occurred in the atmosphere above Lake Tunguska in Siberia, flattening trees for miles around (Figure 8.3). It is now generally thought that this hugely energetic event marked the break-up of an asteroid perhaps ten meters in diameter. Tunguska-scale events are estimated to occur roughly once every 250 years, but the chance of larger impacts falls off rapidly as the size of the impacting body increases. All the same, the consequences of much larger asteroids striking the Earth could be catrastrophic: the impact of an asteroid 10 kilometers across would release more energy than one hundred million one-megaton nuclear warheads – a destructive power many times greater than the world's entire nuclear arsenal – and the effects of such a blast could conceivably wipe out the human race.

The second potential source of meteorites is comets. Although, like the asteroids, comets represent debris that never quite succeeded in aggregating into planets, they come from much further afield. Extending beyond Pluto for a radius of up to two light years or so – halfway to the nearest star – lies the so-called Oort cloud, a shell of objects held weakly by the Sun's gravitational attraction. At these great distances, the bodies in the Oort cloud can feel the gravitational influence of other stars too, and a nudge from a neighboring star can send an object careering through the Solar System in the form of a comet.

Some comets, such as comet Halley, lie on orbits that bring them hurtling through the Solar System and back out again at regular intervals. But very rarely a comet might collide with a planet. At the time of writing, the comet Shoemaker–Levy 9 looks headed for such a collision with Jupiter in July 1994. Events like this are of unimaginable magnitude; but some small meteorites are believed to be lumps of fragmented comets which have passed by the Earth.

In Earth's early days there was much more debris still drifting through the Solar System than there is today, material that had not been swept up when the planets formed. As a result, the bombardment of the Earth by large meteorites was much heavier. The evidence of this heavy bombardment is still visible on the face of the Moon, which is

**Figure 8.3** *In 1908 an explosion in the atmosphere over Lake Tunguska in Siberia flattened trees for hundreds of square kilometers. It is now thought that the explosion was caused by the break-up of a huge meteorite, perhaps 10 meters in diameter. (Photograph courtesy of the American Museum of Natural History, New York.)*

pock-marked with huge impact craters. Most impact craters on Earth have been erased by the deformation, destruction and renewal of the crust that accompanies motions of the tectonic plates; but those that remain bear awesome witness to the enormity of the impacts that shook the young planet. Meteor Crater in Arizona, for instance, is a full two kilometers across (Figure 8.4), while a depression no less than forty kilometers across under the Barents Sea is thought to be the fingerprint of a huge impact about 40 million years ago.

This heavy bombardment ceased about 3,500 million years ago, by which time life may have already appeared. It has long been recognized that giant impacts may have had profound consequences for the evolution of life, but there is now some controversy about whether these consequences were for better or worse.

One's immediate intuition might be that meteorite impacts can do nothing but harm, in view of the amount of energy released in a large impact. The fossil record indicates that the dinosaurs, who had ruled the planet for 150–200 million years, died out rather suddenly at the end of the Cretaceous era, 65 million years ago. The suggestion in 1980 by Walter Alvarez and colleagues from the University of California at Berkeley that this extinction was the result of a giant meteorite impact was based largely on the discovery that buried sediments dating from this time are rich in the element iridium. Iridium is rare on Earth, but is known to be far more abundant in comets. Alvarez and

**Figure 8.4** *Meteor Crater in Arizona bears testimony to the gigantic impacts suffered by our planet early in its history. (Photograph courtesy of David Roddy, U. S. Geological Survey, Flagstaff, Arizona.)*

colleagues suggested that the impact of a comet perhaps 10 kilometers across had scattered iridium-rich dust into the atmosphere, causing major changes in the Earth's climate which wiped out the dinosaurs in a matter of decades. There is now rather compelling geological evidence for this catastrophic event, the favored site of the impact being at Chicxulub in Mexico.

The fossil record indicates that several global extinction events occurred long before the time of the dinosaurs, leading some to suggest that evolution was constantly being frustrated by giant meteorite impacts. It may even be that the origin of life was not a unique event, but had to start again from scratch many times. With all this talk of destruction, it may come as a surprise to hear that meteorites could also have been beneficial to the development of life. The idea that they might even represent the *source* of life could conceivably appear preposterous. But it is not.

### *Organic molecules from meteorites*

It became evident in the 1960s that some meteorites belonging to a class known as carbonaceous chondrites, which are rich in carbon compounds, contain organic compounds. Indeed, certain meteorites were reported to contain fossilized plants, and in 1966 Harold Urey claimed that he had discovered microbes *still living* within meteoritic matter. But these fantastic claims did not stand up to scrutiny: Urey's results, for example, proved to be due to contamination of his samples by terrestrial microbes.

The findings of Sri Lankan chemist Cyril Ponnamperuma and colleagues, working at NASA's Ames Research Center in California, were less easily dismissed, however. They analyzed material from a meteor that exploded in 1969 over the town of Murchison in Australia, and found that it contained tiny amounts of amino acids. Suggestions of terrestrial contamination were countered by the fact that the Murchison amino acids bore hallmarks that distinguished them from those produced by living organisms on Earth. We saw in Chapter 2 that all amino acids in terrestrial organisms (except glycine) are chiral, possessing two possible mirror-image forms (enantiomers); yet in every case only one of the enantiomers (the left-handed or L variant) is found naturally. Ponnamperuma's analysis of the Murchison amino acids, however, indicated that they contained an equal mixture of both the left- and the right-handed (D) forms. Furthermore, the proportion of atoms of the heavy carbon isotope carbon-13 was greater in the meteoritic amino acids than in terrestrial organic matter. Lastly, many of the amino acids in the Murchison fragments were not known to be produced by living organisms: whereas just twenty occur in nature, up to seventy-four varieties have now been identified in the Murchison chondrites.

The implication of Ponnamperuma's results was that amino acids could somehow be formed on extraterrestrial bodies such as asteroids and comets, and delivered to Earth in meteorites. Here, then, was an alternative to the Oparin–Haldane hypothesis and other Earth-based scenarios for the synthesis of life's building blocks. But could large quantities of amino acids really be delivered to the young Earth in this way? Meteorite fragments the size of pebbles simply fall to Earth with a bit of a bump, whereas much larger objects strike with dreadful fury that would surely fry to ashes combustible material like organic compounds. It is conceivable, however, that some might escape if buried deep inside the body of the impactor, or if a large meteorite partly breaks up before impact, as in the Tunguska event.

An intriguing discovery in 1989 by Meixun Zhao and Jeffrey Bada of the University of California in San Diego lends support to the idea that significant amounts of extraterrestrial organic matter could have been brought to Earth by comets. The two geologists were conducting a study of sedimentary clays at Stevns Klint in Denmark, where geological processes have exposed a section of the sediments deposited during the end of the Cretaceous and the start of the following era, the Tertiary.

It was known before Zhao and Bada's work that, like other Cretaceous/Tertiary boundary deposits, the Stevns Klint clays contained a lot of iridium, suggesting a meteorite impact at that time. But Zhao and Bada found that these clays also contained substantial quantities of amino acids – far more than in sediments further from the boundary layer. What was more, the amino acids that they identified were extremely rare in terrestrial organisms but common in carbonaceous chondrites. This suggested to the two researchers that the amino acids had been delivered to Earth by the same giant meteorite that is thought to have wiped out the dinosaurs.

Many geologists now believe that meteorites could have supplied substantial amounts of organic matter to the early Earth. The supply would have begun to dry up once the heavy bombardment ceased, but by this time life seems to have gained a foothold on the planet. Meteorite impacts may also have had an indirect influence on the inventory of organic compounds on the early Earth by supplying energy for Urey–Miller-type reactions in the atmosphere. (Experiments have shown that amino acids can be formed from appropriate mixtures of simple gases heated by explosive shocks.) Christopher Chyba and Carl Sagan of Cornell University have estimated that both terrestrial and extraterrestrial sources of organic material are likely to have contributed significantly to the chemical inventory for the origin of life, but there is much uncertainty in these estimates. Perhaps the only thing we can be sure of is that there will never be any lack of ideas about how organic compounds could have appeared on Earth to seed the beginning of life.

## Why is life not ambidextrous?

### *Hand-picked for handedness*

There is, however, one puzzling aspect of the chemical origin of life that none of these scenarios seems able to explain: why are all the amino acids in the proteins of living organisms of the "left-handed" variety? Similarly, why are all the chiral sugars in carbohydrates right-handed? How did life develop a handedness?

The handedness of certain natural substances, demonstrated by their ability to rotate the plane of polarized light, was recognized by chemists in the early nineteenth century; in particular, quartz crystals were known to have this property. But a perplexing puzzle was identified by the Dutchman Eilhard Mitscherlich, who reported in 1844 that the chemical and physical properties of the sodium ammonium salts of tartaric and racemic acid (both by-products of wine fermentation) indicated adamantly that both represented the same chemical compound; yet the salt of tartaric acid rotated polarized light whereas that of racemic acid didn't.

The French biochemist Louis Pasteur set out to resolve this puzzle. Pasteur found that, when crystals grown from solutions of the two salts were observed under the microscope, differences could be identified in their shapes. Both crystals were asymmetric, but the salt of racemic acid contained two forms that were mirror images of each other whereas the tartaric salt contained just one (Figure 8.5). Pasteur then performed an extraordinary feat of manipulation: he took a pair of fine tweezers and, squinting into the microscope, separated out by hand the two kinds of crystal in the racemic salt. He found that solutions in which each type of crystal were dissolved separately were

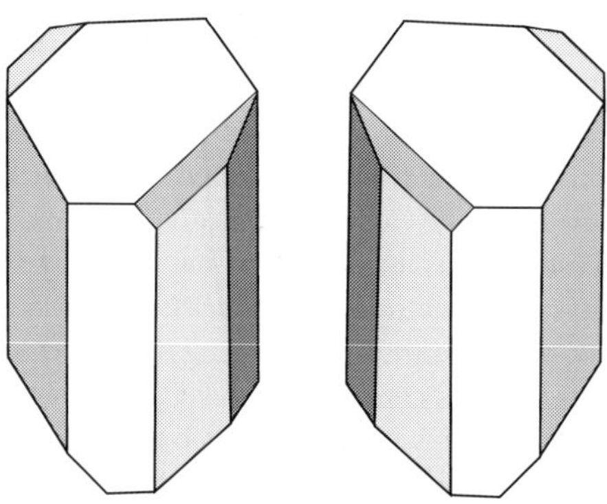

**Figure 8.5** *The sodium ammonium salt of tartaric acid forms two kinds of crystals which are mirror images of one another. Louis Pasteur's separation of the two types by hand in the nineteenth century enabled him to deduce that they contain molecules of opposite optical activity, which we now recognize as enantiomers.*

optically active, like the tartaric salt. Although lacking today's understanding of chemical bonding and molecular shape, Pasteur concluded correctly that racemic acid contained two mirror-image forms of the same compound, which we now recognize as chiral enantiomers (see page 77). In optically active tartaric acid, just one of these enantiomers is present.

Pasteur went on to show that living organics can show a high degree of selectivity towards "optically active" (that is, chiral) molecules. When he fed the two separated enantiomers of tartaric acid to a species of mold, only one of them was metabolized; the mold would not digest the other. Pasteur deduced rightly that this preference reflected some asymmetry in the biochemistry of the mold itself, and went on to infer that chiral discrimination was a characteristic of living things in general, while the inorganic world showed no such preferences. This suggested to Pasteur that optical activity is a distinguishing characteristic of the chemical products of living organisms. He developed the conviction that an explanation of the handedness of the living world would hold the key to an understanding of life itself.

We now know, however, that enantioselectivity does not necessarily require the mediation of living organisms, but only of a chiral environment of some kind, such as a chiral catalyst (see Chapter 2). But far from eliminating the apparent mystery surrounding the enantioselectivity of the natural world, this understanding serves simply to displace the pertinent questions back in time to the chemical origin of life.

## *The origin of chirality*

Chirality begets chirality. The particular handedness of the helical structures adopted by proteins, for example, is the consequence of the chirality of their constituent amino acids, all of which are present uniquely as the L enantiomer. The interaction of enzymes with D sugars is significantly different to their interaction with L sugars (which, in general, known organisms cannot metabolize). Once enantioselectivity appeared in primitive replicating systems, it is therefore not hard to understand how it might have propagated. Because of the central role of amino acids in the origin of life, explanations for life's chiral preferences have been directed towards providing a mechanism by which L amino acids could have gained absolute predominance in prebiotic chemistry. As we shall see shortly, however, the earliest replicating systems were probably not protein-based at all. Nevertheless, most of the proposed scenarios for the origin of enantioselectivity in amino acids can be applied equally well to other chiral molecules.

Urey–Miller syntheses of amino acids use as starting materials methane, ammonia, water and hydrogen. A favored candidate process for their formation on extraterrestrial bodies, meanwhile, is some variant of the so-called Strecker synthesis, involving hydrogen cyanide and ammonia (both of which can plausibly be formed in space). The reaction could take place on the surface of rocky extraterrestrial bodies in the presence of water, perhaps driven by ultraviolet radiation from the Sun. In neither of these processes – terrestrial or extraterrestrial – is there obviously any chirality in the reaction environment, and so they should both produce completely racemic mixtures of amino acids.

Indeed, you may recall that it was the racemic nature of the amino acids found in the Murchison meteorites that provided one of the critical pieces of evidence for their extraterrestrial origin. But in 1990, M. H. Engel and colleagues from the University of Oklahoma reported that their careful re-examination of the Murchison amino acids showed the extract of the amino acid alanine to be slightly *non*-racemic: there was about an 8 per cent excess of the L enantiomer over the D form. This claim is highly controversial; in view of the past history of false trails resulting from terrestrial contamination, some researchers are unwilling to believe that the same had not occurred here, especially as the preference for L-alanine was exactly what one would expect from contamination. But even if the results of Engel and colleagues are robust, they do not provide a fundamental explanation for the L preference found on Earth: one still has to account for this preference on the meteorites themselves. In other words, the origin of chiral selectivity is simply shifted to outer space.

It is a classic chicken-and-egg problem: chiral discrimination seems to be possible only when it already exists. From amongst the several suggestions that have been advanced to circumvent this impasse, we can identify two themes. Either the choice of left- or right-handed molecules was made at random, or some influence (which must then be nonchemical) tipped the balance in the specific direction that we now observe. If the choice was a random one, chemistry alone may furnish a sufficient explanation.

In 1953, Charles Frank of Bristol University described a hypothetical reaction scheme for the formation of a racemic mixture of chiral molecules which could be swayed irreversibly by small random fluctuations so as to yield one or other enantiomer essentially exclusively. In Frank's scheme, the formation of each enantiomer was envisaged as being autocatalytic – that is, each enhanced the rate of its own formation – while actively inhibiting the formation of the other. Within a certain range of conditions, this system is potential unstable in the face of small, random fluctuations, because if one or other enantiomer is formed in slight excess by chance, that excess is rapidly amplified by the combined effects of autocatalysis and inhibition. If there is no difference in stability between the two enantiomers, which way the instability tips is entirely down to chance, so that the D form is just as likely as the L form to end up dominating the system.

If life began in this way in several different locations more or less at the same time, one would then be faced with the possibility of colonies of primitive organisms of different handedness that would eventually encounter each other and perhaps have to do battle for predominance. Such a situation might make a good scenario for a science fiction tale, but it is rather hard to credit.

But might the choice between right and left in prebiotic chemistry have been nonarbitrary? One suggestion along these lines invokes the rotation of the Earth, which exerts a "twist" called the Coriolis force on the oceans and atmosphere. This twist operates in opposite senses in the northern and the southern hemisphere, and so it might, according to this theory, have been sufficient to generate a left- or right-handed preference to life, depending on which hemisphere it happened to originate within. Quite how the Coriolis force might be felt at the molecular scale is not clear, however, and this theory has few advocates. Scenarios involving rotational "stirring" have gained a little more credence, however, from the observation by Dilip Kondepudi and colleagues of Wake Forest University in North Carolina that the optical activity of chiral crystals of sodium chlorate forming within a stirred solution can be determined by the direction of stirring (here the chirality is a consequence of the spiral packing of ions, which can in principle twist either way).

Another proposal for the nonarbitrary origin of handedness ascribes it to the asymmetry of sunlight. At certain times of the day, the Sun's rays exhibit a kind of helical, or "circular," polarization (as distinct from the plane polarization that we have encountered earlier). Sunlight is very slightly circularly polarized in one sense at sunrise and in the other sense at sunset. This could have been sufficient, suggest some researchers, to tip the balance via photochemical interactions with chiral molecules. Calculations show, however, that the size of this effect is extremely small. Related to this idea is the interaction of chiral molecules with circularly polarized high-energy particles such as cosmic rays or beta particles produced in natural radioactivity. High-energy beta particles may be sufficiently energetic to break molecules apart, and for chiral molecules the strength of this interaction depends very slightly on the relative handedness of the beta particle and the molecule. Again, however, it is not clear that the difference is big enough to have a significant effect.

## *The left-handed Universe*

One of the most enticing explanations for the origin of life's handedness links the specific preference for left-handed amino acids with the discovery in the 1950s that the Universe itself has a certain kind of left-handedness. This discovery came as a great shock to physicists, to whom it seemed to undermine a commonsense intuition that the fundamental laws of nature should recognize no such distinction. The labeling of right and left was considered at this time to be entirely arbitrary and impossible to define by an experiment.

This presumed lack of any distinction between left and right became embodied in a fundamental physical "law" called the conservation of parity. The value of the parity of a subatomic particle can be thought of crudely as a measure of its handedness; but it is really a property of the quantum-mechanical equations that describe the particle. The conservation of parity held that in any physical process whatsoever, the sum of all the parity values of the system's constituents before the process takes place must always equal their sum after the process. This is a formal way of saying that there is no physical process for which its mirror image does not produce the same result.

In the 1950s Wolfgang Pauli, one of the fathers of quantum mechanics, expressed a willingness to bet a large sum on the inviolability of parity conservation. Yet the possibility that situations might exist in which parity is not conserved, proposed in 1956 by the Chinese physicists Chen Ning Yang and Tsung Dao Lee, led another

Chinese-born physicist, Chien-Shiung Wu, then at Columbia University in New York, to devise a way of testing the idea experimentally. Wu investigated the decay of the radioactive element cobalt-60, which decays by ejecting beta particles from the atomic nucleus. Beta decay is the result of interactions between particles in the nucleus due to the so-called "weak" force, one of the four fundamental forces of nature. (The others are the electromagnetic and gravitational forces, and the "strong" force which binds together protons and neutrons in atomic nuclei.)

Cobalt-60 nuclei are rather like tiny magnets with a north and south pole. It was known that when a cobalt-60 nucleus decays, the beta particle is ejected preferentially in the direction of the magnetic poles. If interactions involving the weak force observe conservation of parity, there should be an equal chance of the beta particle coming out from either of the two magnetic poles. Wu tested whether this was so by first cooling down a piece of cobalt almost to absolute zero and then placing the sample in an electromagnetic field. This ensured that all of the nuclear magnets lined up in the same direction, with their poles aligned with the applied field; and that this alignment was not disturbed by the random thermal motions which would occur at higher temperatures. She then counted the number of beta particles ejected from each end of the sample. The results showed clearly that more particles appeared from one pole than from the other. Thus the two poles of the cobalt-60 magnets are not equivalent; by implication, the labeling of north and south magnetic poles is not arbitrary, and parity is not always conserved.

How, if at all, does this help us to solve the mystery of enantioselectivity in the origin of life? Even if in particle physics left and right are not always equivalent, is there a way of transferring this distinction to chemical processes? The difficulty of doing so lies with the fact that, while the parity-violating weak force is, in spite of its name, stronger than the electromagnetic forces that are ultimately responsible for chemical interactions, it is of very short range. Although it plays a central role in the interactions of particles within atomic nuclei, the weak force exerts almost no influence beyond these confines; beta decay is just about the only manifestation of the force outside of the nucleus. Any influence that the weak force might have on chemical reactions will therefore be exceedingly tiny.

Nevertheless, such an influence does exist, and its magnitude can be calculated. The left-handed preference of the weak force conveys a very slight difference between the relative stabilities of enantiomers of a chiral molecule. For amino acids, the L enantiomer is more stable than the D enantiomer by a mere whisker (in fact, by a factor of something like 1.00000000000000001 at room temperature). This difference has been dismissed by some scientists as far too small to provide an explanation for the absolute predominance of L amino acids in nature. However, Dilip Kondepudi and his colleague George Nelson have re-evaluated Charles Frank's scheme for producing instabilities in autocatalytic syntheses of chiral molecules, and have shown that the tiny preference for L amino acids due to the electroweak asymmetry is just about sufficient to make a difference, so that a system like this will yield the L enantiomer with around 98 per cent probability.

Vitalii Goldanski and colleagues of the Institute of Chemical Physics in Moscow have advanced a rather startling scenario in which an abrupt "flip" in handedness could have occurred during the synthesis of organic molecules in the freezing depths of space, which might then have been brought to Earth by meteorites. Goldanski and colleagues suggest that Frank's model, which was made rather more general by the Soviet chemist Leonid Morozov in 1978, could provide one possible cause of the flip.

According to classical thermodynamics, space is too cold for chemical reactions, except in the vicinity of stars. The deep freeze – only a few degrees above absolute zero – means that molecules have too little energy to overcome the free energy barriers to reaction (page 60). But Goldanski and colleagues have shown that quantum mechanics allows molecules to "cheat" by *tunneling* through the barriers rather than surmounting them. Goldanski's team has conducted experiments which show that, even at the temperature of liquid helium (about four degrees above absolute zero), formaldehyde molecules can join up into polymer chains a few thousand molecules long. Classical theories predict that this reaction should be extremely slow at such temperatures, whereas quantum mechanics allows it to proceed at a respectable pace. Goldanski has outlined scenarios in which "quantum chemistry" in the bitter cold of space, coupled with an enantiospecific flip in favor of L amino acids owing to the tiny advantage conferred to these isomers by nonconservation of parity, could have led to an extraterrestrial "cold origin of life."

Whether any of these ideas can really provide a satisfactory explanation for the handedness of life is not yet clear. An answer would, however, surely help us to pin down exactly where and how life arose. But sadly it may be that no unequivocal answer will ever be found, since most of the proposals remain hard, if not impossible, to test. We may never know whether the protein chemistry of our bodies is left-handed by chance or by grand design.

## The jigsaw of life

### *Origin of the gene banks*

So much for amino acids, the constituents of proteins. What of the origin of the other principal ingredients of living organisms: the nucleic acids DNA and RNA? We have seen that these are polymers of units called nucleotides, of which there are four varieties in DNA and four in RNA. Each nucleotide contains a purine or pyrimidine base, a sugar and a phosphate group (Figure 5.3; page 149).

On the face of it, nucleotides are more complicated beasts than amino acids, and to synthesize them from simple precursor compounds poses no mean challenge. Nonetheless, chemists have devised plausible schemes to explain how these compounds might have been formed on the early Earth, and have in many cases shown experimentally that such schemes will work. For the purine and pyrimidine bases, the building block is hydrogen cyanide (HCN), which contains in a compact form all three of the elements

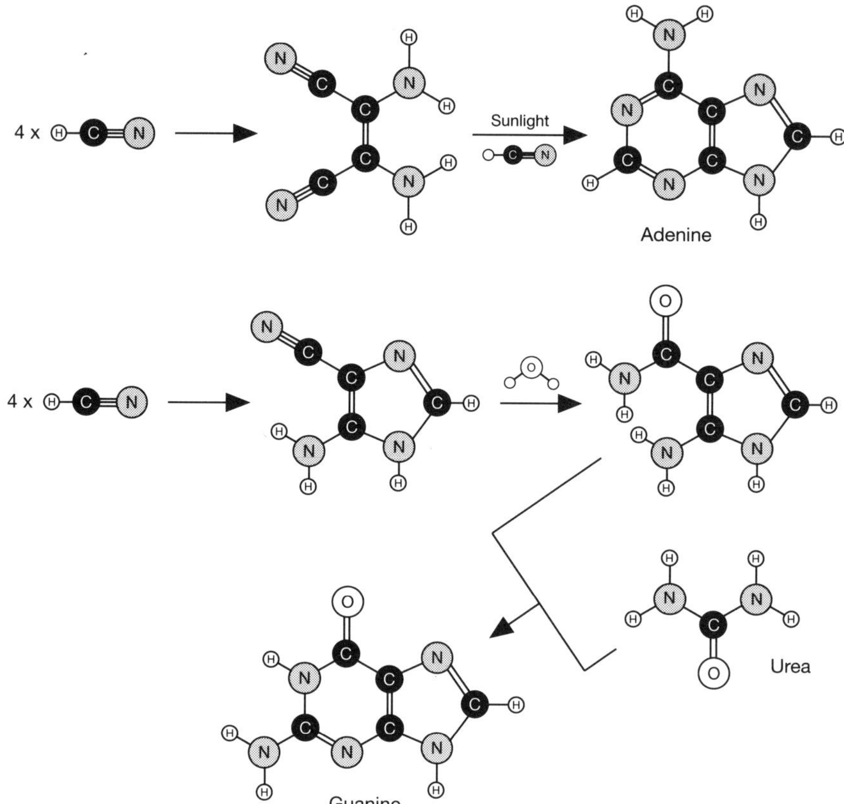


**Figure 8.6** *The purine bases in nucleic acids, adenine and guanine can be synthesized from simple organic molecules. Adenine can be produced simply from the reaction of five molecules of hydrogen cyanide (HCN), although the final step is assisted by sunlight. Guanine can be created by the linking of four HCN molecules into a monocyclic precursor; this reacts with urea to give the base.*

found in all nucleotide bases. Adenine can be produced from HCN alone, five molecules of which will join up in several stages to give the double-ring compound (Figure 8.6). It might appear to be a rather delicate matter to get the five molecules to link in just the right way; but John Oró of the University of Houston showed in 1960 that merely heating a solution of ammonia and HCN in water would produce adenine in small quantities. Cyril Ponnamperuma found in 1963 that adenine appeared in a Urey–Miller mixture of methane, ammonia, water and hydrogen when it was irradiated with a beam of electrons.

Guanine is a more awkward customer, but can be formed from HCN and urea; the latter is not overly difficult to create from simple precursors (Figure 8.6). Urea is also one of the reagents in a suggested prebiotic route to the pyrimidines: first to cytosine,

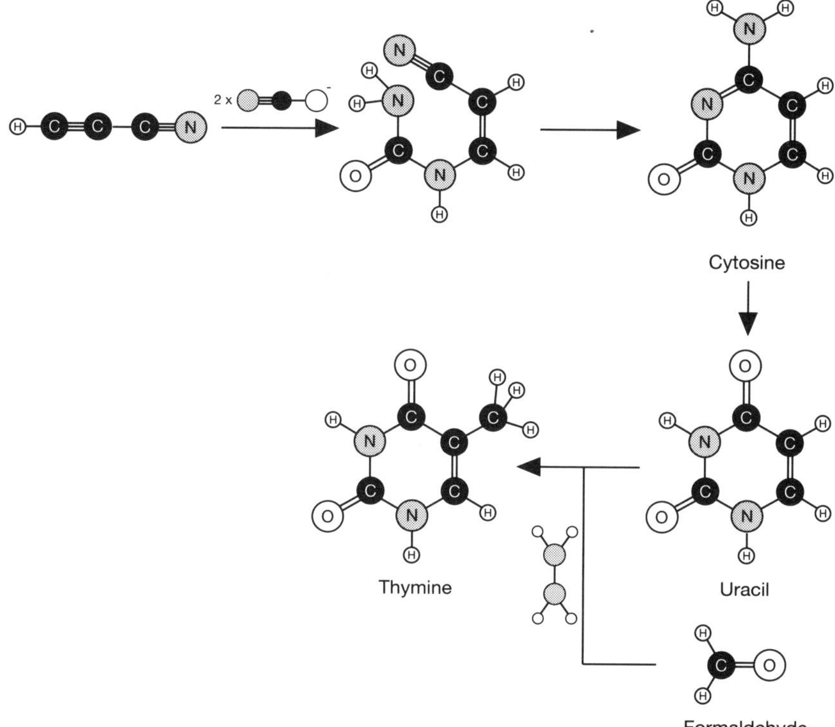


**Figure 8.7** *The pyrimidine bases cytosine and uracil are produced by the reaction of cyanoacetylene with either cyanate ions (shown here) or urea. Uracil itself reacts with formaldehyde, in the presence of hydrazine ($N_2H_4$), to give thymine.*

which then undergoes reactions with other simple organic molecules to form uracil and thymine. Cyanate ions can substitute for urea in this scheme (Figure 8.7).

These reactions may seem rather complicated, but remember that we are not concerned with neat laboratory syntheses that produce the desired products in high yield; we want to establish merely whether it is *feasible* to create these bases from a mixture of simple organics. In other words, we want to demonstrate that *in principle* there is no reason to believe that they could not be formed on the prebiotic Earth. All of the processes above involve reagents or conditions that we might reasonably expect to be accessible in this environment.

The sugar molecules that are found in RNA (ribose) and DNA (deoxyribose) are likewise plausible products of crude syntheses using simple reagents. Both sugars contain a five-membered ring of carbon and oxygen atoms (Figure 5.2; page 148, and both can be formed from formaldehyde (HCHO) alone, which polymerizes into five- and six-membered sugar rings (pentoses and hexoses) in alkaline solution (Figure 8.8). But

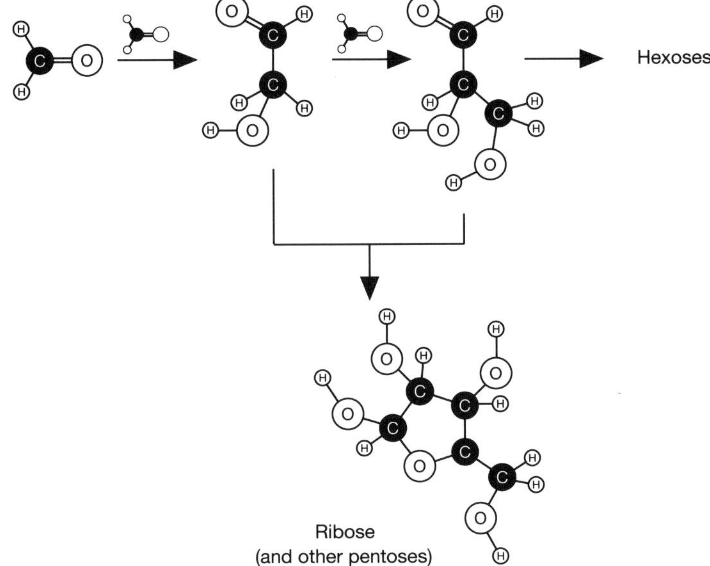


**Figure 8.8** *Sugars such as ribose can be formed by the polymerization of formaldehyde. The reaction produces a variety of other pentose and hexose sugars too.*

life may not have been so simple, so to speak; for ribose does not survive long under alkaline conditions, decomposing quite rapidly into acidic compounds. How it would have managed to accumulate in the prebiotic environment therefore remains something of a puzzle. So, indeed, does the fact that only ribose and deoxyribose, out of the fifty or so pentoses and hexoses that can be formed by formaldehyde polymerization, came to be selected for inclusion into nucleic acids.

One might think that the phosphate components of nucleic acids would be the least problematic to produce, since these groups are not organic at all but are simple inorganic ions which are plentiful in minerals such as apatite. But phosphate minerals are generally extremely insoluble, which has raised a long-standing question as to how phosphate ions could have become involved in the chemistry presumed to be going on between dissolved organic compounds in the oceans. Phosphates *can* exist in a soluble form, however, consisting of chains and rings of the basic $PO_4$ units (Figure 8.9). These polymeric structures, called polyphosphates, can be formed from phosphate-containing rocks under the kind of conditions that they experience in volcanoes.

Although they are inorganic compounds, phosphates may have played a crucial role in the development of more complex prebiotic molecules. Some organic molecules can be linked together more readily into long chains when they have phosphate groups attached. Phosphates will, for example, promote the formation of peptide bonds between two amino acids. Moreover, because polyphosphates are built up from units with

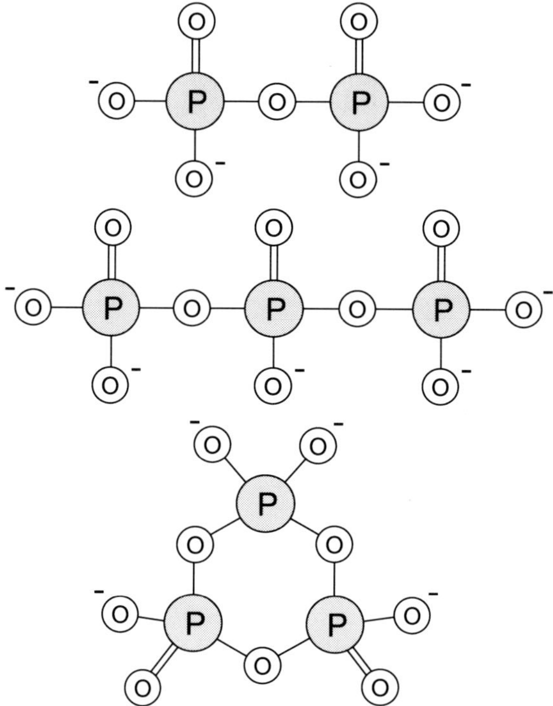


**Figure 8.9** *Polyphosphates – polymeric forms of the phosphate ($PO_4^{2-}$) ion – are relatively soluble in water, and may have been the source of phosphorus in prebiotic synthesis of the precursors to nucleic acids.*

negative charges, which therefore repel each other, a lot of energy resides in the bonds between these units. Polyphosphates therefore represent a kind of chemical battery which can store energy for subsequent use in driving reactions. Living cells store energy in the form of adenosine triphosphate (ATP), an adenosine nucleotide with a string of three phosphate groups attached. ATP or related molecules could have provided the power source for the first living organisms.

## *The first cells*

Having now addressed the problem of generating the principal building blocks of living organisms, the daunting prospect remains of linking them together in the highly organized manner that characterizes real proteins and nucleic acids. If it is hard to imagine how this delicate and sophisticated construction process could have taken place at all with nothing but the principles of chemistry to direct it, it is harder still to conceive of this happening in the storm-tossed waters of the primordial ocean, where in all likelihood organic molecules would be diluted and dispersed rather than conjoined. The reactions that take place in our own cells would not stand a chance in such an environment. But these biochemical processes have a safe, cosy and isolated microcosm within which to unfold – the cell membrane. It seems likely that any chemistry more complex than that which we have considered already would require a similar protective shell. How might such shells have arisen?

We saw in Chapter 7 that amphiphilic molecules such as phospholipids can assemble themselves spontaneously into hollow vesicles with double-layer membranes. Alexander Oparin recognized in the 1920s that something of this sort takes place with certain kinds of oily organic molecules. His "protocells," however, were not formed from amphiphiles but from a combination of natural polymers, usually gum arabic and a protein such as gelatin. Together in solution, these substances combine to form an insoluble compound that coalesces into tiny droplets, which Oparin called coercevates. Unlike bilayer vesicles, coercevates are not hollow shells but oily blobs not unlike the microscopic droplets suspended in well-shaken French dressing. Looked at in this light, coercevates do not appear to be very good models for cells, and it might seem odd that Oparin got so excited about them. They do, however, show some very interesting properties.

In particular, Oparin was able to create coercevates with a crude sort of metabolism that allowed them to grow and divide. Into coercevates made from gum arabic and the protein histone, Oparin incorporated an enzyme that links together glucose molecules to form polymeric carbohydrates called starches. These compounds are used as food stores by plants – they can be broken down again into their glucose components when required. When Oparin added glucose to a suspension of the enzyme-containing coercevates, the droplets absorbed it and transformed it into starch, swelling in the process. Eventually the droplets split, and the smaller progeny proceeded to ingest more glucose and to grow. There is clearly some similarity between this process and that by which cells metabolize, grow and divide. But the resemblance is a superficial one; the coercevates do not pass on genetic information to their progeny, they have no mechanism for producing the crucial enzyme themselves (it has to be added ready-made as an extract from living cells), and indeed the coercevates cannot even produce their basic constituents, the histone protein and gum arabic. Oparin felt that a few million years of evolution might be sufficient to allow coercevates to build proteins, but with no real knowledge at that time of exactly how genetic information was stored and transmitted, he did not fully appreciate the barriers to making a coercevate-based origin of life plausible.

Oparin's work did, however, provide the inspiration for a later variant of the "protocell" scenario, advocated by the biologist Sidney Fox of the University of Miami. Fox's protocells are tiny globules of randomly polymerized amino acids formed by heating mixtures of amino acids in the dry state; Fox calls these polymers protenoids. Attempts to polymerize amino acids by heating usually produces a black tar that is of no use to anyone. But one particular amino acid, aspartic acid, *will* polymerize into a polypeptide by this route. Fox found that in the presence of aspartic acid, other amino acids can also be induced to polymerize, or more correctly to *co*polymerize, forming

**Figure 8.10** *Sidney Fox's protenoid microspheres, generated by the polymerization of amino acids, can grow, reproduce and exhibit primitive forms of metabolism. Fox proposes that they can be regarded as living "protocells." (Photograph by Steven Brooke and Sidney Fox, courtesy of Sidney Fox, University of Southern Illinois.)*

polypeptides that include some aspartic acid too. Another amino acid, glutamic acid, also has this ability to induce polymerization into protenoids.

When dissolved in water, protenoids will spontaneously form spherical structures typically a few thousandths of a millimeter across. Unlike Oparin's coercevates, these microspheres are not droplets but are hollow protenoid vesicles (Figure 8.10). Moreover, they are all of similar size and can coalesce or develop buds that split off into new microspheres, all of which suggests a far greater similarity with real cells. Fox's protocells can even show a degree of catalytic activity for some biochemical reactions, vaguely reminiscent of the behavior of enzymes but without the specificity. This behavior might conceivably be seen to suggest a plausible role for protenoids in prebiotic evolution as protective reaction vessels for chemical reactions. But Fox has gained many antagonists by wishing to go further, proposing that the protenoid microspheres might themselves have developed into primitive systems showing some of the characteristics of life. This idea is very difficult to sustain, since even though the protocells are made of protein-like material, like coercevates they do not obviously contain any basis for replication and transmission of genetic information.

Perhaps more appealing as a candidate for prebiotic protocells are the self-replicating micelles of Pier Luigi Luisi and colleagues described on page 229. Not only are these made of amphiphilic molecules like those that constitute real cell walls, but they can be transformed into bilayer vesicles which really do look something like the shells of living cells. Luisi's experiments require rather specific starting materials, however, and it is not easy to see how these might have formed spontaneously on the early Earth.

### *DNA and proteins: which came first?*

The prebiotic assembly of both amino acids and nucleotides into polymeric forms is possible in principle, if difficult in practice. Amino acid polymerization is not easy in a watery environment. The formation of a peptide bond between amino acids involves the expulsion of a molecule of water; but the bond can also be split again by water (a process known as hydrolysis). In the presence of a lot of water, hydrolysis will be more favorable than peptide formation. It is conceivable that amino acids could have become linked up into polypeptides by being concentrated in evaporating pools or in hot, dry environments close to volcanoes; but experiments that attempt to mimic such processes give only tiny yields of polypeptides. The linking together of amino acids can be assisted, however, by certain kinds of reactive molecules, called condensing agents, which can "soak up" the water that is produced. The compound cyanamide, for example, can induce linkage of the amino acids glycine and leucine (Figure 8.11). Cyanamide can itself be formed from hydrogen cyanide under plausibly prebiotic conditions.

John Oró, Cyril Ponnamperuma and others have shown that condensing agents such as cyanamide will also assist bases, sugars and phosphate to combine into nucleotides, while Leslie Orgel of the Salk Institute in California has found that metal ions such as zinc will help nucleotides to link up into oligonucleotides (short polymers containing a few nucleotides). Primed with polymerized nucleotides and polypeptides, we seem to be well on the road to a world containing the nucleic acids and proteins of primitive living organisms. Is our quest for a chemical origin of life near an end? Not a bit of it. We have just reached one of the most difficult hurdles of all.

So far we have been relying pretty much on chance. That is to say, we have supposed our raw materials to be created amidst a whole mess of other compounds from very crude reactions. And we have seen that it is possible for these compounds to link up in random fashion to give polymers. We now have to face the fact that life is anything but a random process; it is in fact just about the most impressive feat of molecular organization that we know of. Chapter 5 suggested that a reasonable definition of life can be specified in terms of three functions: self-replication, self-repair and metabolism. Moreover, it is generally recognized that living systems must be bounded systems – they must have boundaries of some kind. All of these characteristics require a high degree of organization and cooperation at the molecular scale. Where does the organization come from? Surely not from the insensate, random world of prebiotic chemistry that has been described so far?

Well, it must, of course, unless we are going to shrug our shoulders and accede to divine intervention. Having got this far, I think that we would be advised to reserve that option for when we get really desperate!

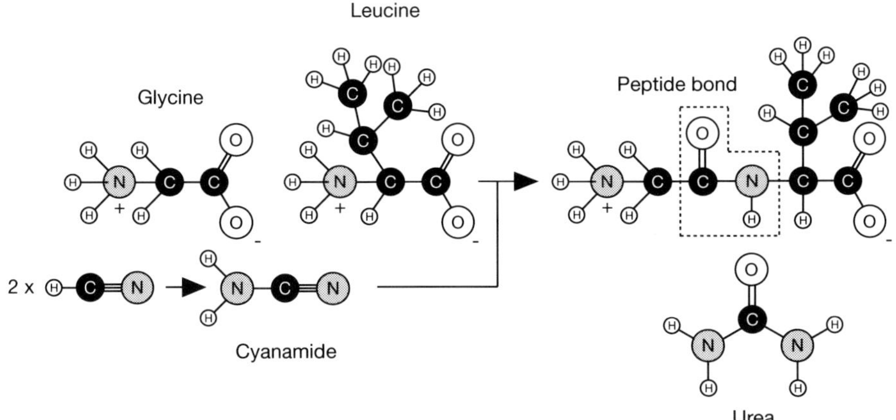


**Figure 8.11** *The amino acids glycine and leucine can be linked into a dipeptide using cyanamide as a "condensing agent."*

The organization within an organism stems ultimately from its genetic makeup – its genome – which encodes the information for construction of the organism's molecular machinery. Most of this machinery is in the form of proteins, while the information itself is encoded on the nucleic acid DNA and is translated into proteins via the agency of RNA. Here we run up against one of the conundrums that defeated origin-of-life researchers for nearly three decades after the ground-breaking experiments of Miller and Urey. Assembling DNA-like oligonucleotides by the random joining of nucleotides has a vanishingly small chance of ever producing anything that resembles the blueprint for an organism, which is to say, for the protein enzymes that are essential to life. In just the same way, we can forget about the chance of making efficient enzymes from the random assembly of amino acids. Both DNA and proteins can be considered replete with *meaningful* information – they are preprogrammed for specific functions. But preprogrammed how?

The problem can be posed in another way. Proteins require DNA for their formation, the protein plan being coded in the four-symbol DNA alphabet. But DNA – as distinct from random polynucleotide chains – cannot be created without the assistance of protein enzymes, which assist the assembly of new strands of nucleic acid on the templates of existing ones. If we can imagine a world with DNA but no proteins, we can see how the latter could be put together from the information encoded in the former. And if we imagine a world with protein enzymes but no DNA, it is conceivable that the enzymes could act together to synthesize nucleic acids from prebiotic nucleotides. But until we have one, we cannot have the other – and vice versa. The problem of the chicken and the egg, far from being a trivial philosophical paradox, provides a singularly apt metaphor for the origin of life.

## *The RNA world*

In the early 1980s, a possible way was discovered to break the protein–DNA circle. So far neglected in the discussion is the go-between in the process of translating DNA to proteins – the humble RNA molecule. But this function as a middle man makes RNA in many ways the ideal candidate for the basis of the first truly replicating molecules. RNA is able both to store genetic information (recall that messenger RNA, or mRNA, carries the information encoded in genes) and to act as a template for the formation of proteins. In other words, it acts both as a carrier of the genetic plan of an organism (the genotype) and as a means for the external expression of that genetic information (the phenotype), via the formation of proteins.

The idea that RNA might have been the first molecular replicator dates back to the 1960s, but such early speculations floundered on one crucial point: the replication of RNA seemed to suffer from the same limitation as that of DNA, in that it required help from enzymes. In the 1980s, however, the molecular biologists Sidney Altman and Thomas Cech discovered that this was not necessarily the case. They found that some RNA molecules can catalyze the assembly of other RNA molecules, thereby acting as "nonprotein enzymes." The implication is that such catalytic RNA molecules might be able to facilitate their *own* replication. Cech and Altman called these catalytic RNA molecules ribozymes.

The discovery of catalytic RNA, which won Altman and Cech the Nobel prize for chemistry in 1989, revitalized interest in the suggestion of a prebiotic world populated by replicating RNA molecules – not exactly living systems, but well on the way towards them. The biologist Walter Gilbert of Harvard University has christened this scenario the "RNA world." The first inhabitants of the RNA world would have been simple RNA-like oligonucleotides with some ability to catalyze their reproduction. Mutations in the nucleotide sequences would occasionally arise from imperfect replication on the RNA template, and mutant forms that turned out to be better at replication would dominate over the others by Darwinian selection. These RNA replicators would become ever more efficient, and could eventually learn how to put together proteins, perhaps in a crude imitation of the codon-based translation process that goes on in our own cells (Chapter 5). Some of these proteins might turn out to assist in RNA replication, making them primitive enzymes. RNA molecules that could produce their own enzymes would gain a tremendous evolutionary advantage over their less capable fellows, and every advance made in this ability would produce a new dominant strain. Only quite late in the day would DNA appear – a double-stranded version of RNA in which the base uracil had become replaced by thymine. As it is a more stable database for storing genetic information, DNA would have gradually taken over as the central component of a replicating system, subverting RNA to the role of an intermediary in protein synthesis.

The catalytic behavior of RNA provides one of the crucial links required for this picture to seem plausible, but it is by no means the only good reason for believing that life, in its earliest manifestation, arose from an RNA world. While DNA is generally a passive memory bank for genetic information, the various forms of RNA play a very diverse and active part in the biochemical processes of the cell. In particular, they are central to processes that are thought to be of most ancient origin. Many coenzymes – molecules that assist enzymes in their various tasks – are based either on true RNA nucleotides or on related compounds, suggesting that RNA-like species may have had to exhibit a wide range of talents before proteins became the dominant biochemical catalysts.

Nevertheless, the scenario of the RNA world is not without its problems, not the least of which is the question of how RNA-like molecules evolved in the first place. We have seen that the basic components of nucleotides can be created from simple organic molecules, albeit with some considerable artifice. Nucleotides can be linked at random in the presence of condensing agents, but the oligonucleotides produced will represent gobbledigook in genetic terms, a far cry from information-laden RNA. It is likely that life did not really start with RNA at all, but with some simpler, albeit similar, kind of molecule that also had some ability to replicate and carry crude genetic information. Perhaps sugars other than D-ribose were incorporated into these pre-RNA replicators, and truly RNA-like molecules were then gradually singled out from this jumble by virtue of their particular aptitude for reproducing themselves reliably. If this was so, what might these earliest of replicating molecules have looked like?

## *The first replicators*

We have already encountered some examples, real and imagined, of molecular assemblies far cruder than RNA that can nevertheless show some degree of self-replicating ability. The (hypothetical) clay minerals of Cairns-Smith are one such; Luisi's autocatalytic micelles represent another. Gerald Joyce, a molecular biologist at the Research Institute of Scripps Clinic in La Jolla, California, has suggested that systems like these might have paved the way for RNA or related precursors, either by simply altering the environment so as to make the evolution of more complex replicators less troublesome, or by developing an ability to catalyze the formation of RNA-like molecules themselves.

One putative RNA precursor that has been studied in some detail is a "pseudonucleoside" formed from glycerol and a purine base (a nucleoside is simply a nucleic acid base attached to ribose or deoxyribose – a nucleotide without phosphate, if you like). Unlike the nucleic acid sugars, glycerol is not cyclic and the nucleoside-like molecule it can form with a purine is not chiral. While ribose–purine link-ups may produce a bewildering array of possible isomers and enantiomers, the number of glycerol analogs is therefore much more limited and their chemistry much simpler. Molecules of this sort can join into oligomers (short polymeric chains) which can act as templates for the linking together of true RNA-like nucleotides.

The German chemist Gunter von Kiedrowski has demonstrated replication in synthetic molecules based more explicitly on DNA-style chemistry. He showed in 1986 that short sequences of DNA-like oligonucleotides can act as templates for the assembly of copies of themselves. On a molecule consisting of six linked nucleotides containing the complementary cytosine (C) and guanine (G) bases, von Kiedrowski brought together two three-nucleotide fragments of this molecule; when held together in this way, the two fragments could link up to form a copy of the six-unit template (Figure 8.12*a*). This template assembly process is an example of replication because the six-unit molecule is *self*-complementary – it binds to an identical molecule head-to-tail.

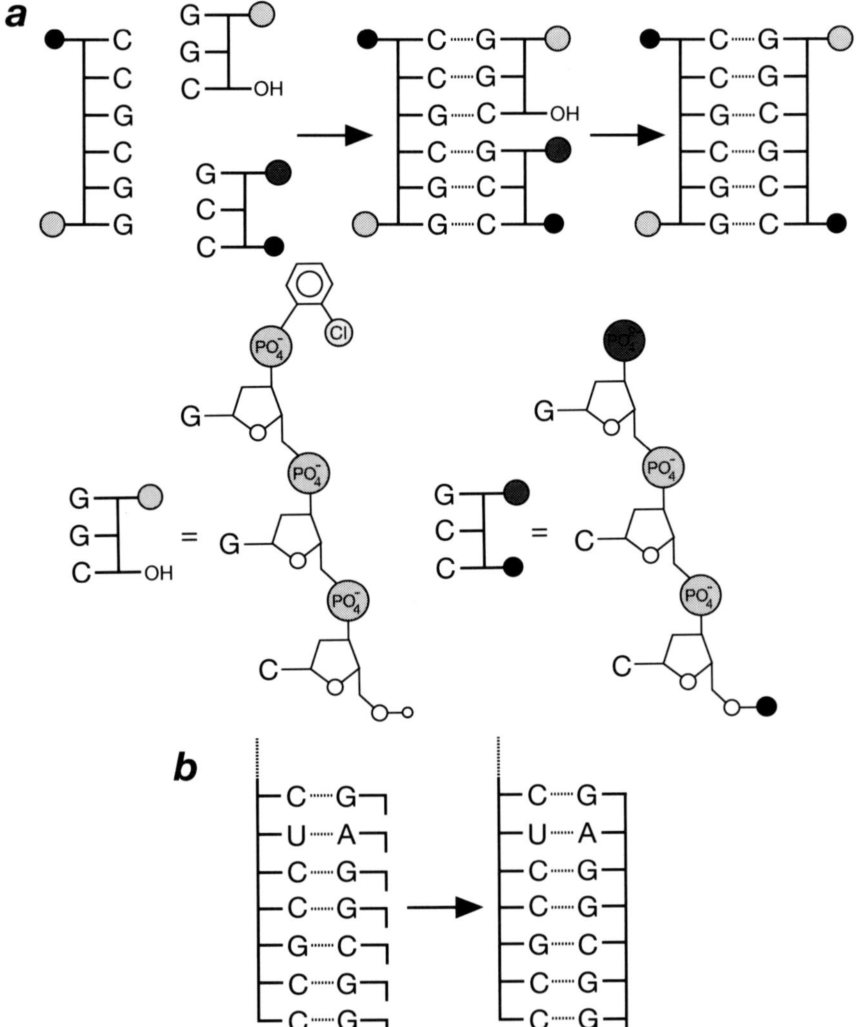


**Figure 8.12** *Synthetic strands of DNA-like nucleotides can act as templates for the assembly of copies or of complementary strands. Gunter von Kiedrowski found that the six-nucleotide strand in (a) would catalyze the formation of copies from its two three-nucleotide fragments. Leslie Orgel found that longer strands could be built from individual nucleotide units assembled one by one on a complementary template (b). As the strands get longer, however, copying "mistakes" become increasingly likely. In DNA replication, such mistakes are identified and corrected by proof-reading enzymes.*

Leslie Orgel's group has gone a stage further in showing that oligonucleotides can act as templates for the nucleotide-by-nucleotide assembly of a complementary strand, rather than being presented with preformed chunks as in von Kiedrowski's experiments (Figure 8.12*b*). This is really a feat of copying rather than replication, as the assembled strand is complementary to, rather than identical to, the template.

One of the difficulties in extending these approaches to longer strands, however, is that once a copy is formed on the template, it tends to stick there in a stable double helix rather than separating so that the molecules can catalyze further replication. And in experiments like Orgel's, the fidelity of the copying process starts to decline once the oligonucleotides become longer than a dozen or so units. On the other hand, these results show that a limited amount of information encoded in nucleic-acid-based chains can be reproduced without the need for the complex enzyme machinery that DNA employs.

The few studies that have been conducted on compounds like these have provided little more than vague hints as to how RNA-like molecules might have created a true RNA world. If the earliest replicators were ever going to turn chemistry into biology, they would need some way of *evolving*. They would need to be able not only to store and pass on information, but also to undergo mutations that might confer evolutionary advantages, allowing them to become better adapted to survival. Evolution and natural selection at the molecular scale would have been responsible for replicating molecules getting *clever*.

But the very fact that it now seems to be possible to conduct experiments to address these problems and uncertainties is tremendously exciting. As I suggested at the end of Chapter 5, it is really rather remarkable that researchers are now becoming able to think about prebiotic chemistry in evolutionary terms – to apply to mere molecules the kinds of concepts, such as mutation and natural selection, that Darwin developed from studies of highly evolved living creatures. As well as having the considerable virtue of generating some ideas accessible to experimental tests, the concept of the RNA world illustrates one of the attractive aspects of studies into the chemical origin of life: one is no longer faced with the prospect of having to work doggedly and without gaps from start to finish, from inorganic chemistry to biology. The puzzle is beginning to resemble a jigsaw in which several areas are starting to take on definite shape, even though substantial blank regions remain between them. We are not yet sure how to get into the RNA world, nor how to get out into a DNA world, just as we cannot solve the riddle of enantioselectivity or give clear answers regarding the relative importance of different proposed sources of primordial amino acids. But the overall picture is starting to make sense nevertheless, and we can concentrate on issues for which answers seem within reach, with the understanding that we can go back and plug gaps later. We are no longer helpless in the face of one of the world's biggest mysteries: where we all came from.

# 9

# Far from Stable

## Fractals, chaos, and complexity in chemistry

*... the physical phenomena which meet us by the way have their forms not less beautiful and scarce less varied that those which move us to admiration among living things.*

D'Arcy Wentworth Thompson

It was during the seventeenth century that mathematics came to be adopted as the official language of science – indeed, to be celebrated as the "Queen of Sciences." For the central figures in this new era of rational philosophy, such as Isaac Newton, Rene Descartes and Gottfried Leibniz, the proper way to express one's investigations into the machinations of the natural world was in the form of mathematical equations. Leibniz in particular considered that mathematics should provide the universal language with which to describe all forms of human enquiry, be they scientific, historical, philosophical or economic. While this is perhaps claiming for mathematics an overstated generality, it remains still the agreed mode of expression for fundamental scientific principles. From Newton's inverse square law of gravitation to Einstein's $E = mc^2$, the language provides a common tongue to scientists of every nationality. In some scientific papers the words of the text scarcely matter: the equations say it all.

This convention exists, of course, because it is such an effective one – mathematics turns out to be appropriate for describing the way that nature behaves to a degree that the Hungarian–American physicist Eugene Wigner could only regard as "unreasonable." Why is this so? Some scientists consider mathematics to be a human invention, an arbitrary formalism that merely serves our purposes. Others cannot but conclude that mathematics is as much an integral part of nature as are subatomic particles or the fundamental forces that act on matter.

Mathematics is a very formal language. To the unversed, a mathematical treatise might as well be written in ancient Sanskrit. And as any student of geometry will testify, it appears to be a paradigm of orderliness and predictability. The mathematical world seems to be inhabited by forms of perfect, simple symmetry: circles, squares, straight lines. The aesthetic appeal of perfect geometrical shapes was not lost on the ancients, who saw them as an expression of divinity. Said Plato, "geometry will draw the soul towards truth, and create the spirit of philosophy." Plato proposed that the fundamental particles of the four Greek elements, earth, air, fire and water, possessed the shapes of the perfect three-dimensional solids identified by Euclid. To Pythagorus and his followers, numbers were not a means of quantifying the natural world; they were the very essence of which the world was made.

This essentially mystical faith in the perfection of mathematics as an underlying principle of nature can be seen too in the work of the sixteenth century astronomer Johann Kepler, who attempted to show that the orbits of the six planets then known could be rationalized by inscribing within them the perfect solids of Euclid. One of the principal objections to Galileo's heliocentric (Sun-centered) Solar System was that it necessitated planetary orbits that were not circles but ellipses, with the Sun at one of the two "focal points" of the ellipse. How, it was argued, could a celestial body follow any other path through the heavens but one that formed a perfect circle?

Since the nineteenth century, however, scientists of all persuasions begin to acquire the uneasy suspicion that there was something amiss with this insistence on the geometrical perfection of the natural world. After all, how many natural objects are truly perfect circles, spheres, cubes, hexagons, tetrahedra or whatever? On the contrary, the evidence suggests that nature cares not one whit for geometry. Instead we have trees, clouds, flowers, mountains and living creatures, demonstrating all manner of irregularity of shape and form. So how, if mathematics is the language of science, can science ever hope to account for this bewildering variety of complex forms?

That question constitutes one of the principal themes of this chapter. To supply an answer is a task that frequently falls to chemists, since it is their business to explain the way in which molecules come together in the shapes and forms that we see around us. D'Arcy Thompson, the unorthodox polymath whose words begin this chapter, tells us that "Of the chemistry of his day, Kant declared that it was a science, but not Science ... for that the criterion of true science lay in its relation to mathematics." Thompson's eccentric yet influential book *On Growth And Form,* which made an almost quixotic attempt to explain the complexity of natural forms through an erudite but sometimes incompatible amalgam of physics, mathematics and mechanics, succeeded at the very least in providing some intimation that geometry and the organic world need not be mutually exclusive.

Today's chemistry is as mathematical as you could like (some would say more so). But the "reductionist" approach to many-molecule systems, which tries to provide a description based on a few simple equations representing the interactions between molecules, becomes rapidly intractable as the numbers grow larger. The world therefore abounds with systems that have in the past been regarded as far too complex to permit any obvious means of realistic mathematical analysis. One of the astounding revelations of recent decades, however, is that complexity does not necessarily imply disorder, intractability or general messiness. Rather, it has become apparent that complexity can itself give rise to the abrupt manifestation of order, often in the form of *patterns* of startling richness that are a far cry indeed from the geometric sterility generally exhibited by simpler systems. The real surprise is that there may be nothing in the fundamental elements of the mathematical description to warn us that these structures might appear; a reductionist sees only the most prosaic of interactions between the system's individual components, but the "holist" who considers the system as a whole discovers within it an unguessed capacity for intricate organization.

Many of the patterns that arise out of complexity resemble the delicate and often beautiful "organic" forms of the natural world; moreover, similar patterns may be found in systems that are ostensibly unrelated. There is a common thread, however: these systems tend to be undergoing processes of rapid or unstable transformation. This has led to hopes for a unified description of pattern formation in systems that are far from attaining equilibrium.

## Bringing crystals to life

### *An alien microcosm*

Much of the beauty and fascination of natural minerals lies in the symmetrical shapes of their crystals. One would hardly call these prism-like forms complex, however. We saw in Chapter 4 that the atoms or molecules in crystals are stacked into orderly layers, with each atom surrounded by a regular, geometrical arrangement of neighbors. This kind of stacking produces smooth, flat faces and abrupt, angular corners, accounting for the faceted shape of the macroscopic crystals.

In order that these regular stacks be formed during crystal growth, each atom added to the growing crystal must have the opportunity to find an appropriate place in which to fit. It is no good atoms just sticking onto the crystal wherever they first strike it; they must be able to jump around on the surface until they find a free site in the regular lattice. This generally requires that the crystal grows slowly. But what will happen if the crystal is made to grow so quickly that there is no chance for "defective" atoms to rearrange in this way, so that each atom simply stays where it first hits, regardless of whether or not this happens to be a site on a regular lattice? This kind of growth process can be induced, for example, by suddenly dropping the temperature of a liquid far below its freezing point, producing a "supercooled" liquid. Crystallization is then a nonequilibrium process. Nonequilibrium crystallization can also be induced for a solid growing from its solution rather than from its melt, by abruptly cooling the solution when it is saturated with the dissolved material. A saturated solution is one that can dissolve no more (many people saturate their coffee with sugar, which is why some lies undissolved at the bottom). The amount of dissolved material that a saturated solution can hold usually decreases as it gets colder; so rapid cooling of a saturated solution will cause some to precipitate out.

The shapes of crystals grown under nonequilibrium conditions can be very far from faceted, geometric prisms. Figure 9.1 shows some of the exquisite, almost organic forms adopted by alloys of iron, chromium and silicon grown by rapid crystallization from a hot vapor. An electron microscope shows that these materials adopt all manner of bizarre shapes, resembling nothing so much as an alien landscape in a science fiction

**Figure 9.1** *When prepared by depositing the atoms from a vapor onto a cold surface, metal silicides such as* $(Cr,Fe)_5Si_3$ *crystallize into a variety of bizarre and dramatic shapes. (Photographs courtesy of Seiji Motojima, Gifu University, Japan.)*

film. Yet these structures are not simply land scape random or disorderly – they are complex, to be sure, but there is a certain symmetry, a certain pattern to them. Something in this far-from-equilibrium process is imposing a degree of orderliness.

## *Tenuous tales*

To try to gain some insight into nonequilibrium crystallization, scientists have focused much attention on a mode of growth called aggregation, in which a cluster of particles grows by their random collision. When a particle encounters the growing cluster, it sticks where it first strikes. This process mimics rapid crystal growth, during which the atoms do not have time to rearrange. Aggregation is a common process in nature: it is encountered, for instance, in the formation of large soot particles in smoke by the clumping together of smaller fragments, or the "flocculation" of tiny particles of organic material in rivers, which turns the waters opaque and muddy.

In many aggregation processes, the rate at which a cluster grows depends on the length of time it takes for the aggregating particles to drift through the surrounding

**Figure 9.2** *A cluster of particles formed by diffusion-limited aggregation (DLA) takes on a tenuous, ramified shape. This cluster is the result of a computer model, in which particles are allowed to drift around at random until they encounter another particle, whereupon they stick together. (Image kindly prepared and supplied by Thomas Rage and Paul Meakin, University of Oslo.)*

medium and encounter the surface of the cluster. The random motion of particles through the medium is called diffusion, and aggregation controlled by these motions is said to be diffusion-limited. Clusters grown by diffusion-limited aggregation (DLA) develop delicate, branching tendrils (Figure 9.2). Once a tendril starts to sprout, new particles are likely to encounter and stick to it before they can find their way deeper into the cluster's core; and so the "valleys" inside the cluster are never filled.

On close examination, DLA clusters turn out to have some strange characteristics. Imagine taking a close look at a large cluster under a microscope. We see an object with highly irregular, ramified branches. Now we step up the magnification, zooming in on one particular region of the cluster. The bumps and tendrils that we could just about make out initially now become easier to distinguish, but we will find that they too bear smaller bumps; the picture in the microscope's eyepiece looks much the same as it did at lower magnification (Figure 9.3). Increasing the magnification still further gives the same again: the irregular surface turns out to have still finer tesselations that were not previously visible, so that again the image remains much the same.

An object that appears to have an unchanging structure when viewed at different levels of magnification is said to be self-similar. It is impossible to define a meaningful length scale for self-similar structures. We were reminded in Chapter 4 that the "block" provides a convenient measure of distance in New York City, since the city is divided by criss-crossing streets into units which share much the same size. The same is true of regular crystals, in which lengths can be expressed in terms of a number of unit cells. But in self-similar structures one cannot pick out blocks of a distinct size by which to measure distances, because it is in the very nature of self-similarity that these will turn out to be made up of still smaller blocks, and will, furthermore, be components of larger ones. There is no natural length scale at all; self-similar structures are "scale-invariant."

The branches of a DLA cluster are thin chains of particles – string-like, one-dimensional entities. Yet as these grow outwards, what we see in Figure 9.2 is a delicate

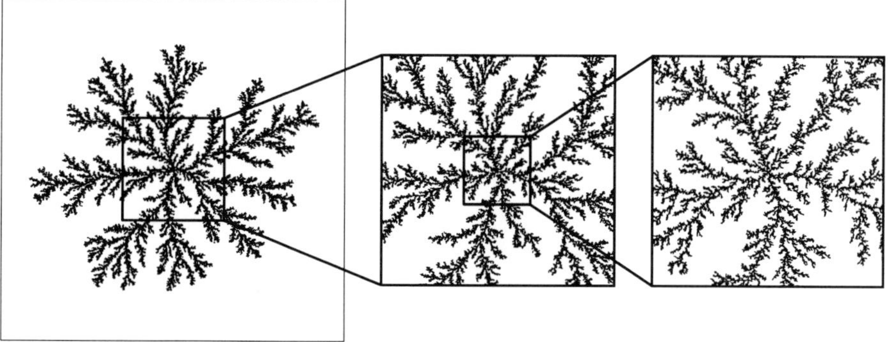

**Figure 9.3** *As one looks ever more closely at a DLA cluster, increasingly fine levels of structure become apparent. The cluster therefore looks the same at many levels of magnification; it is said to be self-similar. (Image kindly prepared and supplied by Thomas Rage and Paul Meakin, University of Oslo.)*

mass of branches that largely fill a two-dimensional area. (Real-world DLA clusters such as soot particles send out their tenuous arms in all three dimensions, but computer-generated "flat" images like that in Figures 9.2 are easier to visualize.) Is a flat DLA cluster one- or two-dimensional?

There is a simple way of determining the dimensionality of an object, and that is to look at how its mass increases as it grows. The amount of material in a line-like entity – say, the amount of ink required to draw it – grows in direct proportion to its length. As the radius of the tenuous star-like object in Figure 9.4*a* grows, the mass of the object increases at the same rate. In mathematical terms, the object's mass is proportional to its radius. An object for which this relationship holds is one-dimensional. The amount of material in (and hence the mass of) the solid object of Figure 9.4*b*, meanwhile, depends on its area, which is related to the square of the radius (that is, to the quantity radius × radius, or $(\text{radius})^2$). This is characteristic of a two-dimensional object. For a three-dimensional object such as a sphere, the mass depends on the volume, which grows as the radius *cubed*: radius × radius × radius, or $(\text{radius})^3$.

The general relationship between dimensionality, radius (or width, if you like) and mass should now be evident: the mass increases at the same rate as the radius multiplied together as many times as the object has dimensions. So we can determine the dimensionality of a DLA cluster by measuring how its mass increases as its size increases. This is done most easily by simulating the DLA process on a computer.

The result of this computer experiment seems to defy all intuition. The mass does not grow in proportion to the radius, nor in proportion to the radius squared, but in proportion to something in between: the radius multiplied by itself 1.7 times, or $(\text{radius})^{1.7}$. (What is meant by multiplying something by itself 1.7 times may not be obvious, but there are mathematical ways to calculate the result – all you need is a book of logarithms.) According to our rule of thumb, this implies that the cluster is 1.7-dimensional. What can this mean? We are used to thinking only in terms of whole number of dimensions. A particle confined to a line can move in just one dimension; a particle confined to a surface moves in two; while our everyday world is three-dimensional. But how can a particle be confined to 1.7 dimensions?

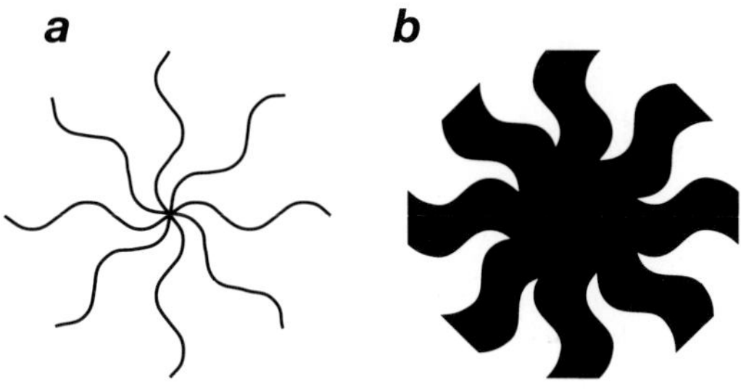


**Figure 9.4** *The dimensionality of an object can be deduced by determining how its mass changes with increasing size. For a one-dimensional object (a) the mass is proportional to the size; for a two-dimensional object (b) the mass increases in proportion to the size squared. A "flat" DLA cluster like that in Figure 9.2 lies somewhere between these two extremes, indicating that it has a nonintegral dimension between 1 and 2. This is a characteristic property of a fractal shape.*

**Figure 9.5** *Many natural forms, such as plant roots and trees, possess self-similar structures. The island of Socotra, off Africa, produces these particularly spectacular examples, called dragon's blood trees. Compare the branching patterns seen here with that in Figure 9.7. (Photograph courtesy of J. E. D. Milner/Acacia.)*

## *Nature's geometry?*

When he discovered objects whose dimensionality is not a whole number, the mathematician Benoit Mandelbrot named them fractals; but privately he thought of them as "monsters," since they seemed to defy commonsense notions about geometry. It is now clear, however, that fractals are not at all grotesque invaders from the abstract world of mathematics – objects exhibiting the characteristics of fractals can be found all around us. Plant roots and trees have a fractal character, forking repeatedly at ever smaller scales (Figure 9.5). Clouds too often have a fractal structure, and likewise the natural topography of mountainous country and the patterns of river networks (Figure 9.6). Coastlines are fractal shapes: from satellite images of a continental coastline to small-scale maps of local bays and inlets, the boundary of land and sea remains tesselated on ever smaller scales.

Indeed, fractals seem to crop up so commonly in the natural world that one is forced to wonder how they could have been overlooked for so long. Perhaps their past neglect stems from the fact that fractals demand a totally new way of thinking about shapes. We are used to describing objects in terms of geometrical outline: a square house, a round apple and so forth. But fractals cannot easily be described in these terms. Their outlines are highly complex and their scale invariance prevents the identification of

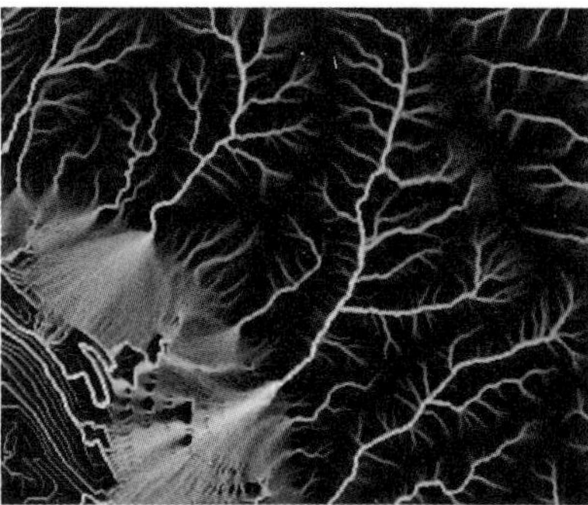

**Figure 9.6** *Many natural landforms exhibit self-similarity and fractal geometry. Mountain ranges, for example, undulate over a wide range of length scales, and river networks commonly display a complex branching structure, as shown here in a pattern of streams (in white) from northeast Nevada. (Picture courtesy of Colin Stark, University of Oxford.)*

characteristic structural units from which the overall pattern is built. Instead, the natural way to describe fractal patterns is in what we might call "algorithmic" rather than geometric terms – rather than giving a pictorial description of the shape, a fractal object can be characterized by providing a set of "rules" that generate it. To describe a tree-like fractal, we might say "Start with a line; branch off at a specific angle into two lines after a distance $d$, then make each of those two branch off in the same way after a distance $\frac{1}{2}d$, then do the same again after $\frac{1}{2} \times \frac{1}{2}d$ (that is, $\frac{1}{4}d$), $\frac{1}{2} \times \frac{1}{2} \times \frac{1}{2}d$, and so on." The pattern that results will look like that shown in Figure 9.7. The basic rule is this: draw each line half as long (and half as wide) as the one it sprouts from, then branch again into two new lines. A set of sequential instructions like this is called an algorithm; the word appears also in computer science, where it describes the sequence of steps that a program follows in order to solve a problem. An algorithm is essentially a strategy for carrying out a particular task.

Figure 9.8 shows another example of a fractal object, called a Sierpinski gasket. The algorithm for generating this structure runs as follows: divide each black triangle into four smaller ones of equal size and remove the central triangle. Each time this operation is performed it leaves three new triangles, each with one-fourth the area of the initial triangle; the operation is then performed again on each of these. To obtain a perfectly fractal object, the algorithm must be repeated an infinite number of times. This means that the black triangles get smaller and smaller, and one might think that an infinite number of repetitions would cause them to vanish altogether, leaving just a blank, white space. But it is clear that, each time the algorithm is repeated, we leave behind three times as much area as we remove, so we can never remove all the black entirely. The ideal Sierpinski gasket is therefore a highly tenuous sponge-like creation. And while we start off with a black triangle which "fills" the two-dimensional space it encloses, after an infinite repetition of the algorithm we find that the object is no longer two-dimensional but has a fractional (which is to say, a fractal) dimension – in this case, it is about 1.58-dimensional.

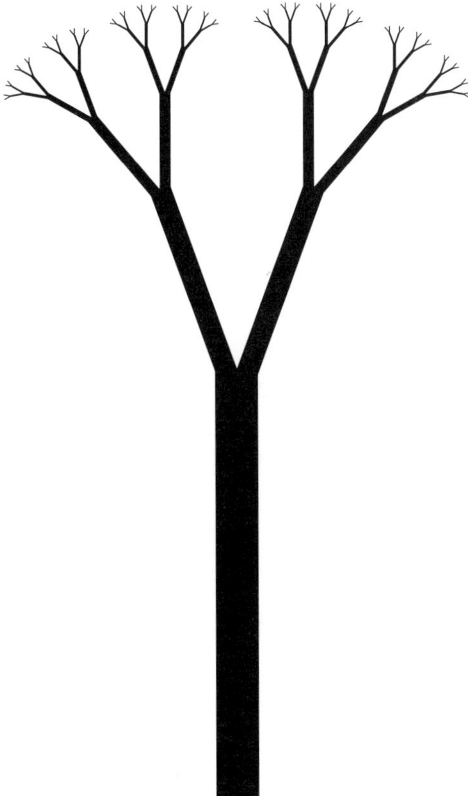

**Figure 9.7** *A simple algorithm in which columns branch repeatedly in a well-defined manner gives rise to a fractal "tree." (The width of the branches is also varied so as to ensure self-similarity.) A variety of natural-looking shapes can be generated by using simple rules such as this.*

The illustration in Figure 9.8 does not show a perfectly fractal Sierpinski gasket. The algorithm has been stopped after just six repetitions, because it is not easy to draw the details finely enough after this. And in nature too, objects that have a self-similar character do so only over a limited range of scales, since ultimately new factors (such as the cellular or molecular structure) start to control the shape.

Fractals crop up in so many different natural systems that Mandelbrot has proclaimed fractal self-similarity "the geometry of nature." Regardless of whether this grand claim can be justified, fractal geometry does go some way towards showing how unifying principles may lie behind some of the complex patterns found in nature. And the example of DLA clusters illustrates another important consideration: fractal geometry is commonly a consequence of a process of change occurring out of equilibrium.

## *Fingers and flakes*

Fractal DLA-type clusters may be generated in a crystal growth process known as electrodeposition. This involves the growth of metal clusters from a solution of dissolved

**Figure 9.8** *The Sierpinski gasket, formed by repeatedly chopping out the central portion of equilateral triangles. The end product is a tenuous, sponge-like object with a dimensionality of about 1.58.*

metal ions by applying a voltage to an electrode dipped in the solution. When the voltage at the electrode is relatively low, the deposition process is leisurely and a smooth film of metal develops – this is the basis of electroplating. But if the voltage is much larger the process no longer proceeds under conditions close to equilibrium. Then the metal deposit acquires a highly irregular shape (Figure 9.9). Robin Ball and Robert Brady of the University of Cambridge pointed out in 1984 that in many ways the electrodeposition process provides an experimental realization of diffusion-limited aggregation, and indeed the metal cluster shown in Figure 9.9 turns out to have a fractal dimension of about 1.7, more or less identical to the dimension of the computer-generated DLA cluster in Figure 9.2.

By altering the voltage of the electrode one has a convenient way of tuning the deposition process further from or closer to equilibrium. This kind of tuning reveals that the DLA-type fractal patterns are not the only alternative to smooth, near-equilibrium deposition. At certain voltages one can observe changes in the growth mode, giving deposits with a variety of shapes (or "morphologies"). Figure 9.10 shows two non-DLA growth modes. The shape in Figure 9.10*a* is called the dense-branching morphology; the filaments are no longer fine thread-like shapes, but fatter fingers that split and branch at their tips. The growth mode in Figure 9.10*b* looks somewhat more regular; the

**Figure 9.9** *Electrodeposition can produce metal deposits that resemble DLA clusters. The one shown here has a fractal dimension of about 1.7, just like the DLA cluster in Figure 9.2. (Photograph courtesy of John Melrose, University of Cambridge.)*

growing fingers branch in a more symmetrical way, and the main fingers themselves do not split at the tips but instead develop side branches. This mode is called dendritic growth.

Both of these new morphologies can be found in other systems too. The dense-branching morphology can be observed, for example, when a liquid is injected under pressure into another, more viscous (thicker) liquid with which it cannot mix – water into oil, for instance. The branching of the boundary between the two liquids is called viscous fingering. Injection of water into oil is the means by which oil is commonly squeezed out of porous, oil-bearing rocks in petroleum reclamation. In this situation viscous fingering is a nuisance because it makes the process inefficient: rather than the oil being pushed out by a uniform, advancing bubble of water, the oil and water become intimately entwined by the fingering process, even though they do not actually mix: Understanding the conditions under which viscous fingering occurs might therefore help to improve methods of oil recovery.

Viscous fingering can be studied using an apparatus devised in the nineteenth century by the British naval engineer Henry Hele-Shaw. The Hele-Shaw cell consists of two flat plates (one transparent) between which is enclosed a layer of the more viscous liquid. The less viscous liquid is injected through a hole in the center of one plate, pushing the other liquid outwards. The force with which the liquid is injected is analogous to the voltage in the electrodeposition experiment: the greater the injection pressure, the further the system is driven from equilibrium. By altering this pressure, one can induce a change in the mode of growth of the injected "bubble," just as cluster growth can be controlled in electrodeposition. All of the growth modes shown above for elec-

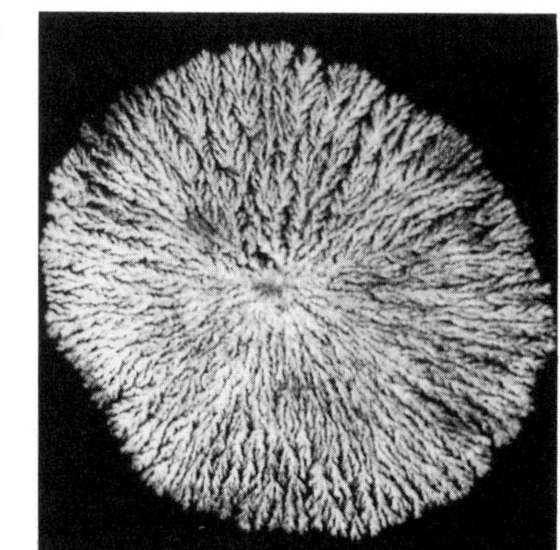

**Figure 9.10** *Various growth modes can be observed in electrodeposition simply by modifying the growth conditions, such as the voltage of the electrode. Shown here are the dense-branching (a) and dendritic (b) growth morphologies. (Photograph (a) courtesy of John Melrose, University of Cambridge; (b) courtesy of Peter Garik, Boston University.)*

trodeposition – the dense-branching morphology, DLA-type fractal growth, and dendritic growth – can be generated in the Hele-Shaw cell (Plate 14).

To obtain symmetrical dendritic patterns in the cell (Plate 14*c*), it is necessary to induce a preferred direction of bubble growth; this is generally achieved by engraving the base plate with grooves to form a regular lattice. These dendritic patterns may look familiar – they resemble the ice crystals found on window panes on a frosty morning. When ice crystals like this grow outwards from a central "seed," the result is a snowflake (Figure 9.11).

**Figure 9.11** *The classic example of dendritic growth is the beautifully symmetric form of a snowflake. These delicately patterned ice crystals display sixfold symmetry, reflecting the symmetry in the crystal structure of ice.*

The directional preference in the growth of such dendritic crystals arises from their microscopic structure – that is, the way that the atoms or molecules are stacked together. Because it may cost more energy to develop a bulge on one kind of crystal facet than on another, growth and branching are easier in some directions than in others. The most favorable growth directions tend to reflect the symmetry of the crystal structure: in crystalline ice, for instance, the water molecules are locked together in an array that displays hexagonal (sixfold) symmetry, so this symmetry is preserved in snowflakes. The crystal structure of solid carbon dioxide, meanwhile, has "square" (fourfold) symmetry, so carbon dioxide snowflakes (which one might encounter on Mars) would also have this symmetry, possibly looking like the pattern in Plate 14*c*.

## Waves and patterns in chemical reactions

### *Flux and change*

The scientific study of processes occurring far from equilibrium is a relatively young discipline, but it should be clear by now that there is nothing unusual or unnatural about such things. Quite to the contrary, we are surrounded by phenomena of this sort. The skies are perpetually in a state of flux: clouds, winds and storms come and go, driven by patterns of atmospheric circulation that are never stilled. The oceans likewise ebb and flow, their surface ruffled into waves ranging in size from the invisible to the awesome. The very face of the planet – the arrangement of seas and continents – has been changing through continental drift ever since dry land first appeared. Land masses collide and break apart, seas open up or are swallowed.

How strange it would seem if all this activity were suddenly to cease, if the oceans were to become calm, smooth mirrors or the weather to crystallize into a pattern that repeated day after day. Yet this is how chemists have long viewed the process of chemical change. Change occurs, they would grant, but it is an ephemeral thing. Two compounds come into contact, there may be a dramatic reaction accompanied by clouds of smoke, a flash of light, a detonation; but eventually a new equilibrium is reached. The smoke clears, and there sit the end products, idly content with their new state. The assumption was that nonequilibrium situations do not persist.

But we now know that there are some chemical reactions for which the "fizzing" does not stop – for which, to be more precise, the chemical components do not settle down into a final equilibrium state but appear instead to keep changing their mind about which state they prefer. Provided that we keep supplying these peculiar reactions with reactants, they will not simply regurgitate a given product but will oscillate between one state and another, often generating complex spatial patterns in the process.

The classic example of an oscillating reaction was discovered in 1951 by the Soviet chemist Boris P. Belousov. Belousov had a hard time convincing anyone that the oscillations were real, rather than an artifact induced by incomplete mixing of the reactants – critics maintained that a reaction that could proceed spontaneously in either direction contravened the Second Law of thermodynamics (Chapter 2), which prescribes a directional preference to all processes of change. Only after the diligent work of Anatol Zhabotinsky at Moscow State University in the 1960s was the oscillating behavior of Belousov's chemical system accepted as an intrinsic property. The Second Law is in no danger from this reaction – free energy is always decreasing, but the concentrations of the compounds involved rise and fall with time.

The oscillations of the Belousov–Zhabotinsky (BZ) reaction are all the more striking because the two states between which it switches can be made brightly colored by adding a chemical "indicator" which makes them red or blue. Mixing the brew of BZ ingredients will produce initially a red solution. Provided that the mixture is well stirred, it will turn blue fairly abruptly as the reaction proceeds. But the chemical transformation is not over: moments later the red color reappears. As time passes, the system has yet another change of mind, and we are back with a blue solution. And so it continues, the solution changing from red to blue and back again like an unconventional traffic light. After several hours, however, the oscillations will cease if the system is left alone.

If the reaction mixture is poured into a shallow dish and not stirred, the red–blue transformation occurs in a remarkable way. The solution does not change color throughout the dish all at once; rather, the new color appears at isolated points, or cores, where perhaps the mixture is nonuniform or where there are impurities such as dust particles. The blue color then radiates outwards, while the core produces subsequent

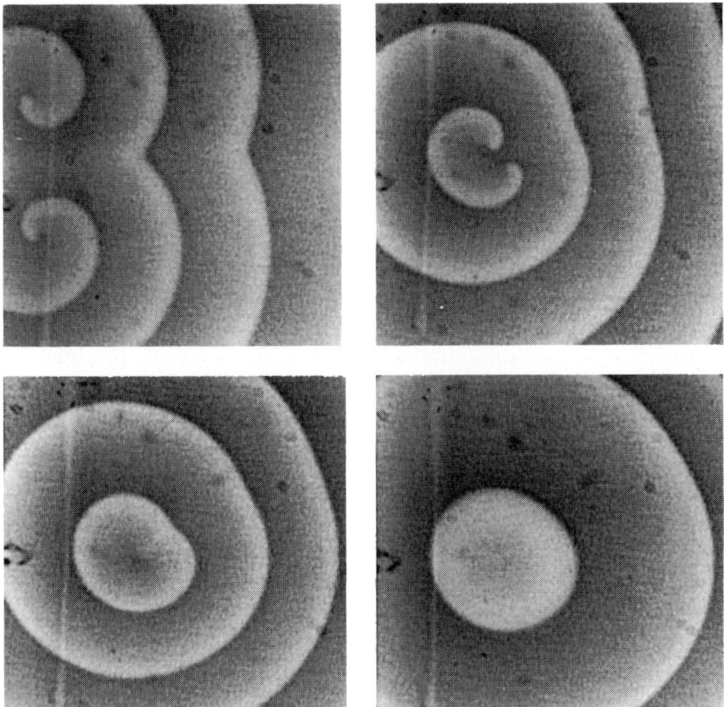

**Figure 9.12** *Colliding spiral waves in the BZ reaction. When the spirals have opposite twists, they annihilate one another. (Photographs courtesy of Stefan C. Müller, Max-Planck-Institut für Molekulare Physiologie, Dortmund.)*

blue pulses at periodic intervals. The result is a series of concentric propagating waves, like ripples on a pond (Plate 15).

On occasion the target patterns may become deformed into spirals that twist outwards into the surrounding medium. When the arms of these spirals collide, they may merge or annihilate one another (Figure 9.12). These target and spiral patterns are known as "chemical waves" – chemical reaction fronts which move through the reaction medium rather like the crests of ocean waves or the pressure fronts depicted on weather maps. The swirling patterns are irresistibly reminiscient of the spiraling vortices characteristic of hurricanes in satellite images of the atmosphere, or of the whirlpools that form and disperse in turbulent water. Is this perhaps another example of a "universal" nonequilibrium pattern?

## *Feedback and oscillations*

We saw in Chapter 2 that chemical reactions tend to be downhill processes. When the reactants are brought together, bonds are broken and made, yielding products that have a lower free energy. This seems to dictate that all chemical reactions must be unambiguously one-way affairs: a process that is downhill in one direction must necessarily be uphill in the opposite direction. If we want to drive a reaction uphill, we will need to expend some free energy to do so. How, then, is it possible for a reaction to go in either direction, and furthermore to do this in regularly repeating cycles?

Chapter 2 also showed that chemical change can be accelerated by the use of a catalyst, a substance that lowers the free energy barrier to a reaction's progress. Catalysis holds the key to the BZ reaction; but what gives it such an unusual character is the fact that one of the catalysts is produced by the reaction itself. The amount of catalyst therefore varies as the reaction proceeds.

This kind of behavior, called autocatalysis, arises when the rate at which the reaction takes place depends on the concentration of one of the products. In other words, this product, rather than remaining passive once formed, interacts with those starting materials that have not yet reacted in a way that encourages them to form still more of the product. The result of this "back-action" is that, as the product appears, it accelerates the reaction through a feedback loop that causes the reactants to be consumed even faster.

Feedback of this sort, in which the effect amplifies its cause, is said to be "positive." Positive feedback has a tendency to send a system's behavior out of control. The opposite phenomenon – negative feedback – has the contrary effect of bringing things back under control, of maintaining a "steady state." In a negative feedback loop, the effect diminishes its cause, so that disturbances are damped.

Feedback implies that the way a system will behave depends on how it has been behaving previously: the *result* of its behavior is the *cause* of its subsequent behavior. In mathematical parlance, this kind of system exhibits nonlinear dynamics. Both positive and negative feedback are characteristic properties often displayed in systems that show erratic or unpredictable behavior – systems that are now commonly described by the term "chaotic." We will return to chaos later, but we should note that the spiral and target patterns in the BZ reaction are not in themselves chaotic: they are complex, to be sure, but there is also a regularity, a periodicity, to them.

Positive feedback provides a general mechanism by which small fluctuations can become amplified into large ones, and thus in principle we can imagine that the positive feedback of autocatalysis might carry a chemical reaction far away from equilibrium. But to provide a recipe for oscillations, there must be another, competing process which acts in opposition to the autocatalytic cycle. To illustrate this, let's first consider an autocatalytic reaction far simpler than the BZ reaction, in which the product (call it compound B) produced spontaneously from a starting material (compound A) interacts with A to accelerate the formation of more B. According to the formalism for chemical reactions introduced in Chapter 2, the spontaneous formation of B from A can be written as:

$$A \rightarrow B \quad (1)$$

In the crucial autocatalytic step, a molecule of B interacts with A to give more B. We start off with a molecule of B and a molecule of A, and end up with the initial B molecule still around (having acted as a catalyst) but with the A converted to a second molecule of B:

$$A + B \rightarrow 2B \tag{2}$$

In one sense, this second step is no different from step (1), since the net result is simply the conversion of a molecule of A to a molecule of B; the only difference is that a molecule of B mediates the transformation. But this difference is critical, since we are proposing that the molecule of B on the left-hand side of the equation speeds up (catalyzes) the conversion of A to B. So step (2) occurs more quickly than step (1).

To obtain oscillatory behavior we can add a third step in which a molecule of B is converted to a new molecule, C. Let's assume that this process *requires* there to be some of compound C present already, which assists the transformation of B to C. The step is then much like step (2), but with B in place of A and C in place of B:

$$B + C \rightarrow 2C \tag{3}$$

This third step is therefore also autocatalytic: as more C is produced, the conversion of B to C becomes accelerated, since C itself assists this. In the absence of step (3), steps (1) and (2) between them would simply produce B at an ever-increasing rate (because of the positive feedback) until all the A is consumed. Step (3) is the crucial *competing* process that prevents this.

What now happens if we start off this reaction scheme with compound A and a little of C? First, step (1) will produce B. The molecules of B so formed could then follow either step (2), reacting with A to create more B, or step (3), reacting with C to consume B. As there is initially much more A around than C, step (2) predominates, and because it is an autocatalytic step that speeds up the formation of B, there is a surge in the concentration of this compound. Nevertheless, step (3) cannot be ignored, since it too is autocatalytic. Only a few molecules of B go down this route to begin with, generating a little more C; but this C reacts with the increasing amount of B from step (2) to generate even more C. After the initial surge of B, therefore, there follows a surge in concentration of C, as step (3) becomes increasingly important. Step (3) consumes B, however, so as C increases, B declines. If we monitor the progress of the reaction by adding an indicator that turns red in the presence of B and blue in C, we would see a red solution change to blue.

Can we get the reaction to go red again? To do this, we have to be able to increase the amount of B and decrease C. The simplest way in which we can imagine the latter occurring is to add to our hypothetical reaction scheme a step in which C changes spontaneously to a different compound, say D, just as A changes spontaneously to B.

But we will stipulate that D takes no further part in the reaction:

$$C \rightarrow D \tag{4}$$

While step (3) produces C, step (4) consumes it. But C can be replenished by step (3) only while there is still B around, whereas step (4) happens regardless. So the rise of C at the expense of B, which turns the indicator blue, cannot persist indefinitely; once there is scarcely any B left, there is no way to replace the C lost by step (4), so the concentration of C will start to fall again. This gives step (2) the chance to regain control, and B prospers once more – the mixture turns red.

If things are to get much further, it is necessary to keep supplying A to the mixture, since this is the only source of B. We will also need to get rid of D, so that it doesn't clog up the whole mixture (it does nothing but sit there once formed). We must therefore set up the reaction vessel in such a way as to allow a constant supply of A and constant removal of D. Then only the concentrations of B and C, which determine the color of the mixture, vary significantly with time. The archetypal reaction vessels of this sort are called continuous flow reactors.

Our reaction mixture has now changed from red to blue and back to red. Once step (2) takes over again, the system is back to where it was previously: a high level of B and a low level of C. The whole cycle then repeats again, a surge of B being followed by a surge of C and so forth. The mixture oscillates in color between red and blue, so long as we continue to supply A and remove D. Note that these influx and removal processes are essential for keeping the system away from equilibrium.

Thus, the color of this simple four-step scheme oscillates under the influence of two autocatalytic but competing steps. But we saw earlier that the color of the BZ reaction mixture can be made to vary in *space* as well as time – that is, it differs from place to place in the reaction vessel. To produce the spatial patterns we must simply ensure that the reactants are not well mixed. Because of the feedback loops, the reaction is very sensitive to the small, random variations in concentration that are likely to arise from place to place in an unstirred mixture – they can induce a switch from the dominance of B to that of C, and vice versa. These imbalances spread out from their point of initiation in the form of colored chemical waves.

### *Belousov's oscillator*

The BZ reaction itself is significantly more complicated than this idealized four-step process. But the basic principle remain the same: several autocatalytic steps provide feedback loops which cause the reaction to swing first one way and then another. The starting materials for the BZ reaction are an organic compound, malonic acid ($HOOC—CH_2—COOH$) and salts containing bromate ($BrO_3^-$) and bromide ($Br^-$) ions. In the course of the reaction, malonic acid is converted to bromomalonic acid ($HOOC—CHBr—COOH$). The reaction requires a catalyst, for which purpose cerium ions are generally used. The important feature of these ions is that they can change back and forth readily between two states that differ in charge – $Ce^{3+}$ and $Ce^{4+}$. The number of charges on the ions is called their oxidation state, and the change from one oxidation state to the other involves the loss or gain of an electron. The colors in Plate 15 are produced by an indicator called ferroin, which turns red in the presence of $Ce^{3+}$ ions and blue where $Ce^{4+}$ ions dominate.

The bromination of malonic acid might appear to be a simple enough process, but the reaction in fact involves a great many steps in which several intermediate chemical species are formed and consumed. This sequence of steps was deduced in 1972 by Richard Field, Richard Noyes and Endre Körös of the University of Oregon. The major processes involve the conversion of bromate into a variety of other species containing bromine and oxygen, such as $HBrO_2$, $BrO_2$ and HOBr. Some of these conversions are catalyzed by the cerium ions, which pick up or lose electrons in the process and thereby switch back and forth between their two oxidation states.

The sequence of steps can be represented as two cyclic processes, reactions A and B, coupled together by the interconversion of $Ce^{3+}$ and $Ce^{4+}$. The starting materials, bromate and bromide ions, react together to form $HBrO_2$ and HOBr. In reaction A, $HOBr_2$ reacts with bromate to form two molecules of $BrO_2$, which are both converted to $HBrO_2$ by $Ce^{3+}$; the metal ion is transformed into $Ce^{4+}$ in the process, turning the indicator blue. Because this reaction starts with one molecule of $HBrO_2$ and ends up with two, it is autocatalytic, like reaction (2) on page 307. Reaction B is still not understood completely, but in the first stage of this reaction $HBrO_2$ and bromide ions conspire to brominate malonic acid. The interaction of bromomalonic acid (denoted BrMA below) with $Ce^{4+}$ then regenerates both bromide ions and $Ce^{3+}$, restoring the red color. Thus in both reactions A and B the products are themselves reactants, giving rise to autocatalytic feedback. The whole scheme can be thought of as a system of rotating cogs, with the $Ce^{3+} \leftrightarrow Ce^{4+}$ cycle linking the two reactions:

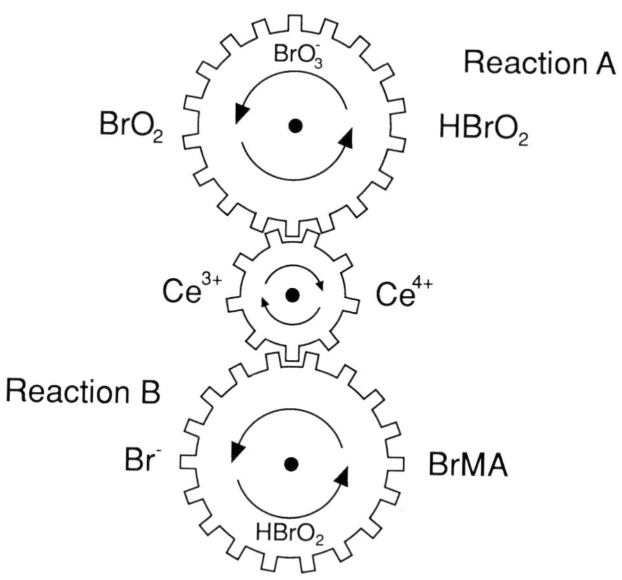

The oscillations are caused by periodic bursts of reaction A. Initially $HBrO_2$ reacts mainly with bromide to form HOBr; when there is little bromide left, this process slows down and is overtaken by reaction A, in which $HOBr_2$ reacts with bromate instead. But this reaction consumes the $Ce^{3+}$ ions, converting them to $Ce^{4+}$, and so by itself it will eventually run down. Reaction B, however, restores both bromide ions and $Ce^{3+}$ ions. Provided that it doesn't produce *too much* bromide, this process will eventually allow reaction A to start up again, generating another blue wave.

## *Spiral waves in life and elsewhere*

Our implicit quest at the beginning of the chapter was to look for signs of "universality" in pattern formation – to identify types of pattern that crop up in seemingly disparate systems. The targets and spirals of the BZ reaction represent such a case. Figure 9.13 shows snapshots of a reaction between two gases, carbon monoxide (CO) and oxygen ($O_2$), taking place on the surface of a platinum metal catalyst. Bright areas correspond to regions covered by $O_2$ molecules. This reaction, which converts CO to carbon dioxide ($CO_2$), was encountered in Chapter 2 as the principal process taking place inside catalytic converters in automobile exhaust systems. Clearly, there seems to be an unguessed complexity to this important reaction! The need to understand how these patterns form therefore goes beyond mere intellectual curiosity, since their appearance may have profound consequences for the efficiency of the catalytic reaction.

The spiral waves of the BZ reaction are mirrored in the patterns that appear in colonies of the amoebae *Dictyostelium discoideum*, a slime mold (Figure 9.14). The patterns appear when the mold is "starved" (of heat or moisture, for example). Under these conditions, the response of the amoebae is to aggregate into a multicellular body which can then move in search of a more favorable habitat. Aggregation is triggered by the release of the compound cyclic adenosine monophosphate (cAMP) from some of the cells, called "pioneer" cells. The pioneer cells synthesize cAMP in an autocatalytic reaction, causing them to release it in a pulsed, periodic fashion. When the cAMP reaches other cells, they start to move towards regions of higher cAMP concentration – that is, towards the pioneer cells. This phenomenon is known as chemotaxis. The periodic release of cAMP gives rise to highly organized target and spiral patterns of cell

**Figure 9.13** *Target patterns are evident in the distribution of reactants during the catalytic reaction of carbon monoxide and oxygen (bright regions) on the surface of platinum metal. (Photograph courtesy of G. Ertl, Fritz-Haber-Institut, Berlin.)*

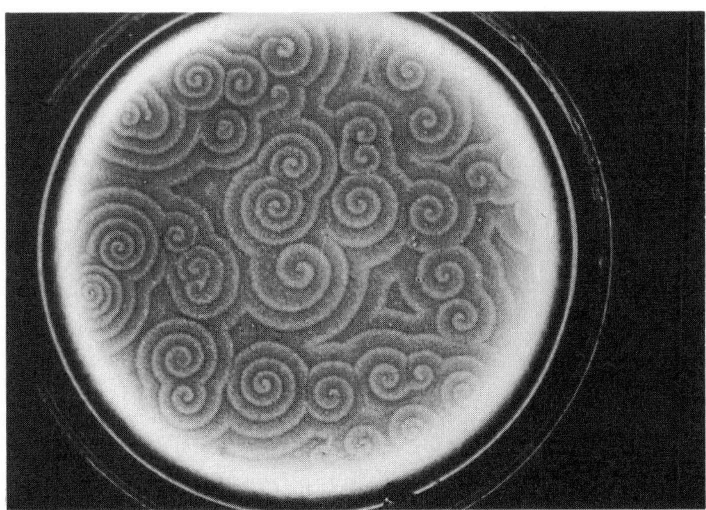

**Figure 9.14** *Spiral waves in colonies of the mold* Dictyostelium discoideum. *These patterns are created when the mold forms aggregates in response to some external "stress," such as a lack of moisture or nutrients. (Photograph courtesy of P. C. Newell, University of Oxford.)*

aggregation. Chemotaxis in other biological systems can create an extraordinary variety of delicately structured modes of aggregation (Figure 9.15) in a manner that is only partly understood at present.

### *Stationary patterns and the leopard's spots*

The chemical patterns that we have seen so far are "dynamic" ones – they change as time passes. Some of the patterns observed in natural systems, on the other hand, are immobile or long-lived. An example is to be found in the very early development of larvae, of which the best studied is that of the fruit fly. The embryos evolve stripes (Figure 9.16), each segment representing a part of the body that will thenceforth follow a different developmental pathway – to become, say, the upper thorax or the head. The appearance of these segments in fruit fly embryos is thought to be related to the concentration of a protein called bicoid, which increases gradually from one end of the developing egg to the other. The concentration of the bicoid protein provides a signal that "switches on" genes at different points along the body axis, dividing the embryo into a sequence of segments which will develop into different parts of the body. This process of differentiation of regions in an initially uniform embryo is called morphogenesis – literally, the genesis of shape – and the bicoid protein is said to act as a morphogen, which is to say, a determinant of body plan. Exactly how the striped pattern arises from a simple gradient in bicoid concentration is, however, a story that is still being unraveled.

In the 1950s, the mathematician Alan Turing proposed a mechanism by which stationary patterns might arise in chemical systems. Turing is one of the legendary

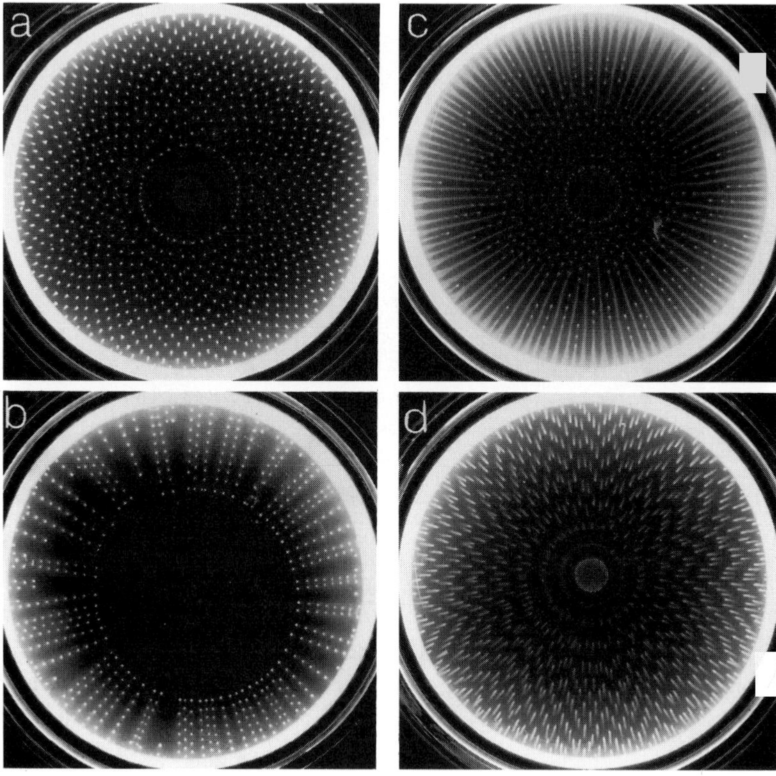


**Figure 9.15** *Colonies of the bacterium* Escherichia coli *grown in agar can aggregate into a fantastic variety of stationary patterns. These structures are created by "chemotactic" signaling between the bacteria, which release chemical attractants towards which neighboring cells are drawn. Chemotaxis provides a very general mechanism by which bacterial colonies can self-organize in complex ways; in other bacterial systems, fractal-like and dendritic patterns have been observed, like those in Figures 9.9 and 9.10. (Photographs courtesy of Elena Budrene and Howard Berg, Harvard University.)*

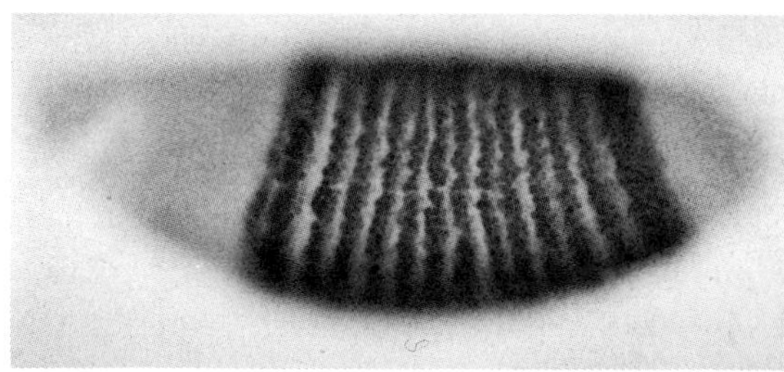

**Figure 9.16** *In the early stages of development, fruit fly larvae develop striped patterns. The regions of different shade represent parts of the embryo that will subsequently follow different developmental pathways. A gradual variation in the concentration of a morphogen called the bicoid protein from one end of the embryo to the other provides the chemical signal that stimulates this patterning. (Photograph courtesy of Peter Lawrence, Laboratory of Molecular Biology, Cambridge.)*

characters of modern science – a mathematician of genius, he gained much prominence for his work in cracking German codes during World War II. His vision of the archetypal computer, now called a Turing machine, has helped mathematicians to decide which problems are solvable by mathematical analysis (and which are not), and now provides one of the cornerstones for the modern theory of information and computation. Turing's contribution to the understanding of morphogenesis, meanwhile, although stimulating to many developmental biologists, lacked widespread acceptance because no one was able to find a real chemical system in which Turing's stable patterns – Turing structures – were actually observed.

What Turing showed theoretically was that these structures could form spontaneously in an autocatalytic system in which the reactant molecules diffuse at different speeds through the reaction medium – a so-called reaction–diffusion system. Specifically, if the faster-traveling reactants inhibit the reaction whereas the slower ones catalyze it, then within a certain range of diffusion rates of the reactants the system can transform suddenly from a uniform mixture into one in which the chemical composition varies in a regular manner from place to place. In effect, the system turns into a kind of crystal, with periodic structure – but a very peculiar kind of crystal, because all the molecules within it are free to move around yet the patterned structure remains. In a normal crystal, this freedom of movement would destroy any periodicity.

The discovery of autocatalytic chemical reactions that could display oscillatory patterns led researchers to wonder whether they might contain the ingredients necessary for Turing structures to be formed. But in general autocatalytic reactions like the BZ reaction tend to produce moving chemical waves, not stationary structures. It was not until 1990, nearly 40 years after Turing presented his ideas, that patterns identified as Turing structures were seen for the first time by Patrick de Kepper and colleagues from the University of Bordeaux. De Kepper's group studied a variant of the BZ process, called the chlorite–iodide–malonic acid (CIMA) reaction, in which the reactants were mixed within a gel to slow down the diffusion rates so that stable patterns might form. The researchers used a starch-based indicator to reveal the changes in chemical composition throughout the system: the indicator turns either yellow or blue, depending on whether the concentration of the ion $I_3^-$, an intermediate species in the reaction, is high or low. They observed a few rows of yellow dots within a certain region of the otherwise blue gel, which they identified as Turing structures (Figure 9.17). Harry Swinney and Qi Ouyang of the University of Texas at Austin were later able to "grow" large patches of these Turing structures. At first the reaction produces radiating target patterns like those of the BZ reaction, but over a period of an hour or so these patterns break up into regular hexagonal arrays of yellow dots which slowly come to rest (Plate 16). By changing the temperature of the system, Ouyang and Swinney were able to transform their dot pattern into a stripe pattern – this kind of transformation too had been predicted by Turing's theory.

Turing's prediction has inspired searches for stationary structures in systems that might be more immediately relevant to developmental biology. Some have suggested that this kind of process might be responsible for the appearance of striped and spotted

**Figure 9.17** *Turing structures – stationary patterns corresponding to regions that differ in chemical composition – can be produced in the chlorite-iodide-malonic acid oscillatory reaction. (Photograph courtesy of Patrick de Kepper, Université Bordeaux I, France.)*

patterns in animal pelts. These patterns are controlled at the cellular level by pigment-producing cells called melanocytes in the epidermis, which generate light-absorbing molecules that determine the color of the hair in that region of the pelt. The activity of the melanocytes is itself controlled by "signaling" compounds which become distributed within the epidermis in complex patterns. The striking patterns observed on shells (Figure 9.18) are also under chemical control; the shell is essentially a mineral crystallized within a matrix of organic matter, biochemically orchestrated by the organism within. Theoretical reaction–diffusion schemes can be formulated that create complex stationary patterns in the reaction products, which are highly reminiscent of those seen on shells. However, our understanding of the chemistry of shell formation is insufficient to permit much more than a qualitative comparison between these theoretical models and the ornate creations of nature.

## Chemical chaos

### *Into the maelstrom*

Turing structures provide an elegant demonstration of how orderly, symmetrical patterns can appear quite suddenly and spontaneously in systems far from equilibrium. I now want to take us to the opposite extreme: to show how these systems can also

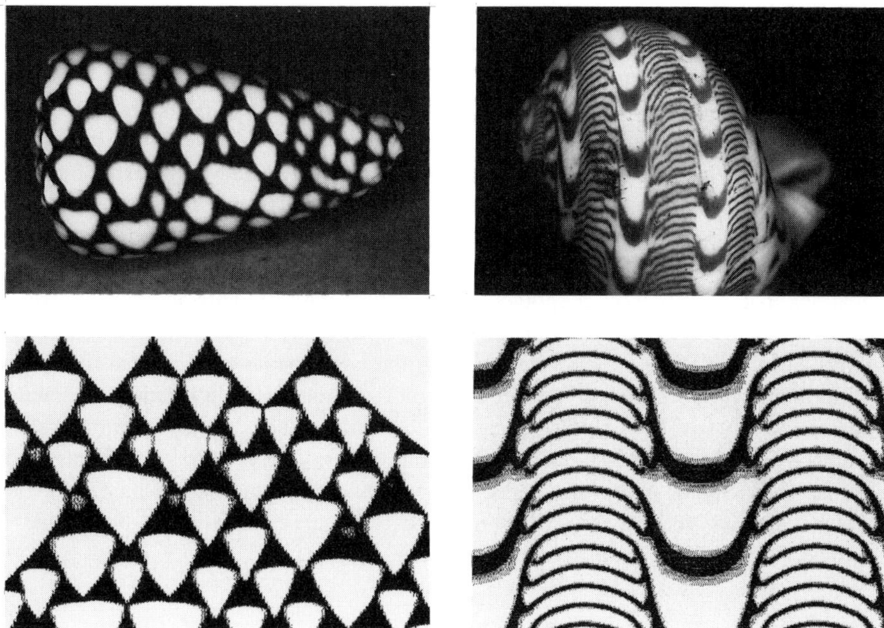

**Figure 9.18** *The complex patterns in shells can be mimicked by those that appear in a reaction–diffusion chemical system. Here the real patterns shown on the top bear a striking resemblance to those calculated for a theoretical reaction scheme at the bottom. (Pictures courtesy of Hans Meinhardt, Max-Planck-Institut für Entwicklungsbiologie, Tübingen, Germany.)*

get so far out of control that we essentially lose all ability to predict their behavior. You might imagine that once this happens, the system would become inaccessible to any kind of scientific study, and indeed that was the point of view taken by most researchers until just a couple of decades ago. But the investigation of unpredictable systems is now one of the biggest growth areas in the whole of science, and is one that brings together scientists from an astonishingly broad range of disciplines. It is the study of chaos.

Chaos is surely one of the most familiar scientific buzz words of recent years. Chaotic behavior is displayed by a tremendously varied range of physical systems, including the world's weather system, patterns of flow in fluids, lasers, electronic circuits, heart tissue, animal populations and economic markets. In crude terms, the hallmark of chaos in all of these cases is complete unpredictability of behavior. More precisely, chaos is manifested as an extreme sensitivity of a system to its initial state or to small perturbations. If they show chaotic behavior, two systems that differ in only the tiniest detail at some initial point in time will rapidly evolve into very different states. The smallest disturbance to a chaotic system can change completely the way that it develops subsequently: the size of the effect may not reflect that of the cause. This is the principle embodied in the now cliched "butterfly effect," whereby the flutter of a butterfly's wing in, say, Tokyo could change the pattern of weather in Oklahoma.

When a system becomes chaotic, we lose the ability to predict how it will look or behave at a future time. The revelation that the weather system is chaotic, which was first recognized by the meteorologist Edward Lorenz of the Massachusetts Institute of Technology, seemed to spell doom for the prospects of making long-term forecasts. There is, however, an important distinction to be made between chaos and randomness. Random processes are the result of unpredictable, chance events which permit of only a statistical description, while for a chaotic system one can often write down mathematical equations that describe how it should evolve in time with absolute accuracy. There are no random influences involved at all in this latter case – everything is exact and mathematical. Yet still we cannot deduce from the mathematics exactly what the system will look like at any specified time in the future, without actually running a numerical computation to find out. This characteristic of being able to specify chaotic systems completely in mathematical terms is reflected in the use of the word "deterministic" to describe their chaotic behavior. Deterministic chaos may sound like an oxymoron, but in fact it implies nothing more than that the chaos arises in a well-defined (that is, nonrandom) way from feedbacks inherent in the system.

## *The road to chaos: up the Devil's staircase*

The BZ reaction provides the ideal test bed for looking at how chaos impinges on chemistry. Provided that we keep up a constant supply of reactants and removal of products for the BZ reaction in a continuous flow reactor, it will continue to oscillate periodically. But if we turn up the flow rate through the reaction vessel, new phenomena start to appear. Imagine that we decide to follow the reaction by measuring the way in which the concentration of bromide ions in the well-stirred vessel changes as time progresses. For low rates of flow, we will see regular, periodic changes as the system jumps back and forth between reactions A and B described on page 309. As the flow is gradually cranked up, these oscillations suddenly double up in a so-called period-doubling bifurcation: the concentration of bromide rises and falls twice in every cycle. And as we turn up the flow rate still further, the behavior becomes increasingly complicated. Eventually the regular oscillations degenerate into what appears to be complete unpredictability (Figure 9.19). The BZ mixture has descended into chaos.

A sequence of period-doubling bifurcations is a common sign of the approach to chaos: the periodic oscillations double up again and again within each cycle as the crossover to full-blown chaos becomes imminent. Other precursory signatures of impending chaos have been identified in other systems – fluid flows, for instance, can show short intermittent bursts of chaotic turbulence as the flow rate is increased towards the point at which fully fledged turbulent, chaotic flow develops. As well as simple period-doubling, the BZ reaction can display a more complicated approach to chaos characterized by so-called "mixed-mode" oscillations. These have a rather subtle kind of periodicity: the repeating cycles are marked out by large-amplitude oscillations with

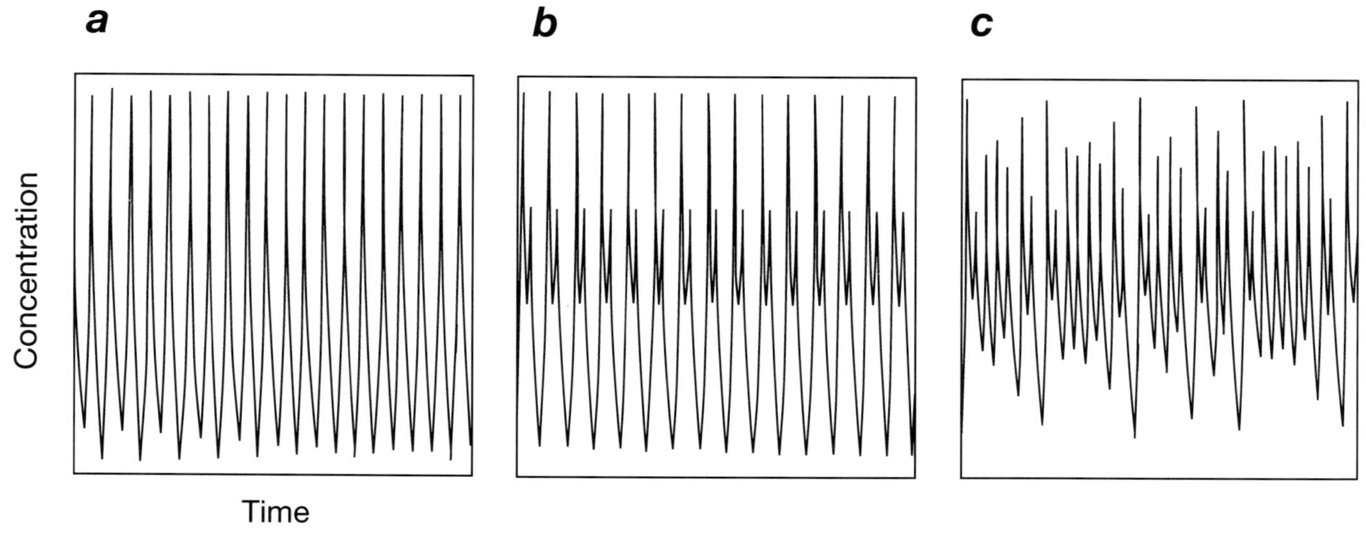


**Figure 9.19** *Oscillations in bromide concentration in the BZ reaction vary from regular (a) to period-doubled (b) to chaotic (c) as the rate of flow of reactants through the reaction vessel is varied.*

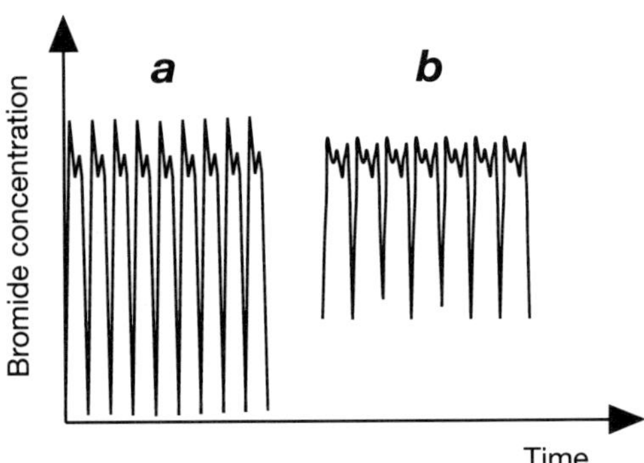


**Figure 9.20** *Within a certain range of flow rates through the continuous flow reactor in the BZ reaction, mixed-mode oscillations are observed. Each cycle is marked out by a repeating sequence of large-amplitude oscillations in concentration, between which smaller-amplitude oscillations are apparent. The pattern in (a) corresponds to the* $1^2$ *mode, and that in (b) is the* $1^3$ *mode.*

smaller ones squeezed between them (Figure 9.20). These patterns are called mixed modes because they incorporate both small- and large-amplitude peaks. Each mode of behavior is assigned a "firing number," which quantifies the number of times the system "fires" a large-amplitude oscillation during each full repeat cycle. The behavior of the BZ mixture as the flow rate is varied can then be depicted by plotting the firing number against the flow rate.

This plot forms a kind of staircase in which plateaus rise from a firing number of zero at low flow rate to a firing number of 1 at high flow (Figure 9.21). The width of

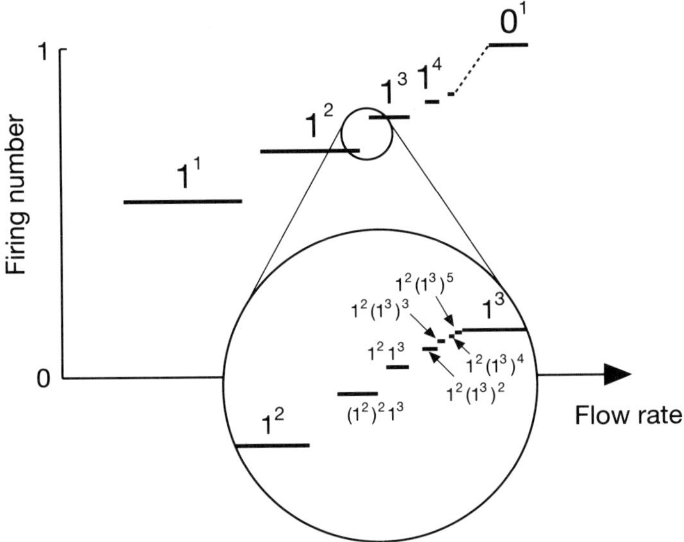


**Figure 9.21** *Transitions between mixed modes of the BZ reaction, as a function of flow rate, delineate a "Devil's staircase" of unequally spaced steps. A closer look at these steps reveals that the jumps occur via a sequence of more complex mixed modes, and that these themselves have still finer structure. The staircase is in fact fractal.*

each step in this staircase, and the drop between successive steps, varies in a highly irregular (in fact, chaotic) way. To descend such a staircase would be a rather perilous exercise, and it is not hard to understand why this structure has earned the label of a "Devil's staircase."

But closer inspection of the region in which one mixed mode switches to another reveals that between each step there is a multitude of smaller steps, on which the system does not seem to have quite made up its mind which of the two mixed modes it prefers. It exhibits periodic patterns in which a single cycle contains varying numbers of the cycles of each of the two "pure" mixed modes to either side. If we look even more closely at the transitions between these intermediate states separating the main rungs of the Devil's staircase, we can identify still more complex oscillations (Figure 9.21). The system never seems to make a clean break, but instead switches its behavior via a complex sequence of patterns. This increasing level of detail at increasing magnifications is of precisely the same nature as that exhibited by fractal structures. Sometimes the oscillations appear to lose all vestige of periodicity, and become simply chaotic. The Devil's staircase of the BZ reaction thus has chaos lurking in its corners.

At first encounter, it is hard to see how one might develop any meaningful description of chaotic behavior. But when depicted in the right way, the behavior of a chaotic system exhibits an underlying structure. For the BZ mixture, instead of plotting concentrations of the reaction species against time, as in Figure 9.19, we can show how the concentration of one component (say bromide) varies with that of another component (say $HBrO_2$). For periodic oscillations, this plot becomes a closed loop, called a limit cycle (Figure 9.22). As time progresses, the variation in the two concentrations traces out a path around the loop. When the flow rate in the system is increased to the point at which a period-doubling bifurcation occurs, this limit cycle develops two loops (Figure 9.22), and the system must traverse both loops in each repeating cycle. Subsequent bifurcations produce an increasing number of loops, so that the diagram begins to resemble a skein of wool. When the system's flow rate is brought into the chaotic region, the behavior loses all periodicity – but this does not mean that the skein simply breaks up. Rather, it retains a characteristic shape, but is very finely structured, so that the concentrations never follow quite the same path on each cycle.

Limit cycles are an example of what chaologists call an "attractor;" they represent a mode of behavior to which the system will inevitably be drawn, no matter what the starting conditions are. If the initial concentrations of the two components lie off the limit cycle, they will soon evolve so as to encounter it, and will thenceforth stay resolutely on the loop. Even a chaotic system may retain an attractor, with more or less well-defined boundaries beyond which the system's behavior will not generally stray. But this attractor is not a single loop, nor indeed can one count the number of loops it has. A chaotic attractor has an infinitely fine structure – as one looks at it ever more closely, new paths will become evident. The attractor is, in other words, a fractal. Because of their peculiar properties, these fractal structures are called "strange attractors." They provide perhaps the clearest demonstration that chaos and fractals are all part of the same new and remarkable science of complexity.

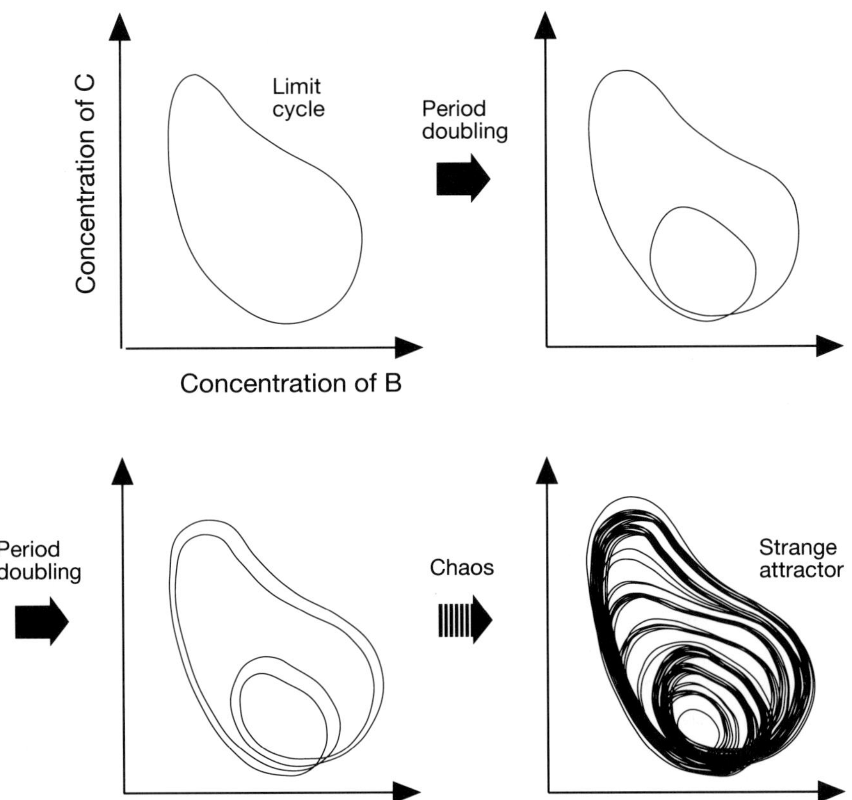


**Figure 9.22** *Regular oscillations in oscillatory reactions correspond to closed trajectories ("limit cycles") on a plot of the concentration of one component against another. If the reaction is started from a point outside of the limit cycle, it is rapidly drawn onto the cycle. At a period-doubling bifurcation, a single cycle splits into two loops. The system can become chaotic via a series of period doublings. In the chaotic state, the "attractor" has an infinitely fine structure, so that the behavior never quite repeats exactly. Yet the plot does not simply break up into a mass of random trajectories – it retains a characteristic structure. The densely spaced set of chaotic trajectories is called a strange attractor.*

## The big picture

I hope now to have persuaded you that the complex behavior and intricate patterns seen in many natural processes, chemical or otherwise, may share common properties. Certain structures are universal; unpredictable, chaotic behavior can have simple, deterministic causes; there is an intimate relation between fractal structures and chaotic dynamics; and complexity can sometimes lead to order and regularity rather than utter confusion. Most importantly, complexity, chaos and the formation of richly structured patterns are characteristics shared by many processes that occur far from equilibrium.

The first concerted attempt to bring some unity to nonequilibrium processes was made in the 1960s and 1970s by Ilya Prigogine from the Institute of Chemical Physics in Brussels, in work that earned him the Nobel prize for chemistry in 1977. Central to Prigogine's studies was the question of how spatial patterns, sometimes of considerable complexity, can arise and survive in a system that has not become frozen in some equilibrium state but is constantly evolving in time.

One of the fundamental properties of many nonequilibrium patterns is self-organization: they are large-scale structures that are not reflected in the microscopic properties of the system. We saw in Chapter 4 that the cubic symmetry of salt crystals formed under equilibrium conditions is a consequence of the cubic packing of the individual ions. But within the interactions between molecules and ions of the chlorite–iodide–malonic acid reaction mixture there is *no* prescription for hexagonally arrayed Turing structures. The spontaneous appearance of this regular pattern in the initially uniform system is an example of what physicists call a symmetry-breaking process. Initially looking the same in all directions, the system switches to a state in which different directions are no longer equivalent – in other words, its symmetry is lowered.

Another property of these nonequilibrium structures is their stability against perturbations. A system exhibiting chaotic behavior is extremely sensitive to perturbations – the smallest push can lead to markedly different subsequent evolution. But if instead the system is trapped on a limit cycle, this acts as an attractor to pull the system's evolution back on course if we disturb it. Dynamic structures that remain stable against perturbations are said to be *dissipative* – they are able to dissipate the extra energy supplied by the perturbation. They present a stark contrast to the behavior of so-called *conservative* oscillating systems such as the simple pendulum. If we give a push to a swinging pendulum, it swings with greater amplitude; a dissipative oscillator, on the other hand, would undergo a short-lived disturbance (called a transient) when "pushed," before returning to exactly the same kind of periodic motion as before. In a sense (which I emphasize is entirely figurative), a dissipative system seems to have a mind of its own which organizes it into a specific mode of behavior and then resolutely maintains it that way. Beyond the set of conditions for which this mode is stable, the system may undergo transitions (such as period-doubling bifurcations) to new steady states.

What is the source of this capacity for self-organization out of equilibrium? To try to answer this question, Prigogine and his coworkers have attempted to construct a system of thermodynamics for nonequilibrium systems just as Gibbs, Helmholtz and others did for equilibrium systems in the nineteenth century. The "Brussels school" draws an analogy between equilibrium phase transitions and nonequilibrium changes of structure, pattern or growth mode (which can also be thought of as transitions).

Many types of equilibrium phase transition are characterized by symmetry-breaking. A system undergoing an equilibrium symmetry-breaking phase transition often develops long-range correlations, which means nothing more than that one part of the system can become very sensitive to what is going on in a distant part. At this point the system as a whole is therefore easily influenced by small, random fluctuations. But we have seen that this is precisely the situation too in a nonequilibrium system such as the BZ reaction. It turns out that analogous long-range correlations develop in nonequilibrium processes when they approach a bifurcation point, and these put remote parts in contact with each other. In this way, the system becomes able to organize itself over large scales, even though the interactions between its individual components may extend only over very short distances. A key difference from equilibrium systems is that, rather than being driven by changes in a thermodynamic quantity such as temperature, nonequilibrium transitions are triggered by changes in the driving force away from equilibrium: the degree of supersaturation or supercooling in crystal growth, the electrode voltage in electrodeposition, the flow rate of reactants in oscillating chemical reactions.

While Prigogine's ideas, and more recently the theory of chaos, have helped to bring some coherence to the rich and often bizarre behavior of systems evolving far from equilibrium, our understanding of such processes remains a very long way from that which has developed for the kind of equilibrium phenomena that are embraced by classical nineteenth-century thermodynamics. There is little question, however, that the beauty of nonequilibrium structures and their importance for the physics, chemistry and biology of a great number of natural systems will continue to motivate the quest to achieve this understanding.

# 10

# Transforming the Globe

## The crises of atmospheric chemistry

*It is not impractical to consider seriously changing the rules of the game when the game is clearly killing you.*

M. Scott Peck

Not so long ago you would have been most unlikely to find in a book about chemistry any mention of the way in which it is relevant to our environment. Today, however, atmospheric and environmental chemistry are no longer regarded as arcane backwaters of science but as matters for immediate and global concern. Atmospheric scientists have rather suddenly found themselves at the focus of public and media attention, and their research determines the policies of governments. For the world has at last awoken to a fact that these scientists have always recognized: the chemical composition of the atmosphere exerts a profound influence on our environment, and upsetting its delicate balance can have grave consequences for the planet.

Foremost amongst the problems to which atmospheric scientists must now address their attention are the threat of global warming (the so-called greenhouse effect), the deterioration of the ozone layer, the deleterious effects of acid rain and the rising levels of pollutants such as lead, mercury and radioactive substances throughout the world. These issues have generated vigorous and sometimes bitter controversy, all the more so because their implications are of so much more than purely scientific interest. Industrial companies are having to face up to the potential damage that can be done to the environment by the chemicals that account for a part of their profits, while the increasing demand for power generation must now we weighted against the harmful effects of the waste gases or hazardous substances that are generally created in the process.

This chapter aims to show how atmospheric chemistry lies at the heart of some of these matters. A discussion of this kind cannot really be comprehended in its true context without some explanation, as far as we can yet provide it, of how our atmosphere got to be the way it is today. As I implied in Chapter 8, the Earth has not always been blessed with a life-sustaining blanket. No, our present atmosphere did not beget life: life begat the present atmosphere. The atmospheric scientist James Lovelock has suggested that it is time to stop thinking of life, the solid Earth, the oceans and the atmosphere as independent systems, and to regard them instead as interconnected aspects of our planet. This point of view, which is central to Lovelock's "Gaia" hypothesis, should help us to appreciate why we cannot divorce our activities from their environmental consequences, or assume that the environment has an infinite capacity to absorb our mess. Our atmosphere is a privilege that should not be taken for granted.

## The chemical control of climate

### *Air fit to breathe*

Imagine that some extraordinarily powerful astronomical instrument were to be invented that allowed us to see the planets that must surely orbit around other stars, and to look at the composition of their atmospheres. Were we to come across one with an atmosphere like our own, we could scarcely conclude other than that here was a planet on which life had evolved. Not *could* evolve, but most definitely *had*. Our atmosphere is a beacon broadcasting our presence to any intelligent beings who might be able to see it.

The reason for this is that, unlike those of the other planets in the Solar System, Earth's atmosphere is in a state of extreme chemical disequilibrium. It is in some sense comparable to a mixture of compounds in a vast beaker that is being maintained in a state far from equilibrium – indeed, rather like those chemical systems we encountered in the previous chapter. What is holding the atmosphere away from chemical equilibrium? Ultimately it is the energy of the Sun, as well as heat from the Earth's interior. But the principal agent that converts this energy into chemical disequilibrium is life itself.

This is to say that it would be quite wrong to view our environment as one miraculously tuned to our needs. It is no coincidence that the atmosphere is suited to the organisms that dwell within and beneath it, for the evolution of life and the attainment of the atmosphere's present composition have not been independent processes.

About 4,600 million years ago, the newly formed Earth was a ball of molten magma which had condensed, along with the Sun and the other planets, out of a primordial gaseous nebula. Within the body of this molten Earth, chemical elements began to separate out. Much of the planet consisted of iron, which sank (together with a smaller amount of nickel) to form a metal core, leaving behind a "scum" of molten rock which contained largely magnesium, silicon, oxygen, some remaining iron, aluminum, sodium, potassium and calcium. This chemical "differentiation" of the Earth is similar to the process that occurs in an iron smelter during extraction of iron from its ores.

By about 3,900 million years ago, much of the planet's heat had been radiated out to space, and the surface was cool enough to solidify into a thin, rocky crust. Two processes now began to contribute to the formation of an atmosphere. The molten rock below the crust contained many dissolved gases, such as water, methane, carbon oxides, nitrogen and neon. These were released from the magma through volcanoes that punctured the solid crust, in a process called degassing. Meanwhile, stray bodies in the Solar System left over from the formation of the planets occasionally collided with the Earth, releasing considerable quantities of volatile gases. It has been suggested that as much as 85 per cent of the water presently on Earth was brought here by impacting extraterrestrial objects.

About 3,800 million years ago, the temperature at the Earth's surface fell below 100 degrees Celsius, at which point water vapor in the atmosphere could condense to liquid. It is hard to picture the rainstorm that ensued: imagine, if you can, the entire contents of the oceans falling from the skies in a deluge lasting for perhaps 100,000 years. With the appearance of the oceans, gases that dissolve to a significant extent in water, such as hydrogen chloride, sulfur dioxide and carbon dioxide, were extracted from the atmosphere into the water. Reactions with minerals would then have precipitated some of these compounds as insoluble salts, such as carbonates and sulfates.

Light gases such as hydrogen, helium and neon, which were abundant in the solar nebula, are too light to be retained by the Earth's gravity, and so they simply rose through the atmosphere and evaporated away into space. Left behind in the early atmosphere were gases such as methane, water vapor, nitrous oxide ($N_2O$) and carbon monoxide (CO). It was under skies such as these that life first appeared.

It is remarkable that the complex chemistry of life may have developed from its raw materials within the space of just 300 million years. Yet this is the implication of the discovery in 1983 by S. M. Awramik and colleagues of 3,500-million-year-old rocks in western Australia that contained evidence of the fossilized forms of bacteria. These organisms appeared to be very similar to some of the very primitive species that still exist today, called blue-green algae or cyanobacteria.

But whereas most algae today obtain their energy by splitting water molecules through photosynthesis, the metabolism of the earliest organisms probably involved much cruder chemical reactions such as those utilized by archaebacteria, the most primitive form of life still extant today. Some of these organisms split apart organic molecules such as acetic acid, releasing energy and forming carbon dioxide and methane in the process. Others convert carbon dioxide to methane, or sulfate ions to hydrogen sulfide.

These resourceful bacteria were quite content living under oxygen-free skies; in fact, oxygen was poisonous to them. But we must assume that one day a species of bacteria made the discovery that the stuff all around them – water – could itself provide a bountiful source of energy when split apart. This was a profoundly antisocial habit, because it yielded as a by-product the toxic gas oxygen. Lynn Margulis, a microbiologist from Harvard University, has described the appearance of photosynthetizing organisms as having heralded a "worldwide pollution crisis" of such magnitude that our present-day industrial emissions are as nothing in comparison. The evolution of life changed the atmosphere beyond recognition.

The time at which this crisis truly took a grip on the planet is open to some debate, but most researchers now place it at around 1,900 to 2,000 million years ago. Oxygen production eventually became overwhelming, presumably because the benefits of using photosynthesis as an energy source were so substantial that bacteria with this capability simply took over, until oxygen was bubbling forth from colonies of algae throughout the world. Inevitably, this polluting activity led to the extinction of many microbial populations, but at the same time mutant strains evolved that were resistant to the poison. Some of these showed still greater adaptability: rather than stoically tolerating the unhealthy new environment, they found a way to thrive in it. The metabolic pathways of these organisms developed so as to actually utilize the oxygen in the atmosphere. They learned to breathe the air of the new world.

These single-celled oxygen-breathing organisms, called protozoa, were the first animals. They made their appearance about 800 million years ago, when the concentration of oxygen in the atmosphere had reached about 5 per cent of its present-day value. Oxygen has probably been maintained at its present proportion of about one-fifth of the atmosphere more or less steadily during at least the past 300 million years, although before this there is evidence of substantial fluctuations: at one time as much as 35 per cent of the air may have been oxygen.

In the upper atmosphere, sunlight splits apart oxygen molecules into their two constituent atoms, and these undergo subsequent reactions with other $O_2$ molecules to form a new type of oxygen compound containing three atoms: ozone ($O_3$). This molecule absorbs ultraviolet light strongly, and so filters this part of the spectrum out of the sunlight impinging on the atmosphere. As ultraviolet light is damaging to organic matter, it was not until the ozone layer had formed, about 400 million years ago, that living creatures could entertain the notion of leaving the protecting blanket of seawater and venturing onto dry land.

## *Recycling the world*

Today's air has an oxygen content of about 21 per cent; most of the remaining 79 per cent consists of the unreactive gas nitrogen. About 0.05 per cent is carbon dioxide, sufficient to support plant growth. This composition is regulated both by the sum of all life on Earth – the biosphere – and by geological processes involving the land masses, the oceans and the planet's interior, which collectively comprise the geosphere. The biosphere encompasses all living things: the forests and grasslands, the microbes in soil and the communities of the seas: phytoplankton and zooplankton, microscopic marine plants and animals respectively. Photosynthesizers (which is to say, plants) strip water of its hydrogen atoms and use them to convert carbon dioxide to energy-rich carbohydrates, releasing oxygen gas in the process. Consumers (that is, animals) breathe in oxygen and use it to burn up ingested carbohydrates, converting carbon compounds back to carbon dioxide, which is released into the air again. This process, known as respiration, releases energy which the consumers generally store for later use in the form of the compound ATP (see page 281). Without photosynthesizers to regenerate the oxygen used up by consumers, the atmosphere's oxygen content would slowly but steadily decline.

Much of the carbon "fixed" into organic matter by photosynthesizers is eventually released back into the atmosphere as carbon dioxide via respiration of consumers (primarily that of the microbes which decompose dead plant matter). But carbon is also cycled to and from the atmosphere via purely "inorganic" geochemical processes. The reaction between atmospheric $CO_2$ and minerals (known as weathering) binds up the carbon in carbonate compounds, while transformation ("metamorphism") of carbonate-rich rocks, perhaps induced by the deformations caused by collision of tectonic plates, can release $CO_2$. Carbon dioxide dissolves in the oceans to form soluble species such as bicarbonate ions. And carbon-rich sediments on the seafloor – the remains of dead organisms from the upper waters – are dragged down into the Earth's interior when a tectonic plate plunges down under another at ocean trenches. The carbon is converted into new forms by the heat within the Earth's mantle, and is recycled into the atmosphere in the effluent of volcanoes which sit behind the ocean trench (Figure 10.1).

Nitrogen, too, is cycled to and from the atmosphere by processes involving the biosphere and geosphere. Certain kinds of bacteria transform the normally very unreactive nitrogen molecules to ammonia, whence it is converted into nitrogen-containing organic compounds such as amino acids. All organisms require amino acids; plants synthesize theirs directly, but animals obtain them from ingested matter – either plant tissue or that of other animals. The nitrogen in organic compounds is ultimately converted back to inorganic forms. Some may be incorporated into urea and then into ammonia again; some is "oxidized" to nitrite ($NO_2^-$) and nitrate ($NO_3^-$) ions. In a process known as denitrification, bacteria strip nitrate ions of their oxygen atoms, releasing nitrogen gas back into the atmosphere.

These cyclic transformations of oxygen, carbon and nitrogen through the atmosphere, biosphere and geosphere are known as biogeochemical cycles. A wonderful account of a carbon atom's journey through parts of this cycle can be found in Primo Levi's *The Periodic Table*. When processes that remove elements from the atmosphere are balanced by those that replenish them, the atmosphere remains in a "steady state" – never achieving thermodynamic equilibrium, yet always staying the same.

We saw in Chapter 9 that the behavior of systems that are out of equilibrium can be hard to predict – in particular, they can undergo large changes in response to small disturbances. We do not know how stable the present steady state of the atmosphere is, but we do know that there were times early in the planet's history when it was in an entirely different steady state, with a different composition.

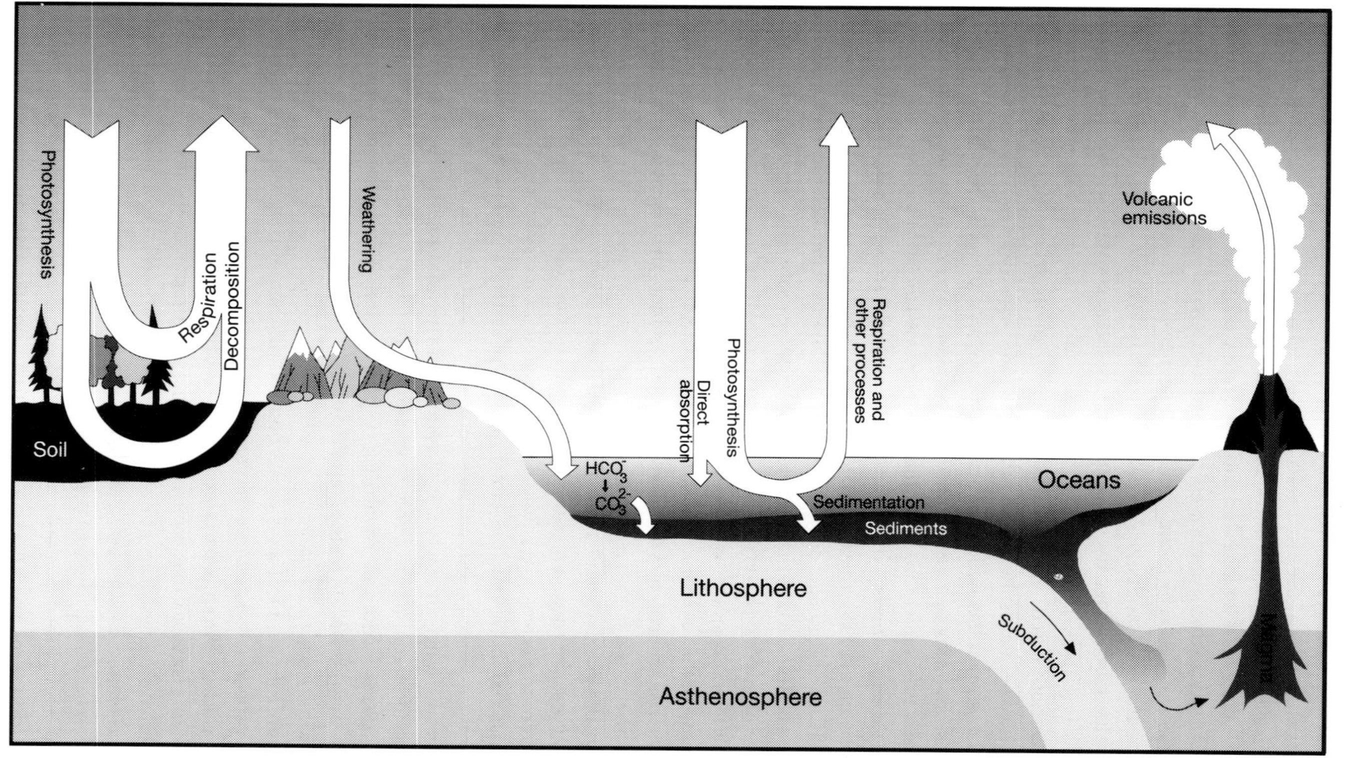


**Figure 10.1** *The major part of the natural carbon cycle involves processes that add carbon dioxide to and remove it from the atmosphere. $CO_2$ is fixed in plant matter by photosynthesis, on land and in the sea. It is released from plants by respiration (particularly at night) and also by bacterial decomposition of dead plant matter. Inorganic reactions between silicate rocks and atmospheric $CO_2$ (weathering processes) release bicarbonate ions ($HCO_3^-$) into the sea, where some are used by marine organisms to build calcium carbonate ($CO_3^{2-}$) shells. Dead organic matter (plant and animal) and the shells of dead creatures sink to the bottom of the oceans and accumulate as carbon-rich sediments. At ocean trenches, the sediment is dragged down into the Earth's mantle by descending tectonic plates, and the carbon is converted back into $CO_2$ and other carbon-containing gases, and released into the atmosphere via volcanism. Other elements of the natural carbon cycle determine the concentrations of atmospheric methane and carbon monoxide.*

*Ice ages: the eternal recurrence*

Climate change is nothing new. The average temperature of the Earth underwent several large, long-term variations before *Homo sapiens* evolved, the most obvious manifestation of which is the irregularly repeating cycle of ice ages. The ice ages are thought to be fundamentally the result of periodic changes in the shape and orientation of the Earth's orbit around the Sun, which bring small but important changes in the seasonal and latitudinal distribution of heat received by the planet. The effect of these changes in the Earth's orbit was first calculated by the Yugoslavian astronomer Milutin Milankovitch in the nineteenth century. Milankovitch suggested that the resulting changes in solar heating are sufficient to trigger changes in the Earth's climate, causing the onset and termination of the ice ages. Periodic climate fluctuations of about 100,000, 44,000, 23,000 and 19,000 years, called Milankovitch cycles, can be identified in the geological records of global climate, reflecting the periodic changes in the planet's orbit.

But the variations in the global distribution of solar heating as a result of the Milankovitch orbital cycles are rather small – insufficient in themselves to cause the world to freeze over or, conversely, to melt the ice sheets. Moreover, Milankovitch's theory predicts very slow and gradual changes in climate, whereas the geological records show much more rapid variations in global temperature, accompanied by changes in the amounts of minor ("trace") gases in the atmosphere. It is thought that the small climate variations induced by the Milankovitch cycles trigger changes in natural processes that affect climate, such as the patterns of ocean circulation and, perhaps most significantly in the long run, the biogeochemical cycles that determine natural levels of trace gases (primarily carbon dioxide and methane) in the atmosphere. These processes then serve to amplify and accelerate the change in climate.

By drilling deep down into ancient ice sheets such as those that cover the Antarctic continent, scientists can extract ice columns (or "cores") which contain bubbles of air trapped when the ice was formed (Plate 17). Extremely sensitive chemical analysis of these bubbles allows the chemical history of the atmosphere to be deciphered. The amount of carbon dioxide in bubbles from a core drilled at the former Soviet Union's Vostok station in Antarctica reveals that atmospheric $CO_2$ levels have been far from steady over the past 160,000 years (Figure 10.2). Sometimes they rose to values comparable to those of today; at other times, $CO_2$ levels dropped to less than two-thirds of those in the pre-industrial modern world. During the last ice age, which lasted from about 120,000 to about 10,000 years ago, the concentration of carbon dioxide was only about 64 per cent of its "modern" (pre-industrial) value.

The amount of heavy hydrogen (deuterium) in the ice depends on the local temperature when the ice was formed, enabling researchers to reconstruct how these temperatures have varied in the past and thus to obtain an historical record of global climate. In the Vostok core, there is a clear relationship between temperature and carbon dioxide levels: when one is high, so is the other.

The analysis of the Vostok core also reveals that atmospheric methane concentrations shadow temperature changes just as do carbon dioxide levels (Figure 10.2), suggesting

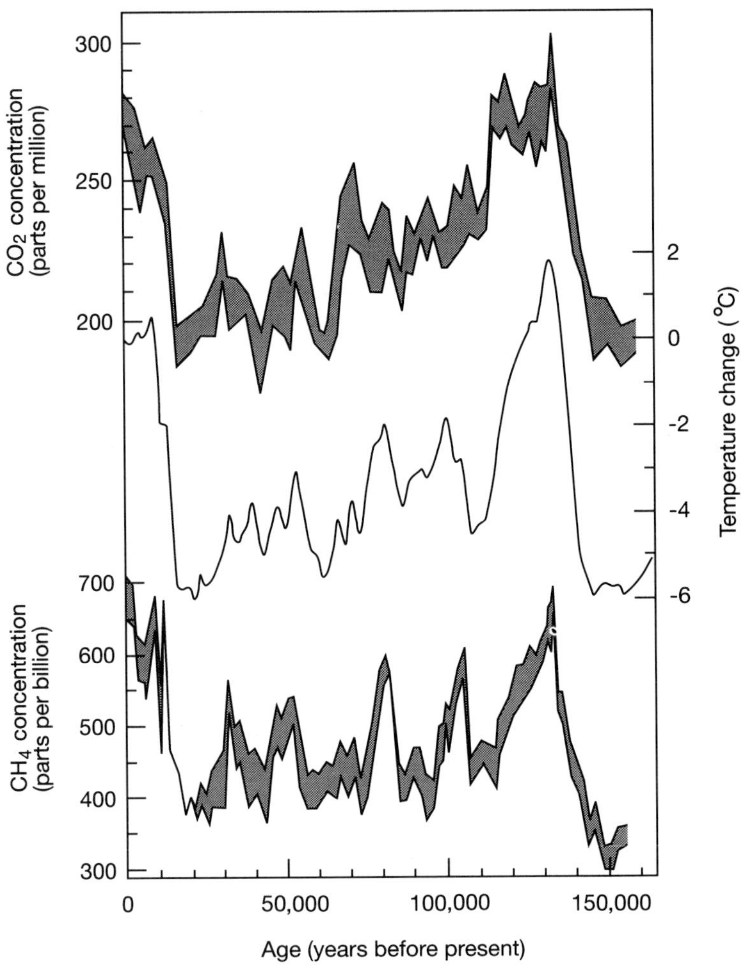


**Figure 10.2** *Records of carbon dioxide (top) and methane (bottom) deduced from analysis of bubbles in an ice core drilled at the Vostok site in Antarctica show that the atmospheric concentrations of these compounds have varied significantly in the past. (The gray areas denote the range of uncertainty in the measurements.) The amount of heavy hydrogen (deuterium) in the ice provides a record of temperature changes (middle). The levels of both carbon dioxide and methane change more or less in synchrony with temperature changes, suggesting that variations in the concentrations of these two gases have influenced the climate of the planet.*

that the natural cycling of this gas too is linked with climate fluctuations. But how can these gases, which comprise less than 1 per cent of the atmosphere, cause changes in global mean temperatures of up to 10 degrees Celsius?

### *Balancing the radiation books*

The Earth receives heat from the Sun in the form of visible sunlight, as well as radiation with wavelengths outside the visible range (particularly in the infrared and ultraviolet). About one-third of the total incident solar radiation is reflected back into space, largely by bright objects such as clouds and ice sheets. The exact proportion of reflected radiation therefore depends to a large degree on the extent and brightness of global cloud cover, and is quantified in terms of the planetary "albedo" – the ratio of the reflected to total incident radiation. An increase in cloud cover or in the area of the ice sheets enhances the albedo and reduces the amount of solar energy absorbed at the Earth's surface.

The solar radiation that is not reflected back into space is absorbed by the atmosphere, the oceans and the land mass, as well as by living organisms such as vegetation and marine plankton. The absorbed energy warms up the absorber, causing it ultimately to re-radiate the energy. But the emitted radiation differs significantly from that absorbed – clearly, the absorbing systems do not start to glow like the Sun! Rather than visible light, they radiate invisible heat – that is, infrared (IR) radiation, which has wavelengths longer than those of light.

While they are transparent to visible light, certain atmospheric trace gases – the so-called greenhouse gases – absorb radiation strongly in the infrared region of the spectrum because they have molecular vibrations with frequencies equal to those of infrared radiation (see Chapter 3). Thus, some of the energy that the Earth receives from the Sun never gets radiated back out into space – it is re-emitted as heat, absorbed by the atmosphere and radiated back once more to the Earth's surface (Figure 10.3). This is how the greenhouse effect works. (Yet it is not how a *real* greenhouse works at all – there the glass simply stops warm air inside from mixing with cold air outside.)

The most important greenhouse gases are carbon dioxide ($CO_2$), methane ($CH_4$), nitrous oxide ($N_2O$) and chlorofluorocarbons (CFCs). In actual fact, atmospheric water vapor has a larger warming effect than any of these because it absorbs IR radiation very strongly, but nevertheless it is not usually classified as a true greenhouse gas because mankind's activities have almost no direct influence on its concentration in the atmosphere. Rather, this is determined solely by natural processes such as evaporation from oceans and precipitation from the skies as rain and snow. Like the elements carbon, oxygen and nitrogen, water is constantly circulated to and from the atmosphere – this circulation is known as the hydrological cycle. In their computer models for predicting the greenhouse effect, researchers include the hydrological cycle as an element of the climate system itself.

It is sometimes implied incorrectly that the greenhouse effect is an unnatural phenomenon arising from the misguided practices of mankind. But as I have already

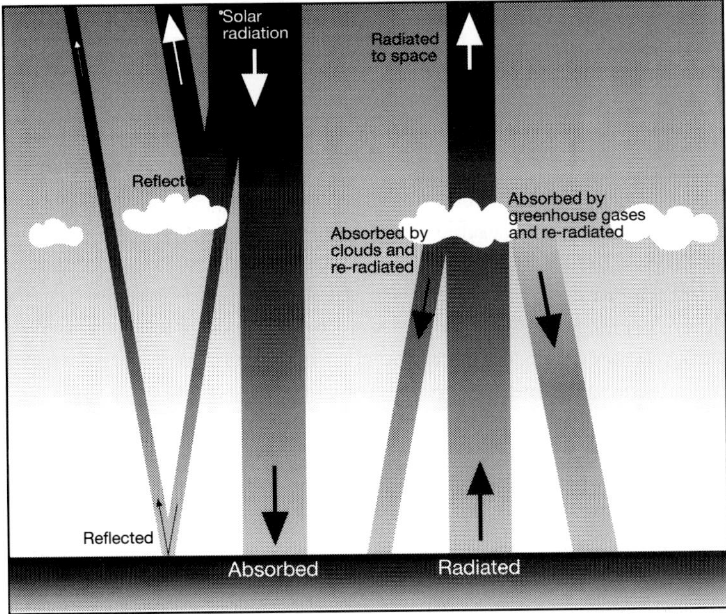


**Figure 10.3** *The greenhouse effect is the result of heat from the Sun becoming trapped in the atmosphere by absorption by greenhouse gases. About 30 per cent of the incident solar energy is reflected back into space; the rest is absorbed by the atmosphere and planetary surface, and then re-emitted as infrared radiation. Some of the re-emitted radiation is trapped by clouds and greenhouse gases; the remainder escapes back into space.*

hinted, all of the greenhouse gases (other than the CFCs) have been present in the atmosphere to some degree since the outgassing of the primordial Earth; human influences are merely increasing the atmospheric concentrations of these gases above the levels that are present naturally. If today's $O_2/N_2$ atmosphere was devoid of greenhouse gases, the average temperature of the planet would be a frigid minus 18 degrees Celsius. Thus the planet probably could not sustain life were it not for a *natural* greenhouse effect (due mainly to small amounts of water vapor, carbon dioxide and methane), which warms the planet by an additional 33 degrees Celsius on average.

## *The global warmers*

Of all the greenhouse gases, carbon dioxide is the most significant. Human activities have increased the amount of atmospheric $CO_2$ by about 26 per cent since the Industrial Revolution, mainly through the burning of fossil fuels (coal, oil and natural gas). The destruction of vast areas of forest, particularly in South America, also has an important impact on atmospheric $CO_2$: forests act as natural "sponges" which soak up this gas from the atmosphere and fix the carbon in plant matter. If the trees are chopped down and burnt or left to decompose, the carbon is released again as $CO_2$. Natural fluctuations of the carbon cycle can bring about substantial changes in concentrations of atmospheric $CO_2$, as the Vostok ice-core record indicates; the important question is how the natural sources and sinks (removal pathways) of $CO_2$ will change in the face of man-made (anthropogenic) inputs.

Although there is roughly two hundred times less methane than carbon dioxide in the atmosphere today, its potential role in global warming is significant. Molecule for molecule, $CH_4$ has a greater greenhouse warming effect than $CO_2$, as it is a stronger absorber of infrared radiation. Since the Industrial Revolution, the amount of methane in the atmosphere has doubled. A variety of human activities have contributed to this rise, largely connected with agriculture and land management. The cultivation of rice in paddy fields makes the biggest contribution: rice plants produce methane as they grow. Rice production, nearly all of which is practised in Asia, has approximately doubled since 1940. Ruminant animals such as cattle and sheep brew up considerable quantities of methane in their digestive systems, expulsion of which represents the second most important anthropogenic source. Burning of vegetation in tropical forests and savannahs releases methane, while other sources include the fermentation and decay of organic wastes in rubbish tips and landfills and the leakage of natural gas during coal mining and gas drilling and transmission along pipelines.

But there are substantial natural sources of methane too. Microbial processes in wetlands such as bogs, swamps and tundra produce roughly as much methane as paddy fields now do, while termite communities are thought to emit as much as is produced by burning of vegetation. Biological processes in oceans, lakes and rivers also produce small quantities of methane.

The most important sink for atmospheric methane is its chemical destruction in the atmosphere. The troposphere – that part of the atmosphere extending from ground level to an altitude of about 10–15 kilometers – contains significant quantities of reactive hydroxyl (OH) species. These attack methane to form a variety of products including carbon monoxide and water (both of which, however, themselves contribute to global warming).

Chlorofluorocarbons are primarily a source of environmental concern because of their ozone-destroying properties (discussed later). But they are also very strong absorbers of infrared radiation, and so make a small but significant contribution to global warming despite being present in the atmosphere in far smaller concentrations than carbon dioxide or methane. The presence of CFCs in the atmosphere is entirely mankind's doing – there are no natural sources. These gases are manufactured industrially for use as aerosol propellants, refrigerants, solvents and foam-blowing agents, uses that rely on their extreme inertness towards chemical reactions. But this unreactivity means that there are no effective removal pathways of CFCs in the atmosphere until they find their way up to the stratosphere and destroy ozone. The good news, however, is that CFCs are now so clearly a bad thing that the pressure on industries to replace them with less harmful compounds is considerable, and has resulted in agreements to phase out their use in the coming decades. It can be expected that the importance of CFCs as greenhouse gases will therefore decrease in the future. Ironically, the ozone-destroying capacity of CFCs diminishes their net global-warming effect, because ozone too is a greenhouse gas.

Nitrous oxide is the product of a wide range of biological processes in both the oceans and soils, although the details of these processes and the precise magnitudes of the natural sources and sinks are not well understood. Human activities have boosted the atmospheric concentration of nitrous oxide by about 8 per cent since pre-industrial times, primarily by the burning of fossil fuels and forests and by the use of nitrogen-rich fertilizers (nitrates and ammonium salts).

Amongst the minor greenhouse gases are other nitrogen oxides, ozone and carbon monoxide. In contrast to its beneficial, UV-filtering role in the stratosphere (10–50 kilometers above the Earth's surface), at ground level ozone is not at all friendly to the environment – it is a hazardous, poisonous pollutant, harmful to the eyes and lungs and damaging to plants. Levels of tropospheric ozone appear to have increased by as much as two- or threefold relative to the past century as a result of fossil-fuel burning and industrial activity.

### *Feedbacks and uncertainties*

Biogeochemical cycles involving the natural greenhouse gases (especially carbon dioxide and methane), as well as the hydrological cycle, introduce feedback mechanisms to climate change. These include both positive feedbacks, which act to accelerate the process of change, and negative feedbacks, which retard it. For example, increases in global temperature could disturb the ecological systems in the oceans and on land, altering the balance between uptake and emission of $CO_2$ and methane. A report compiled in 1990 by an international team of top climate researchers, the working group of the Intergovernmental Panel on Climate Change (IPCC), confessed ominously that "the possibility of unexpected large changes in the mechanisms of the carbon cycle due to a human-induced change in climate cannot be excluded."

The feedbacks involved in the hydrological cycle are dependent primarily on the climatic effects of clouds, formed by the condensation of atmospheric water vapor into tiny liquid droplets. How clouds influence the Earth's radiation budget is still unclear; indeed, this issue represents one of the major uncertainties in predictions of future global warming. There is not even a consensus over whether clouds produce a positive or negative climate feedback – that is, whether their net effect is to amplify global warming or to alleviate it. On the one hand, clouds increase the planet's albedo, reflecting incoming radiation back into space, and in this way greater cloud cover reduces the total amount of radiation reaching the surface. But clouds also absorb infrared radiation from the surface and radiate it back into the atmosphere, just as greenhouse gases do. The overall effect of clouds on the present-day climate seems to be one of cooling – that is, the infrared radiation trapped in the atmosphere by clouds is more than compensated by the solar radiation reflected back into space from cloud tops. But this finding is not necessarily a good guide to the nature of cloud feedback in a warmer climate, because global warming could change the distribution and structure of clouds, and thereby their radiative properties.

We saw in Chapter 9 that feedbacks can make a system acutely sensitive to perturbations by allowing small fluctuations to be amplified into larger ones. The existence of climate feedbacks makes the threat of global warming all the more serious, because it means that we cannot necessarily expect climate to change smoothly and predictably as levels of greenhouse gases increase. We must recognize the possibility that temperature changes induced by mankind may push the climate system just far enough out of balance for some *natural* positive feedback to take over, blowing up the changes far beyond what might be expected on the basis of the human influence alone. On the other hand, negative feedbacks might come into play that act as thermostats to restrict the amount of warming experienced by the planet. Identifying positive and negative feedbacks and estimating their relative effects on climate is proving to be a tremendously difficult task, which is limiting severely our ability to make accurate predictions about future climate change.

It is in the very nature of feedbacks that they make the climate system hard to predict and model, since they imply that small uncertainties in the models can produce large uncertainties in the forecasts. This poses a major problem not just for the science but also for the politics of global warming – most nonscientists have come to expect clear-cut and precise answers from science, and the message that there are many things about which we cannot be sure or which are hard to predict accurately often gets interpreted as a sign that climate modelers do not really understand what is going on. It also means that those who choose, for whatever reasons, to take extreme views (in either direction) about possible future climate change do not experience much difficulty in finding arguments to support their preferred interpretations. It is a common refrain from industries reluctant to reduce greenhouse-gas emissions that such measures are not warranted in the face of all the uncertainties, whereas I hope it is now becoming clear that those uncertainties in fact represent one of the strongest arguments in favor of restrictions on emissions. Of course, by the same token, alarmists and prophets of doom are able to construct highly dramatic and disturbing scenarios of climate change which may turn out to bear little relation to reality.

Because the social, economic and industrial changes that may be required to avert the potentially disastrous consequences of global warming are difficult and costly, those who stand to lose by them are demanding proof that such changes are really necessary. One might imagine that, given all the difficulties of making accurate predictions, the simplest approach is instead to go out into the real world and measure whether any changes are discernible. But here too, the scientific community is hard pushed to provide definitive answers. Most scientists are agreed that the threat of global warming is a real one, but what they cannot yet say for sure is that the signs of global warming are already evident.

We know that concentrations of greenhouse gases have increased dramatically since the Industrial Revolution, and that the global mean temperature has been rising since the start of the century (albeit with a blip from 1940 to 1970) (Figure 10.4). Furthermore, over the past hundred years or so sea levels have risen on average by about 1 to 2 millimeters per year; such rises are an expected longer-term consequence of global

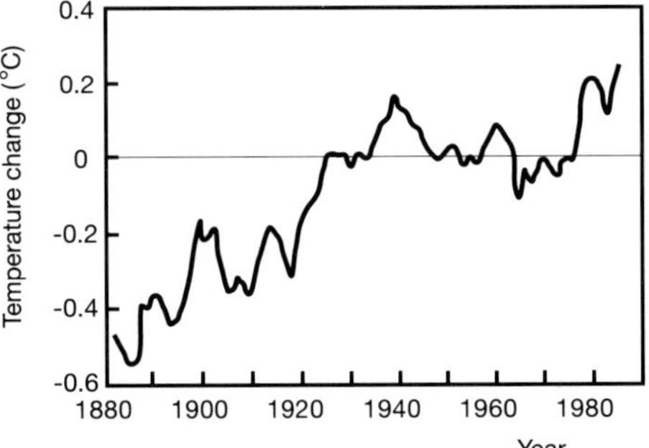


**Figure 10.4** *Global mean temperatures since the beginning of the century show a general (and statistically significant) increase, although on occasion (such as from 1940 to 1970) the trend is broken. This tendency towards a warmer climate suggests, but does not prove, that we are already experiencing the consequences of a man-made greenhouse effect.*

warming, since this should cause melting of polar ice sheets and mountain glaciers. For a scientifically rigorous proof, however, such circumstantial evidence will not do. For one thing, global averages do not tell the whole story. Predictions of the effects of greenhouse warming using computer models indicate that warming will not be equal across the entire planet, and that in fact changes in weather patterns may lead to *cooling* in some locations over the short term. To be sure that greenhouse warming is happening, we require an unambiguous link (not just a coincidence) between rising levels of greenhouse gases and temperature changes, and a correspondence between observations and climate model predictions of the global *distribution* of temperature changes: the so-called "greenhouse fingerprint." No such link has yet been firmly established.

Forecasts of future warming are plagued by these uncertainties and gaps in our understanding. They also depend on the degree to which greenhouse gas emissions will be controlled in the coming decades. If no restrictions on emissions are imposed, so that industrialized societies continue to conduct their business as usual, the predicted temperature rise is between about 1 and 2.5 degrees Celsius by the year 2025, and between 3 and 6 degrees Celsius by 2100. The latter upper bound takes temperatures higher than they have been at any time during the past 150,000 years, and we have little experience in predicting what the consequences of such a change might be. A slightly less dramatic picture appears if we assume that some kinds of reductions on emissions will be enforced in the coming decades. Under a variety of such scenarios, temperature increases of between 2 and 3 degrees Celsius are predicted by 2100. While these changes may not seem very large, the consequences for sea-level rise, weather variations, agricultural productivity and frequency of extreme climate events such as storms and hurricanes may be severe. Of course, we could turn out to be mistaken totally – some negative feedback mechanism might appear, for instance, that limits temperature changes to less than a degree. But to use our current uncertainties as a basis for inaction would be wishful thinking at its most reckless. As some commentators have pointed out, no sensible person demands absolute proof that they will be burgled, will crash their car or will fall seriously ill before they take out an insurance policy. And if the worst happens, it will happen to all of us; we can be sure that there will be no one to bail us out.

## The planetary sunscreen

### *Hole in the sky*

We have seen that concomitant with the development of an oxygen-rich atmosphere, the Earth's stratosphere became enriched in oxygen's sibling gas, ozone ($O_3$). Because ozone absorbs ultraviolet radiation strongly, the stratospheric ozone layer acts as a filter to prevent much of the Sun's UV from reaching the surface of the planet. Ultraviolet light carries energy in considerably larger packets than does visible light, potentially enough to disrupt the structure of delicate biological molecules. It can cause damage to living tissue, giving rise to skin cancer and cataracts as well as causing harm to land plants and to plankton, a crucial part of the oceanic food chain.

This was why the results announced in 1985 by Joe Farman and colleagues from the British Antarctic Survey were greeted with consternation. They found that between 1977 and 1984 the ozone concentrations in stratospheric air above Halley Bay in Antarctica had decreased to about 60 per cent of the normal level (Figure 10.5). It later transpired that atmospheric scientists had for some time been observing similar results from the Total Ozone Mapping Spectrometer (TOMS) on NASA's Nimbus 7 satellite; but because the measurements were so low, they had been assumed to be the result of instrument malfunction. The measurements of Farman's team left no doubt, however, that serious depletion of stratospheric ozone was occurring over most of the Antarctic continent at altitudes of between 12 and 24 kilometers.

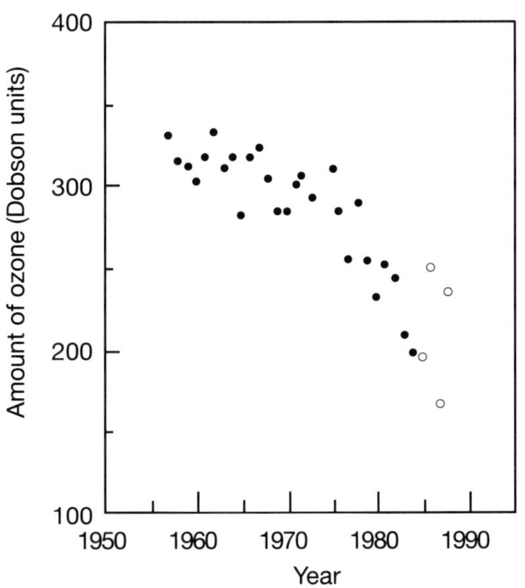


**Figure 10.5** *Joe Farman and colleagues of the British Antarctic Survey reported in 1985 that concentrations of ozone in the springtime over Halley Bay, Antarctica, had been decreasing steadily over the previous decade and a half. Farman's data are shown here in black; the white circles denote later measurements, which follow the same trend. Ozone concentrations are measured in Dobson units, after G. M. B. Dobson, a British scientist who pioneered ozone measurements in the early part of this century.*

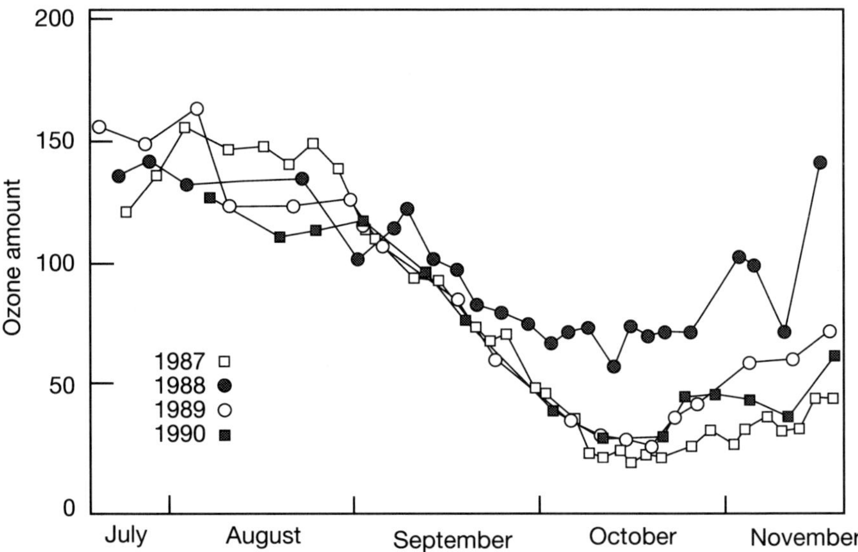


**Figure 10.6** *The "ozone hole" over Antarctica is a regular visitor in the austral spring. It begins in September and lasts until November. Shown here is the total amount of ozone (Dobson units) in the stratospheric layer between 12 and 20 kilometers above South Pole Station for the years 1987 to 1990.*

Subsequent measurements have revealed that, every year since this announcement, ozone concentrations in the south polar stratosphere have started to diminish during the Antarctic spring (which begins around September), remaining low until late October or November when the pattern of atmospheric circulation around the pole changes and the ozone-poor air is dispersed and mixed into that surrounding it. The severity of ozone depletion – the "depth" of the ozone hole – varies from year to year (Figure 10.6).

It is now generally agreed that the ozone hole is caused primarily by compounds derived from man-made chlorofluorocarbons in the atmosphere. These chemicals are essentially hydrocarbons that have some of the hydrogen atoms replaced by chlorine or fluorine. They have been used for decades in the variety of commercial applications mentioned earlier. Their wide range of uses derives from the fact that CFCs are very unreactive and nontoxic. But this property also has the consequence that CFCs are resistant to the chemical reactions that destroy or remove many other trace gases in the lower atmosphere, and so they become dispersed across the entire planet, eventually finding their way up into the stratosphere. At altitudes above about 25 kilometers, the CFC molecules are exposed to the ultraviolet rays that are blocked further down by ozone. The UV radiation breaks apart the otherwise stable molecules, splitting off individual chlorine atoms. In 1974, Mario Molina and Sherwood Rowland of the University of California at Irvine warned of the possible consequences of this process.

Lone atoms are generally very reactive, since they contain electrons that are not paired up either in atomic or in molecular orbitals (the exceptions are the so-called inert

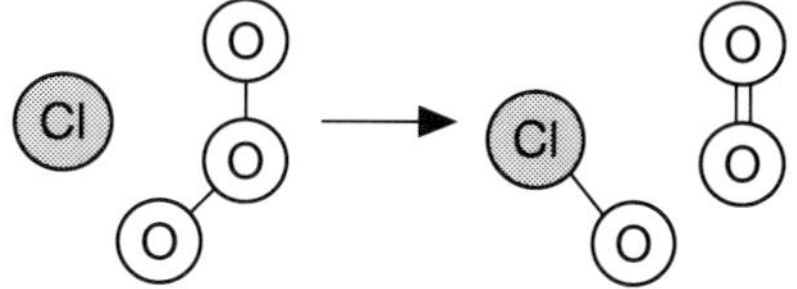


**Figure 10.7** *Free chlorine atoms (radicals) react with ozone to form chlorine monoxide and molecular oxygen.*

gases, such as helium and neon, which have no unpaired electrons). Chemical species with unpaired electrons are called free radicals. Chlorine free radicals are particularly virulent, and laboratory studies show that they will readily consume ozone to form chlorine oxide (ClO) and molecular oxygen ($O_2$) (Figure 10.7). Molina and Rowland pointed out that this reaction could occur in the stratosphere, causing chemical destruction of the ozone layer. The recognition of this danger led to a ban in the United States on the use of CFCs in aerosols during the late 1970s. But in the absence of any evidence at that time to show that CFCs really did have these harmful effects, industries that made extensive use of the compounds resisted with vigor suggestions that they should change their ways, and during the 1970s and 1980s CFCs from various industrial sources throughout the world continued to accumulate steadily in the atmosphere.

### *The cycle of destruction*

If chlorine radicals from CFCs are indeed to blame for ozone destruction, why should it occur only over the Antarctic and only during the spring? In the Antarctic winter, a vast vortex-like column of air forms over Antarctica; the air inside the vortex is effectively isolated from that outside (Figure 10.8). This isolation, and the absence of sunlight during the polar winter, cause stratospheric air temperatures in the vortex to plummet to below minus 80 degrees Celsius. These bitter conditions lead to the freezing of water in the stratosphere, and the resulting ice particles form polar stratospheric clouds (PSCs), whose light-scattering properties render them strikingly visible in the long polar night (Plate 18). The ice particles in PSCs may also contain considerable amounts of nitric acid ($HNO_3$), formed from nitrogen oxides, which are ubiquitous atmospheric trace gases. It is now thought that these ice particles provide the stage on which the key ozone-destroying reactions take place. Elucidating the many interrelated processes involved in that drama has been the aim of extensive studies conducted in the past few years in laboratories and through ground-based, balloon-borne and satellite observations of the chemical composition of the Antarctic stratosphere.

The principal step is the reaction of a chlorine atom with ozone to form chlorine oxide and oxygen, shown in Figure 10.7. But chlorine oxide is itself a very reactive molecule and therefore undergoes further reactions (Figure 10.9). The end result of these reactions is that the chlorine atom in ClO becomes separated from the oxygen atom and is liberated as a free radical once more. Having destroyed one ozone molecule, the chlorine atom is then ready to go to work on another. In other words, chlorine radicals

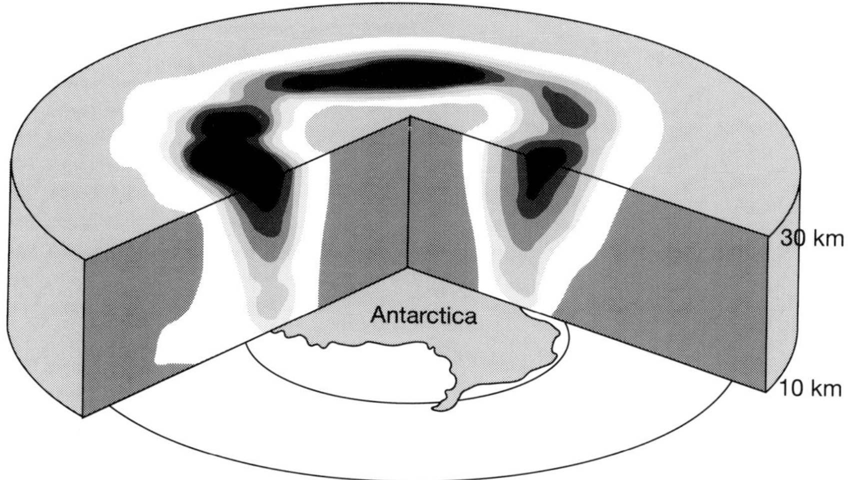


**Figure 10.8** *The pattern of air circulation around Antarctica gives rise to an isolated vortex of cold air every winter. Here I show wind speeds around the South Pole measured at altitudes of between 10 and 30 kilometers in troposphere and stratosphere in October 1990. Increasingly dark regions within the white polar vortex correspond to increasingly higher wind speed. (Picture after an image kindly supplied by Mark Shoeberl, NASA Goddard Space Flight Center, Maryland.)*

act as *catalysts* for ozone depletion. The sequence of reaction that this involves is called the chlorine catalytic cycle.

This malign circle can be broken, however, by competing reactions which consume chlorine oxide or chlorine radicals. The most important of these reactions involves nitrogen dioxide ($NO_2$), which can combine with ClO to form the molecule $ClONO_2$. This reaction can proceed only in the presence of a catalytic surface, like those encountered in Chapter 2. In the Antarctic stratosphere, such surfaces are provided by the ice particles in polar stratospheric clouds. $ClONO_2$ is a relatively stable compound which binds up the chlorine atom and neutralizes its capacity to do harm. Thus nitrogen dioxide in the polar stratosphere moderates the severity of ozone depletion. Another important reaction is that between a chlorine radical and methane to form hydrogen chloride (HCl), which again constitutes a relatively stable, innocuous form of chlorine. But these "inactive" types of chlorine may be split apart by light, or by reactions with other molecules, liberating "active" chlorine once more (Figure 10.10*a*).

Polar stratospheric clouds play a further part in the story by catalyzing a reaction between the two inactive forms of chlorine, HCl and $ClONO_2$. These combine to form $Cl_2$ and nitric acid ($HNO_3$): the former can be split apart by sunlight to yield chlorine radicals once more, while the latter remains in the ice particles. In this way, the $NO_2$ available to tie up active chlorine and thereby alleviate ozone depletion can become

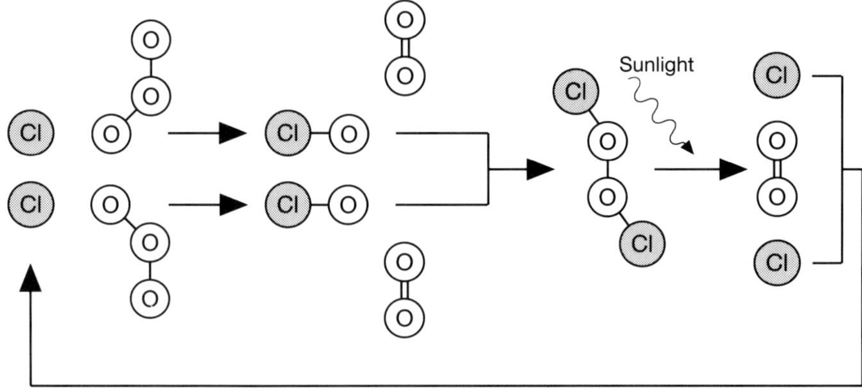


**Figure 10.9** *The chlorine radicals that destroy ozone are regenerated via the chlorine catalytic cycle.*

locked up in nitric acid/ice crystals in the clouds, giving the active chlorine more freedom to do harm (Figure 10.10*b*).

Worse still, the nitric acid/ice crystals may grow so large that they become too heavy to remain suspended in the air, and they will then sink down through the stratosphere. This ensures that the palliative influence of nitrogen is removed permanently from the arena, which is bad news for the ozone layer.

The chlorine catalytic cycle is assisted by bromine compounds. Some of these are injected into the atmosphere by human activities (they are used as fumigants, for example), but most atmospheric bromine comes from natural sources such as marine algae, some of which emit methyl bromide ($CH_3Br$). Light-induced splitting of these bromine compounds leads to the formation of bromine oxide (BrO), which, as well as being able to destroy ozone directly in much the same way as ClO does, also helps to regenerate chlorine radicals from chlorine oxide.

These, then, are the elementary chemical processes involved in ozone depletion. The many pieces of the puzzle fit together to form a picture that gives a convincing explanation of how and why ozone is destroyed over Antarctica. In the dark, cold polar winter, atmospheric circulation patterns create the polar vortex. As temperatures within the vortex fall, water and nitric acid condense in the stratosphere to form PSCs. Chemical reactions on the surfaces of the ice particles convert inactive forms of chlorine (HCl and $ClONO_2$, for example) to active forms (ClO and Cl), thus "priming" the vortex for ozone destruction. When in the spring the Sun reappears, the energy from sunlight sets off the chlorine catalytic cycle. Ozone concentrations decrease rapidly, reaching a minimum in early October.

When the Antarctic vortex breaks up in late October, the ozone-poor air that it contains is mixed with "normal" air from outside. This means that ozone concentrations over Antarctica start to rise again, but by the same token the ozone layer immediately outside of this region becomes diluted. There is now good evidence that the effects of

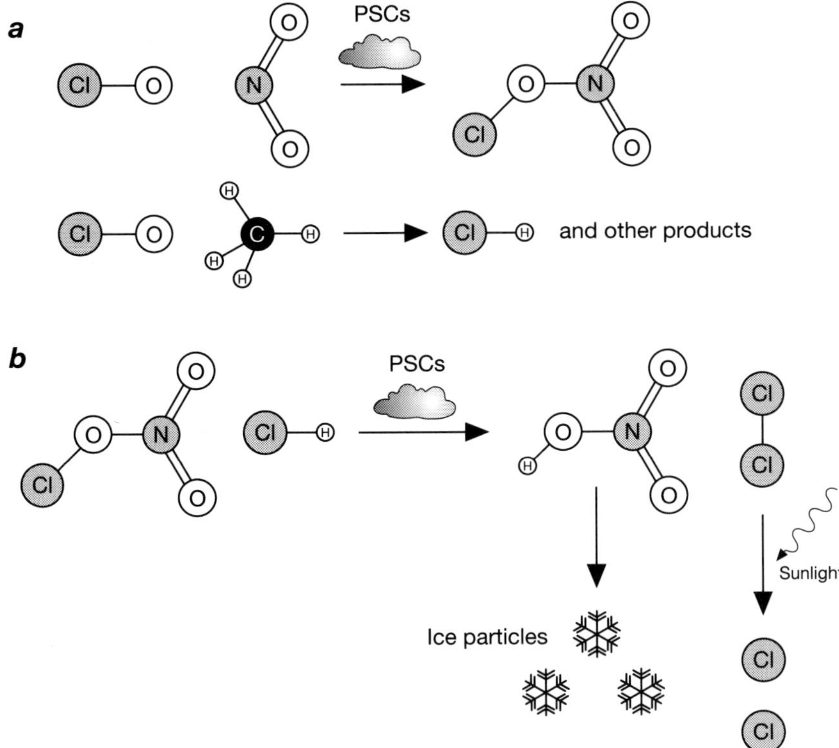


**Figure 10.10** *"Active" forms of chlorine, which have the potential to generate chlorine radicals, can be converted into "inactive" forms by a variety of reactions, such as those involving nitrogen dioxide ($NO_2$), nitric oxide (NO) and methane (a). These reactions disrupt the chlorine catalytic cycle. Active chlorine can be regenerated, however, by the reaction of the inactive forms $ClONO_2$ and HCl on ice particles in PSCs (b).*

*Removal of reactive forms of nitrogen from the polar stratosphere – denitrification – diminishes the ability of these compounds to prevent ozone depletion by binding up "active" chlorine. The freezing of nitric acid in polar stratospheric clouds locks up the nitrogen and prevents it from taking part in chemical reactions; and if the ice particles become very large, they sink down through the stratosphere and are removed permanently from the scene.*

this dilution are felt as far beyond the polar region as Australia. Here the intensity of ultraviolet radiation reaching the ground is found to increase in the late spring as the vortex is dispersed. It is just at this time, of course, that the population starts heading for the beaches, and the health risks of increased UV levels are therefore a cause for considerable concern. Aside from Australia, however, these latitudes of the Southern Hemisphere are populated rather sparsely. But what would happen if a similar phenomenon were to occur over the North Pole, just beyond which lie the far more densely populated regions of Scandinavia, Northern Europe and Canada?

There are now indications that an ozone hole *does* open up over the Arctic. This evidence is, as yet, not incontrovertible, but it seems to suggest that if an Arctic ozone hole is indeed created, it is considerably more shallow than that over the Antarctic. Reasons for this difference are not hard to find. Patterns of atmospheric circulation around the Arctic are not the same as those over the Antarctic, in part because there is more dry land in the Northern than in the Southern Hemisphere. A polar vortex of sorts does appear over the North Pole, but it is nothing like so well defined or so isolated as that in the south. It is partly for this reason that winter air temperatures in the Arctic do not fall as low as those in the Antarctic (the former are typically 15 to 20 degrees Celsius warmer). Therefore, the low temperatures that are necessary for the formation of the polar stratospheric clouds which play such a crucial role in ozone depletion do not occur as a matter of course in the Arctic winter. But seasonal temperatures vary from year to year, and it is conceivable that a particularly cold winter in the Arctic could create conditions appropriate for ozone destruction.

The winter of 1988–89 seems to have been such a case. In January of 1989 the Arctic was colder than it had been during any other January for the past 25 years, with temperatures in the Arctic stratosphere falling below minus 85 degrees Celsius. Polar stratospheric clouds were seen inside the Arctic circle, and ozone levels at some altitudes in late January appeared to be lower, by about 25 per cent, than those measured for the previous three years. The question of Arctic ozone depletion remains contentious, however, in part because ozone depletion and PSC formation don't quite seem to coincide. All the same, there are probably few researchers who would now be surprised if definitive evidence for Arctic ozone depletion turns up.

## *Limiting the damage*

In September 1987, an international agreement was signed by 24 nations in Montreal to restrict production and consumption of CFCs. This treaty, known as the Montreal Protocol, came into force in January 1989. All of the signatory nations agreed to freeze, at 1986 levels, production of the major CFCs in 1990 and then to cut back on production until a reduction of 50 per cent is achieved by 1999. The agreement included compromises in cutbacks for developing nations to lessen the impact of the changes on their economies. The Montreal Protocol was updated by a convention in London in 1990, when the target was revised to that of a complete phase-out of the major CFCs by the year 2000. Without doubt these treaties signify a growing international recognition and acceptance of the global consequences that human activities can have. But it seems likely that ozone depletion will persist for decades even if the London treaty is honored. CFCs remain in the atmosphere for many years before being purged by chemical processes, so that even if not another puff of these gases entered the air from tomorrow, enough is up there already to ensure that the unwelcome effects will remain for some time to come.

Moreover, it now seems clear that ozone depletion need not be limited to the poles. Ozone concentrations at much lower latitudes were significantly lower than normal during late 1982 and most of 1983. At the time these losses went unexplained, but the present understanding of ozone chemistry now provides a possible interpretation. In spring of 1982, the volcano El Chichón in Mexico underwent a massive eruption, exceeded this century only by the eruption of Mount Pinatubo in the Philippines in 1991. El Chichón spewed past quantities of sulfur dioxide into the atmosphere, much of which was converted into airborne particles (aerosols) of sulfuric acid. Scientists believe that the catalytic chemical reactions that take place in polar stratospheric clouds can also occur on these volcanic aerosols. It seems likely that the mid-latitude ozone depletion of 1982–83 was the result of El Chichón's violent outburst. There is evidence to suggest that Pinatubo had much the same effect in 1991–92. Thus volcanic eruptions add a further, unpredictable element into the story of ozone depletion.

There is, however, some cause for hope. For the present time, the unwelcome effects of CFCs are a fact of life; but CFCs will not be with us forever. Replacements are being developed that can fulfill the same industrial role without the atmospheric side effects, or at least with reduced effects. The most popular replacements, called HCFCs, still contain chlorine, as well as fluorine and bromine, but they also contain significant amounts of hydrogen, which makes them slightly more reactive and reduces the length of time that they remain in the atmosphere. It seems entirely possible that the inhabitants of the mid-twenty-first century will be able to look back at ozone depletion as a temporary crisis now averted.

## Brimstone and acid drops

### *Death in Scandinavia*

During the 1980s, the health of large areas of coniferous forest in Scandinavia and the northeastern United States was in decline. Fir, spruce and pine trees were found to be dying in uncharacteristic numbers (Figure 10.11). At the same time, the population of freshwater fish in lakes and streams within these regions were showing signs of equally drastic malaise. Trout and salmon populations were affected particularly badly, while some other freshwater species had actually become extinct. In the Adirondack mountains of New York State, almost half of the lakes contained no fish, compared with just 4 per cent in the 1930s.

So severe were these troubling changes in forest and lake ecology that environmental scientists were forced to suspect that some influence other than natural processes must be at play. Particularly in view of the fact that these effects were seen in or close to regions of intense industrial activity (eastern North America and northern Europe), suspicion fell on pollution from human activities.

Records of the chemical composition of rain water in the affected areas showed evidence of a steady increase in acidity since the Industrial Revolution. Normal rain water is slightly acidic, because it contains dissolved carbon dioxide, which forms carbonic acid. Whereas the pH of neutral water is 7, that of rain water is generally about

**Figure 10.11** *Forest "die-back" became a common phenomenon in northern coniferous forests during the 1980s. (Picture kindly supplied by Richard Wright, NIVA, Norway.)*

5.6 (the more acidic a solution is, the lower its pH). But in the northeast United States, the pH of rain now averages about 4, and values as low as 2.1 have occasionally been reported in the United States and Europe – comparable to the acidity of vinegar. The drop in pH of lakes and streams is smaller than this – acidified lakes in these regions generally reach a pH of little less than 5 – because many natural systems have some capacity to absorb and neutralize acidified rain water. Acidic waters are deposited from the skies not only as rain but also in the form of snow, dew or fog. The term "acid rain" is often used to denote all of these forms of acidic "wet" deposition. Acids in the atmosphere can also be absorbed as gases into streams, lakes, soil and vegetation – this is called dry deposition.

It became rapidly apparent that the deaths of trees and fish were a consequence of wet and dry acid deposition and the consequent acidification of natural water systems. As well as having an impact on ecosystems, acid deposition corrodes stonework, cement and other construction materials, producing potentially hazardous consequences and necessitating hugely expensive protection and restoration schemes.

## *The bitter breath of industry*

Acid deposition is due to the presence of two kinds of trace gas in the atmosphere: sulfur dioxide ($SO_2$) and nitrogen oxides (primarily NO and $NO_2$, denoted collectively as $NO_x$). Both $SO_2$ and $NO_x$ are produced in considerable quantities by industrial processes and other human activities. They are released during the burning of fossil fuels (particularly coal, which contains many sulfur and nitrogen compounds; but automobile exhausts also make a significant contribution). Nitrogen oxides are also produced by the burning of forests and other vegetation. Although acid raid is principally a problem for the industrialized societies of the Northern Hemisphere, burning of vegetation in the tropics releases sufficient $NO_x$ into the atmosphere to affect soil and lake chemistry there too.

Although these pollutants are injected into the atmosphere over centers of industrial activity, they can be transported through the troposphere for hundreds of kilometers before being flushed out in precipitation. This means that regions upwind of and somewhat removed from the areas in which the gases are produced tend to bear the brunt of the pollution. Scandinavia is situated in a particularly unfortunate position, receiving the combined loads of sulfur and nitrogen oxides from Britain, Germany and countries in the Eastern Bloc. Over southern Sweden, more than 70 per cent of the sulfur compounds in the atmosphere is thought to result from human activities, about four-fifths of which has been carried from sources outside Sweden. The extent of the problem is made worse by the trend towards building taller flue chimneys, ironically with the intention of alleviating *local* pollution – this means that the noxious gases become dispersed more widely. Sulfur pollution from Europe and North America shows up in Greenland ice, and contributes to the so-called Arctic haze that sometimes forms a foggy shroud over the north polar region.

$SO_2$ and $NO_x$ are converted in the troposphere to sulfuric and nitric acid respectively, via a series of chemical reactions involving hydroxyl free radicals. These radicals are created when a water molecule is split photochemically after absorbing ultraviolet sunlight. They play a very important role in the chemistry of the lower atmosphere, since their extreme reactivity allows them to react with almost every kind of trace gas. We saw earlier, for instance, that the reaction of hydroxyl with methane provides the main route through which this greenhouse gas is removed from the atmosphere. Because of their ability to purge the air of trace gases, hydroxyl radicals are often described as the "detergent" of the troposphere. The reactions of hydroxyl with $SO_2$ and $NO_x$ provide further examples of this cleansing process. But unfortunately the products are nitric and sulfuric acids, which, being highly soluble in water, readily dissolve in the water droplets of fog or clouds to form acidic solutions.

One might imagine that the effect of acid rain will inevitably be to increase the acidity of lakes, rivers and soils. But while the acidity of precipitation has clearly increased over recent decades, the extent to which this influences soil and lake chemistry has been hotly debated, because some of these natural systems have a considerable ability to absorb moderate inputs of acids or alkalis. Soils are usually slightly alkaline, containing "bases" such as bicarbonate ions ($HCO_3^-$) and ammonia. These species can combine with the hydrogen ions generated by acids to form, respectively, carbonic acid ($H_2CO_3$, which can then fall apart into water and carbon dioxide) and ammonium ions ($NH_4^+$). Many soils contain clay minerals which can act as "buffers" against acidification. Reactions between these aluminosilicate minerals and water can generate compounds containing aluminum and hydroxide ions, such as the mineral gibbsite ($Al(OH)_3$). The action of acids on these compounds frees the aluminum ions, enabling them to become dissolved in and transported by groundwater. A consequence of acid deposition in clay-based soils may therefore be an increase not in acidity but in the amount of dissolved aluminum ions that are flushed into streams and lakes. As aluminum is poisonous to many species of fish, much of the concern over the biological and ecological effects of acid rain has therefore centered around the implications of rising aluminum levels rather than falling pH. There has also recently been increasing concern about the potential effects on human health of dissolved aluminum, which has been implicated as a possible cause of neural degenerative diseases such as Alzheimer's disease.

Some rocks, such as calcium carbonate (chalk), are alkaline and so will react with and neutralize deposited acid; but siliceous rocks such as granite and quartz are weakly acidic themselves and therefore lack this ability. The latter are common in some of the areas most exposed to acid rain, such as Scandinavia and Canada, the Rockies, the Appalachians and the Adirondacks. The streams and lakes of these regions have little natural protection against acidification.

Lake chemistry is rather complicated, being controlled both by the inputs of minerals from the streams that feed it and by the biological processes that take place in the water and lake sediments. How lakes respond to acid inputs is therefore a complex issue that is hard to predict. Some lakes, particularly those in Arctic regions, are rendered alkaline by natural biological processes, and can therefore withstand modest acid inputs without significant changes in pH; but others become rapidly acidified. The effects of increasing acidity are not necessarily the same throughout the food web, so that some species may die off to the benefit of others, giving rise to a shift in the lake's ecology. Generally, however, the number of species adapted to acidic environments is rather small, so acidification has the result of decreasing the diversity of lake ecosystems.

## *Coming clean*

If there is one consolation to the malaise of acid rain, it is that the causes, and thus in principle the solutions, are easy to identify. The fact that the primary causes – human activities that burn fossil fuels – also threaten us with global warming increases the urgency with which reductions in the emissions from these sources need to be imposed. But there is little prospect that the use of fossil fuels will decline significantly in the near future; despite calls for greater energy conservation, emissions are likely to increase, if anything, towards the end of the twentieth century. So solutions may have to lie largely with preventing the harmful gases from reaching the atmosphere.

There exist ways both to lower the sulfur content of fuels before burning and to remove $SO_2$ and $NO_x$ from flue gases afterwards. The use of coals and oils that are naturally low in sulfur is a particularly attractive option, but their sulfur content can also be lowered artificially (albeit somewhat expensively). Sulfur and nitrogen oxides can be removed from effluent gases by passing them through "scrubbers" in which the acidic gases are converted to harmless or nonvolatile forms. One unconventional proposal is to inject the flue gases directly into the sea, where the $SO_2$ will dissolve and be carried to the seafloor. Whether this is economically viable, let alone ecologically safe, is another matter.

The implementation of existing measures, such as installing scrubbing units, is costly

– to cut present global sulfur emissions by half would have a cost running into tens of thousands of millions of dollars each year, which would push up the price of electricity drastically. Nevertheless, the United States Environmental Protection Agency has now decreed that between 70 and 90 per cent of gaseous sulfur compounds must be removed from the emissions from all new coal-burning power plants (the measures do not apply, however, to existing plants.) Now that Europe has finally conceded its share of the blame for Scandinavia's problems, measures are being introduced there to limit sulfur emissions. Ultimately acid rain can last only so long as supplies of fossil fuels themselves hold out. But during that time it seems likely that it will remain with us as one of the least pleasant aspects of an industralized society.

**Figure 10.12** *Is there life on Earth? When the spacecraft Galileo flew past Earth in 1990 for a "gravitational assist" to boost it to Jupiter, it pointed its instruments towards the planet and searched for signs of life. The extreme chemical disequilibrium of the atmosphere (in particular the extraordinarily high concentrations of oxygen and methane) provided the most telling hints that life exists on the planet. No other world in the Solar System has these characteristics. (Picture by NASA, prepared by W. Reid Thompson and kindly supplied by Carl Sagan.)*

## A lonely oasis

As far as we can tell, Earth is the only planet in the Solar System on which life exists or has ever existed. Although both of our neighboring planets – Venus (on the sunward side) and Mars (the first stop outwards) – are of a size comparable to Earth, they are both barren and inhospitable: Mars shivers under a gelid chill of minus 53 degrees Celsius, while the surface of Venus swelters at a steamy 400 degrees. Mars has no ozone layer, and so is unprotected against the ravages of ultraviolet light from the Sun. The result is that the Martian "soil" is coated with a caustic layer of superoxide compounds, which would rapidly burn up organic matter. On Venus, it is thought that a "runaway greenhouse effect" early in the planet's history boiled all of its volatile compounds into the atmosphere, which is now laden with clouds of sulfuric acid. The environmental problems that we are creating by changing the chemistry of our atmosphere therefore have dramatic precedents elsewhere in the Solar System. Even the most fervent of environmental campaigners could not reasonably suggest that the Earth is facing consequences as extreme as these; but in the face of such demonstrations of how the chemical composition of the atmosphere determines the fate of a planet, it would be foolhardy indeed to think that we can continue to tamper at will with the delicate blue skin that separates us from space, without heed to the consequences that such actions may entail.

Seeing the Earth from space has changed our perception of the planet in many ways, not the least of which is that it has demonstrated how life itself determines the face that the planet presents to the stars. From up there, the view is one that commends humility: there are no signs of specifically human life, only a chemical membrane that betokens the presence of living organisms in all their profusion (Figure 10.12). But here at ground zero, we are beginning to realize that we have the capacity to change the appearance of this face. We can only hope that we find the wisdom to contain that power.

# Bibliography

To indicate the audience at which each of these references is pitched, I have used a scale of one to three. One designates that the writing is intended for a non-technical audience, and is similar in level to this book; two denotes articles aimed at a non-specific but scientific readership, or basic text books appropriate for college courses; and three indicates writings intended for a specialist audience. Readers without a scientific training should not, however, be deterred from investigating references in the third category, many of which are not always so difficult as might be feared.

## General chemistry

*General Chemistry*, P. W. Atkins & J. A. Beran (Scientific American Books, W. H. Freeman & Co., 1992). (2)
*The Extraordinary Chemistry of Ordinary Things*, C. Snyder (John Wiley, 1992). (1)
*Molecules*, P. W. Atkins (Scientific American Books, W. H. Freeman & Co., 1987). (1)
*Chemical Evolution*, S. F. Mason (Clarendon Press, Oxford, 1992). (2)
*A Short History of Chemistry*, I. Asimov (Heinemann, London, 1965). (1)
*The World of Physical Chemistry*, K. J. Laidler (Oxford University Press, 1993). (1)
*General, Organic and Biological Chemistry*, J. R. Amend, B. P. Mundy & M. T. Arnold (Sanders College Publishing, 1990). (2)
*Chemistry Imagined*, R. Hoffmann & V. Torrence (Smithsonian, Washington, 1993). (1)

## Chapter 1

### *Structure of the atom*

*Atom*, Isaac Asimov (Dutton, New York, 1991). (1)
*Taming The Atom*, H. C. von Baeyer (Viking, 1992). (1)

## Chemical bonding

*The Nature of The Chemical Bond* (2nd Edition), Linus Pauling (Cornell University Press, 1940). (2)
*Valence*, C. A. Coulson (Oxford University Press, 1952). (2)
*Physical Chemistry* (4th Edition), P. W. Atkins (Oxford University Press, 1990). (2)
*The Chemical Bond*, ed. A. H. Zewail (Academic Press, 1992). (2)

## Carbon molecules

*Organic Chemistry: The Name Game*, A. Nickon & E. F. Silversmith (Pergamon, 1987). (2)
*Fascinating Molecules in Organic Chemistry*, F. Vogtle (John Wiley, 1992). (2)
*Cyclophanes*, F. Diederich (Royal Society of Chemistry, London, 1991). (3)

## Dodecahedrane

"Total synthesis of dodecahedrane," L. A. Paquette, R. J. Ternansky, D. W. Balogh & G. J. Kentgen, *Journal of the American Chemical Society* **105**, 5446 (1983). (3)

## Buckminsterfullerene

"$C_{60}$: Buckminsterfullerene," H. W. Kroto, J. R. Heath, S. C. O'Brien, R. F. Curl & R. E. Smalley, *Nature* **318**, 162 (1985). (3)
"Space, stars, $C_{60}$ and soot," H. W. Kroto, *Science* **242**, 1139 (1988). (2)
"Probing $C_{60}$," R. E. Smalley & R. F. Curl, *Science* **242**, 1017 (1988). (2)
"Solid $C_{60}$: a new form of carbon," W. Krätschmer, L. D. Lamb, K. Fostiropoulos & D. W. Huffman, *Nature* **347**, 354 (1990). (3)
"Great balls of carbon," R. E. Smalley, *The Sciences* **31**(2), 22 (March/April 1991). (1)
"Great balls of carbon," J. Baggott, *New Scientist* **34** (6 July 1991). (1)
"Fullerenes," R. F. Curl & R. E. Smalley, *Scientific American* **256**, 54 (October 1991). (1)
"$C_{60}$: Buckminsterfullerene, the celestial sphere that fell to Earth," H. W. Kroto, *Angewandte Chemie* (English Edition) **31**, 111 (1992). (2)
*Buckminsterfullerenes*, eds W. E. Billups & M. A. Ciufolini (VCH, Berlin, 1993). (3)
*The Fullerenes*, eds H. W. Kroto, J. E. Fischer & D. E. Cox (Pergamon, 1993). (3)
*Perfect Symmetry: The Accidental Discovery of a New Form of Carbon*, J. Baggott (Oxford University Press, 1994). (1)

## Carbon nanotubes and nanoparticles

"Helical microtubules of graphitic carbon," S. Iijima, *Nature* **354**, 58 (1991). (3)
"Down the straight and narrow," M. S. Dresselhaus, *Nature* **358**, 195 (1992). (2)
"Curling and closure of graphitic networks under electron-beam irradiation," D. Ugarte, *Nature* **359**, 707 (1992). (3)
"Carbon onions introduce new flavour to fullerene studies," H. W. Kroto, *Nature* **359**, 670 (1992). (2)
"Single metal crystals encapsulated in carbon nanoparticles," R. S. Ruoff, D. C. Lorents, B. Chan, R. Malhotra & S. Subramoney, *Science* **259**, 346 (1993). (3)

## Chapter 2

### *Thermodynamics*

*Basic Chemical Thermodynamics* (4th Edition), E. B. Smith (Oxford University Press, 1990). (2)
*Chemical Thermodynamics*, M. L. McGlashan (Academic Press, 1979). (2)

### *Surface catalysis*

*Perspectives in Catalysis: A Chemistry For The 21st Century*, eds J. M. Thomas & K. I. Zamaraev (Blackwell Scientific Publications, 1992). (3)
*Catalysis at Surfaces*, I. M. Campbell (Chapman & Hall, 1988). (2)
"Catalysis on surfaces," C. M. Friend, *Scientific American* **268**, 42 (April 1993). (1)

### *Zeolites*

"Synthetic zeolites," G. T. Kerr, *Scientific American* **82** (July 1989). (1)
"Solid acid catalysts," J. M. Thomas, *Scientific American* **112** (April 1992). (1)

### *Zeolite engineering*

"Catalytic aspects of inclusion in zeolites," N. Herron in *Inclusion Compounds* Vol. 5, eds J. L. Atwood, J. E. D. Davies & D. D. MacNicol (Oxford University Press, 1991). (3)

### *Enzyme catalysis*

*Understanding Enzymes* (3rd Edition), T. Palmer (Ellis Horwood, 1991). (2)
*Introduction to the Chemistry of Enzyme Action*, A. Williams (McGraw-Hill, London, 1969).(2)
*The Machinery of Life*, D. S. Goodsell (Springer-Verlag, Berlin, 1993). (1)

### *Enzymes in industry*

"The greening of chemistry," S. Roberts & N. Turner, *New Scientist* **126**, 38 (21 April 1991). (1)

### *Biosensors*

*Biosensors*, E. A. H. Hall (Prentice Hall, 1991). (3)
*Biosensors: Fundamentals and Applications*, eds A. P. F. Turner, I. Karube & G. S. Wilson (Oxford University Press, 1987). (3)
"Biosensors," J. S. Schultz, *Scientific American* **64** (August 1991). (1)

## Chapter 3

### *Light*

*Light*, R. W. Ditchburn (Dover, 1991). (2)

### *Spectroscopy*

*Introduction to Molecular Spectroscopy*, G. M. Barrow (McGraw-Hill, 1962). (2)
*Fundamentals of Molecular Spectroscopy*, C. N. Banwell (McGraw-Hill, 1972). (2)
*Physical Chemistry* (4th Edition), P. W. Atkins (Oxford University Press, 1990). (2)

### *Photochemistry*

*Principles and Applications of Photochemistry*, R. P. Wayne (Oxford University Press, 1988). (2)
*Light, Chemical Change and Life*, eds J. D. Coyle, R. R. Hill & D. R. Roberts (Open University Press, 1982). (2)

### *Ultrafast laser spectroscopy*

"The birth of molecules," A. H. Zewail, *Scientific American* **263**, 76 (1990). (1)
"Laser femtochemistry," A. H. Zewail, *Science* **242**, 1645 (1988). (3)
"Ultrafast reaction dynamics," M. Gruebele & A. H. Zewail, *Physics Today* **43**(5), 24 (1990). (3)
"Real-time laser femtochemistry," A. H. Zewail & R. Bernstein, in *The Chemical Bond* ed. A. H. Zewail (Academic Press, 1992). (2)
"Femtosecond clocking of the chemical bond," M. J. Rosker, M. Dantus & A. H. Zewail, *Science* **241**, 1200 (1988). (3)
"Direct femtosecond mapping of trajectories in a chemical reaction," A. Mokhtari, P. Cong, J. L. Herek & A. H. Zewail, *Nature* **348**, 225 (1990). (3)

### *Bond-selective photochemistry*

*"State- and bond-selected unimolecular reactions,"* F. F. Crim, *Science* **249**, 1387 (1990). (3)

## Chapter 4

### *Crystallography and diffraction*

*Crystallography and Its Applications*, L. S. D. Glasser (Van Nostrand Reinhold Co., 1977). (2)
*Inorganic Solids*, D. M. Adams (John Wiley, 1974). (2)
*Diffraction Methods*, J. Wormald (Oxford University Press, 1973). (2)
"Architecture of the invisible," J. M. Thomas, *Nature* **364**, 478 (1993). (1)
*Fearful Symmetry*, I. Stewart & M. Golubitsky (Penguin, 1992). (1)

### *Quasicrystals*

"Metallic phase with long-range orientational order and no translational symmetry," D. Schectman, I. Blech, D. Gratias & J. W. Cahn, *Physical Review Letters* **53**, 1951 (1984). (3)
*Introduction To Quasicrystals*, ed. M. V. Jaric (Academic Press, 1988). (2)
*The Physics of Quasicrystals*, eds P. J. Steinhardt & S. Ostlund (World Scientific, Singapore, 1987). (3)
"Quasicrystals," D. Nelson, *Scientific American* **255**, 32 (August 1986). (1)
"The structure of quasicrystals," P. W. Stephens & A. I. Goldman, *Scientific American* **264**, 24 (April 1991). (1)

## Chapter 5

### *Biochemistry and genetics*

*The Chemistry of Life* (3rd Edition), S. Rose (Penguin, 1991). (1)
*Biochemistry* (2nd Edition), J. D. Rawn (Carolina Biological Supply Co., 1989). (2)
*Biochemistry*, C. K. Mathews & K. E. van Holde (Benjamin/Cummings, 1990). (2)
*Genetics* (2nd Edition), P. J. Russell (Scott, Foresman & Co., 1990). (2)
*Molecular Cell Biology* (2nd Edition), eds J. Darnell, H. Lodish & D. Baltimore (Scientific American Books Inc., Freeman, 1990). (2)

### *DNA*

*The Double Helix*, J. D. Watson (Penguin, 1968). (1)
"Molecular structure of nucleic acids," J. D. Watson & F. H. C. Crick, *Nature* **171**, 737 (1953). (3)

### *Chemical molecular recognition and supramolecular chemistry*

*The Chemistry of Macrocyclic Ligand Complexes*, L. F. Lindoy (Cambridge University Press, 1989). (2)
*Macrocyclic Chemistry*, B. Dietrich, P. Viout & J.-M. Lehn (VCH, Weinheim, 1993). (3)
"Supramolecular chemistry – scope and perspectives," J.-M. Lehn, *Angewandte Chemie* (English Edition) **27**, 89 (1988). (2)
*Bioorganic Chemistry* (2nd Edition), H. Dugas (Springer-Verlag, 1989). (2)
*Host-Guest Molecular Interactions: From Chemistry to Biology* (John Wiley, 1991). (3)
*Inclusion Compounds* Vol. 4, eds J. Atwood, J. E. D. Davies & D. D. MacNicol (Oxford University Press, 1991). (3)

### *Crown ethers*

*Crown Ethers and Cryptands*, G. W. Gokel (Royal Society of Chemistry, London, 1991). (3)

### *Calixarenes*

*Calixarenes*, C. D. Gutsche (Royal Society of Chemistry, London, 1993). (3)

### *Carcerands*

"Molecular container compounds," D. Cram, *Nature* **356**, 29 (1992). (3)

### *Rotaxanes and catenanes*

"A [2] catenane made to order," P. R. Ashton *et al.*, *Angewandte Chemie* (English Edition) **28**, 1396 (1989). (3)

"Molecular trains: the self-assembly and dynamic properties of two new catenanes," P. R. Ashton *et al.*, *Angewandte Chemie* (English Edition) **30**, 1042 (1991). (3)

"Polyrotaxanes: molecular composites derived by physical linkage of cyclic and linear species," H. W. Gibson & H. Marand, *Advanced Materials* **5**, 11 (1993). (3)

### *Molecular replication*

"A self-replicating system," T. Tjivikua, P. Ballester & J. Rebek, *Journal of the American Chemical Society* **112**, 1249 (1990). (3)

"Molecular recognition with model systems," J. Rebek, *Angewandte Chemie* (English Edition) **29**, 245 (1990). (3)

"Crossover reactions between synthetic replicators yield active and inactive recombinations," Q. Feng, T. K. Park & J. Rebek, *Science* **254**, 1179 (1992). (3)

"Competition, cooperation, and mutation: improving a synthetic replicator by light irradiation," J.-I. Hong, Q. Feng, V. Rotello & J. Rebek, *Science* **255**, 848 (1992). (3)

"Life in a test tube," L. D. Hurst & R. Dawkins, *Nature* **357**, 198 (1992). (2)

"Molecular replication," L. E. Orgel, *Nature* **358**, 203 (1992). (3)

## Chapter 6

### *Solid-state physics*

*Introduction to Solid-State Physics* (6th Edition), C. Kittel (John Wiley, 1986). (2)

*The Solid State*, A. Guinier & R. Jullien (Oxford University Press, 1989). (2)

*The Electronic Structure and Chemistry of Solids*, P. A. Cox (Oxford University Press, 1987). (2)

### *Molecular electronics*

"Molecular electronics," C. A. Mirkin & M. A. Ratner, *Annual Reviews of Physical Chemistry* **43**, 719 (1992). (3)

### *Conducting polymers*

"Plastics that conduct electricity," R. B. Kaner & A. G. MacDiarmid, *Scientific American* **258**, 60 (February 1988). (1)

"New semiconductor device physics in polymer diodes and transistors," J. H. Burroughes, C. A. Jones & R. H. Friend, *Nature* **335**, 137 (1988). (3)

### *Molecular conductors*

"Linear-chain conductors," A. J. Epstein & J. S. Miller, *Scientific American* **241**, 48 (October 1979). (1)

### *Superconductivity*

*Superconductivity – The Next Revolution?*, G. F. Vidali (Cambridge University Press, 1993). (2)
*The Path of No Resistance*, B. Schechter (Simon & Schuster, New York, 1989). (1)

### *Organic and molecular superconductors*

"Organic superconductors," K. Bechgaard & D. Jerome, *Scientific American* **247**, 52 (July 1982). (1)
"Superconductors go organic," D. Carlson & J. M. Williams, *New Scientist* 26 (14 November 1992). (1)
*Organic Superconductors (Including Fullerenes): Synthesis, Structure, Properties and Theory*, J. M. Williams *et al.* (Prentice Hall, 1992). (3)
"Molecular inorganic superconductors," P. Cassoux & L. Valade, in *Inorganic Materials* eds D. W. Bruce & D. O'Hare (John Wiley, 1992). (3)

### *Fullerene superconductors*

"Superconductivity at 18 K in potassium-doped fullerene ($K_3C_{60}$)," A. F. Hebard *et al.*, *Nature* **350**, 600 (1991). (3)
"Superconductivity at 28 K in $Rb_xC_{60}$," M. J. Rosseinsky *et al.*, *Physical Review Letters* **66**, 2830 (1992). (3)
"Superconductivity in doped fullerenes," A. F. Hebard, *Physics Today* **45**, 26 (November 1992). (3)

## Chapter 7

### *Colloid science*

*Introduction to Modern Colloid Science*, R. J. Hunter (Oxford University Press, 1993). (2)
*Introduction to Colloid Science*, W. J. Popiel (Exposition-University Press, New York, 1978). (2)

### *Gels*

"Gels," T. Tanaka, *Scientific American* **244**, 110 (January 1981). (1)
"Phase transitions of gels," Y. Li & T. Tanaka, *Annual Reviews of Materials Science* **22**, 243 (1992). (3)
"Environmentally sensitive polymers and hydrogels," A. S. Hoffman, *MRS Bulletin* **16**, 42 (Materials Research Society, September 1991). (2)

### *Surfactancy, micelles and liposomes*

*The Science of Soap Films and Soap Bubbles*, C. Isenberg (Dover, 1992). (1)

"Molecular architecture and function of polymeric oriented systems: models for the study of organization, surface recognition, and dynamics of biomembranes," H. Ringsdorf, B. Schlarb & J. Venzmer, *Angewandte Chemie* (English Edition) **27**, 114 (1988). (3)
"Micelles and microemulsions," D. Langevin, *Annual Reviews of Physical Chemistry* **43**, 341 (1992). (3)
"Liposomes," M. J. Ostro, *Scientific American* **256**, 90 (January 1987). (1)
*Liposomes: from Physics to Applications,* D. D. Lasic (Elsevier, 1993). (2)

### *Self-replicating micelles*

"Self-replicating reverse micelles and chemical autopoiesis," P. A. Bachmann, P. Walde, P. L. Luisi & J. Lang, *Journal of the American Chemical Society* **112**, 8200 (1990). (3)

### *Langmuir films*

"Seeing phenomena in flatland: studies of monolayers by fluorescence microscopy," C. M. Knobler, *Science* **249**, 870 (1990). (3)
"Phase transitions in monolayers," C. M. Knobler & R. C. Desai, *Annual Reviews of Physical Chemistry* **43**, 207 (1992). (3)

### *Langmuir-Blodgett films*

*Langmuir-Blodgett Films,* ed. G. Roberts (Plenum, 1990). (2)

### *Liquid crystals*

*Liquid Crystals,* P. J. Collings (Princeton University Press, 1990). (2)
*Liquid Crystals* (2nd Edition), S. Chandrasekhar (Cambridge University Press, 1992). (2)
"The world of liquid crystals," R. Templer & G. Attard, *New Scientist* 25 (4 May 1991). (1)

## Chapter 8

### *The early Earth*

*The Young Earth,* E. G. Nisbet (Allen & Unwin, 1987). (2)
*Chemical Evolution,* S. F. Mason (Clarendon Press, Oxford, 1992). (2)
"The nature of the Earth prior to the oldest known rock record: the Hadean era," D. J. Stevenson, in *Earth's Earliest Biosphere* ed. J. W. Schopf (Princeton University Press, 1983). (3)

### *The chemical origin of life*

*The Origin of Life,* M. G. Rutten (Elsevier, 1971). (2)
*The Origin of Life,* C. E. Folsome (W. H. Freeman, 1979). (2)
*Origins of Life,* F. Dyson (Cambridge University Press, 1985). (2)
*Seven Clues to the Origin of Life,* A. G. Cairns-Smith (Cambridge University Press, 1985). (1)

"Chemical evolution and the origin of life," R. E. Dickerson, *Scientific American* **239**, 62 (September 1978). (1)
"The origin and early evolution of life on Earth," J. Oró, S. L. Miller & A. Lazcano, *Annual Reviews of Earth & Planetary Science* **18**, 317 (1990). (3)

### *The origin of chirality*

*The Ambidextrous Universe*, M. Gardner (Penguin, 1974). (1)
*Chemical Evolution*, S. F. Mason (Clarendon Press, Oxford, 1992). (2)
"Origins of biomolecular handedness," S. F. Mason, *Nature* **311**, 19 (1984). (3)

### *The RNA world and ribozymes*

"RNA evolution and the origins of life," G. Joyce, *Nature* **338**, 217 (1989). (3)
*The RNA World* ed. R. F. Gesteland & J. F. Atkins (Cold Spring Harbor Laboratory Press, 1993). (2)
"RNA as an enzyme," T. R. Cech, *Scientific American* **255**, 76 (November 1986). (1)

### *Early evolution of living organisms*

*Earth's Earliest Biosphere*, ed. J. W. Schopf (Princeton University Press, 1983). (3)
"The evolution of the earliest cells," J. W. Schopf, *Scientific American* **239**, 84 (September 1978). (1)
*Microcosmos*, L. Margulis & D. Sagan (Summit, New York, 1986). (1)
*The Emergence Of Life*, S. Fox (Basic Books, New York, 1988). (1)

### *Models of nucleic-acid replication*

"Molecular replication," L. E. Orgel, *Nature* **358**, 203 (1992). (3)
"A self-replicating hexadeoxynucleotide," G. von Kiedrowski, *Angewandte Chemie* (English Edition) **25**, 932 (1986). (3)

## Chapter 9

### *Form in physics and biology*

*On Growth And Form*, D'A. Thompson (Cambridge University Press, 1992). (2)

### *Fractals*

*The Fractal Geometry Of Nature*, B. B. Mandelbrot (W. H. Freeman, 1982). (1)
*Fractals*, J. Feder (Plenum Press, 1988). (2)
*The Beauty of Fractals*, H.-O. Peitgen & P. H. Richter (Springer-Verlag, Berlin, 1986). (1)
*Fractals*, H. Lauwerier (Princeton University Press, 1991). (2)
"Fractal phenomena in disordered systems," R. Orbach, *Annual Reviews of Materials Science* **19**, 497 (1989). (3)

## *Crystal growth and pattern formation*

"Fractal growth," L. M. Sander, *Scientific American* **256**, 82 (January 1987). (1)
"The formation patterns in non-equilibrium growth," E. Ben-Jacob & P. Garik, *Nature* **343**, 523 (1990). (3)
"Pattern formation in materials science," J. P. Gollub & L. M. Sander, *MRS Bulletin,* **12**, 98 (Materials Research Society, August/September 1987). (2)

## *Oscillating chemical reactions*

*When Time Breaks Down,* A. T. Winfree (Princeton University Press, 1987). (2)
*Oscillations and Travelling Waves in Chemical Systems,* eds R. Field & M. Burger (John Wiley, 1985). (3)
"Chemical waves," J. Ross, S. C. Müller & C. Vidal, *Science* **240**, 460 (1988). (3)

## *Morphogenesis*

*The Making of a Fly,* P. A. Lawrence (Blackwell Scientific Publications, 1992). (2)
"The shape of things to come," L. Wolpert, *New Scientist* 38 (27 June 1992). (1)

## *Turing structures*

"Experimental evidence of a sustained standing Turing-type nonequilibrium chemical pattern," V. Castets, E. Dulos, J. Boissonade & P. De Kepper, *Physical Review Letters* **64**, 2953 (1990). (3)
"Transition from a uniform state to hexagonal and striped Turing patterns," O. Ouyang & H. L. Swinney, *Nature* **352**, 610 (1991). (3)
"Crystals from dreams," A. T. Winfree, *Nature* **352**, 568 (1991). (2)

## *Pattern formation in skins and shells*

"How the leopard gets its spots," J. D. Murray, *Scientific American* **258**, 80 (March 1988). (1)
*Models of Biological Pattern Formation,* H. Meinhardt (Academic Press, 1982). (2)

## *Chaos*

*Chaos,* J. Gleick (Sphere, 1988). (1)
"Chaos," J. P. Crutchfield, J. D. Farmer, N. H. Packard & R. S. Shaw, *Scientific American* **255**, 38 (December 1986). (1)
"What is chaos, that we should be mindful of it?," J. Ford, in *The New Physics* ed. P. Davies (Cambridge University Press, 1989). (2)
*Exploring Chaos,* ed. N. Hall (W. W. Norton & Co., 1992). (1)

## *Chemical chaos*

"Chemical chaos," S. K. Scott, in *Exploring Chaos,* ed. N. Hall (W. W. Norton & Co., 1992). (1)
*Chemical Chaos,* S. K. Scott (Oxford University Press, 1991). (3)

### *Non-equilibrium thermodynamics*

*From Being To Becoming*, I. Prigogine (Freeman, 1980). (1)
*Self-Organization in Nonequilibrium Systems*, G. Nicolis & I. Prigogine (John Wiley, 1974). (2)
"Physics of far-from-equilibrium systems and self-organization," G. Nicolis, in *The New Physics* ed. P. Davies (Cambridge University Press, 1989). (2)
*Exploring Complexity*, G. Nicolis & I. Prigogine (Freeman, New York, 1989). (2)

## Chapter 10

### *Origin and evolution of the atmosphere*

*The Chemical Evolution of the Atmosphere and Oceans*, H. D. Holland (Princeton University Press, 1984). (2)
"How climate evolved on the terrestrial planets," J. F. Kasting, O. B. Toon & J. B. Pollack, *Scientific American* **258**, 90 (February 1988). (1)

### *Atmospheric chemistry*

*Chemistry of Atmospheres* (2nd Edition), R. P. Wayne (Oxford University Press, 1991). (2)
*Atmospheric Change*, T. E. Graedel & P. J. Crutzen (W. H. Freeman, 1993). (2)
*Atmosphere, Weather and Climate* (6th Edition), R. G. Barry & R. J. Chorley (Routledge, 1992). (2)
"The changing atmosphere," T. E. Graedel & P. J. Crutzen, *Scientific American* **261**, 28 (1989). (1)
*Gaia*, J. Lovelock (Oxford University Press, 1979). (1)
*The Ages Of Gaia*, J. Lovelock (Oxford University Press, 1988). (1)

### *Biogeochemical cycles*

*Biogeochemistry*, W. H. Schlesinger (Academic Press, 1991). (2)
*Global Biogeochemical Cycles*, eds S. S. Butcher, R. J. Charlson, G. H. Orians & G. V. Wolfe (Academic Press, 1992). (2)

### *Climate records and palaeoclimate*

*Ice Ages*, J. Imbrie & K. P. Imbrie (Macmillan, London, 1979). (1)
"The ice-core record: climate sensitivity and future greenhouse warming," C. Lorius, J. Jouzel, D. Raynaud, J. Hansen & H. Le Treut, *Nature* **347**, 139 (1990). (3)

### *Global warming*

*Global Climate Change*, ed. S. F. Singer (Paragon House, New York, 1989). (2)
*Hothouse Earth*, J. Gribbin (Bantam Press, London, 1990). (1)
"The changing climate," S. H. Schneider, *Scientific American* **261**, 38 (September 1989). (1)
*Climate Change. The IPCC Scientific Assessment*, eds J. T. Houghton, G. J. Jenkins & J. J. Ephraums (Cambridge University Press, 1990). (2)

### *Ozone depletion*

"Large losses of total ozone reveal seasonal $ClO_x/NO_x$ interaction," J. C. Farman, B. G. Gardiner & J. D. Shanklin, *Nature* **315**, 207 (1985). (3)
"The Antarctic ozone hole," R. S. Stolarski, *Scientific American* **258**, 20 (1988). (1)
"Progress towards a quantitative understanding of Antarctic ozone depletion," S. Solomon, *Nature* **347**, 347 (1990). (3)
"Stratospheric ozone depletion," F. S. Rowland, *Annual Reviews of Physical Chemistry* **42**, 731 (1991). (3)
"Polar stratospheric clouds," R. Turco & O. B. Toon, *Scientific American* **264**, 40 (June 1991). (1)

### *Acid rain*

"Acid rain," G. E. Likens, R. F. Wright, J. N. Galloway & T. J. Butler, *Scientific American* **241**, 39 (October 1979). (1)
*Acid Rain*, B. J. Mason (Clarendon Press, Oxford, 1992). (2)

## Other topics

*Advanced Inorganic Chemistry* (3rd Edition), F. A. Cotton & G. Wilkinson (John Wiley, 1972). (2)
*Organic Chemistry* (3rd Edition), R. T. Morrison & R. N. Boyd (Allyn & Bacon, 1973). (2)
*Introduction to Polymers*, R. J. Young (Chapman & Hall, 1981). (2)
*Polymer Chemistry*, M. P. Stevens (Oxford University Press, 1990). (2)
*Electrochemistry*, C. M. A. Brett & A. M. Oliveira Brett (Oxford University Press, 1993). (2)
*Organometallics* (2nd Edition), Ch. Elschenbroich & A. Salzer (VCH, Weinheim, 1992). (2)

# Credits

Quote on page 3 from Geoffrey Willans & Ronald Searle, *Down With Skool,* © Geoffrey Willans & Ronald Searle, 1958, reprinted with permission of the Tessa Sayle Agency. Quote on page 216 from Primo Levi, *The Monkey's Wrench,* reprinted with permission of Simon & Schuster Inc.; translation by William Weaver, © Summit Books, 1986. Quote on page 290 from D'Arcy Wentworth Thompson, *On Growth and Form,* reprinted with permission of Cambridge University Press.

Figure 9.1 from S. Motojima & H. Iwanaga, *Journal of Chemical Vapour Deposition* **1**, 87 (1992). Figure 9.12 from J. Schütze, O. Steinbock & S. C. Müller, *Nature* **356**, 45 (1992). Figure 9.15 from E. O. Budrene & H. C. Berg, *Nature* **349**, 630 (1991). Figure 9.18 from H. Meinhardt & M. Klingler, *Journal of Theoretical Biology* **126**, 63 (1978). Figure 10.2 redrawn from C. Lorius, J. Jouzel, D. Raynaud, J. Hansen & H. Le Treut, *Nature* **347**, 139 (1990). Figure 10.12 from C. Sagan, W. R. Thompson, R. Carlson, D. Gurnett & C. Hord, *Nature* **365**, 715 (1993).

# Index

HAMILTON TOWNSHIP PUBLIC LIBRARY
3 1923 00292877 4